S0-DUU-416

Numerical Methods in Quaternary Pollen Analysis

Numerical Methods in Quaternary Pollen Analysis

H. J. B. BIRKS

Botanical Institute
University of Bergen
Norway

A. D. GORDON

Department of Statistics
University of St. Andrews
Scotland

1985

ACADEMIC PRESS
Harcourt Brace Jovanovich, Publishers
London Orlando San Diego New York
Austin Montreal Sydney Tokyo Toronto

ACADEMIC PRESS INC. (LONDON) LTD.
24–28 Oval Road
LONDON NW1 7DX

United States Edition published by
ACADEMIC PRESS, INC.
Orlando, Florida 32887

British Library Cataloguing in Publication Data

Birks, H. J. B.
Numerical methods in quaternary pollen analysis.
1. Pollen, Fossil
I. Title II. Gordon, A. D.
561'.13 QE993.2

Library of Congress Cataloging in Publication Data

Birks, H. J. B. (Harry John Betteley)
Numerical methods in quaternary pollen analysis.

Bibliography: p.
Includes index.
1. Pollen, Fossil--Statistical methods.
2. Paleobotany--Quaternary--Statistical methods.
I. Gordon, A. D. II. Title.
QE993.2.B57 1985 561'.13 84-21739
ISBN 0-12-101250-6 (alk. paper)

PRINTED IN THE UNITED STATES OF AMERICA

85 86 87 88 9 8 7 6 5 4 3 2 1

Contents

Preface

In the last two or three decades, ecology has developed from a subject that was almost entirely qualitative and observational in outlook to a science in which investigators seek generalisations on the basis of observations that can be quantified and analysed statistically. More recently, the same trend has been evident in Quaternary palaeoecology. Quaternary pollen analysis has always had a quantitative basis, in that the primary data are the numbers of different types of pollen grains and spores counted in a sample, but there has recently been an increasing awareness of the potential contribution of statistical methodology to palynological investigations. There has also been some scepticism about the magnitude of this contribution, possibly fuelled by excessive claims for, and inappropriate uses of, statistical techniques. This book, which has arisen out of collaborative research by the two authors over more than a decade, describes and reviews numerical methods which we believe can assist an investigator in the analysis of Quaternary pollen analytical data. Many of the methods are of more widespread applicability, for example to palynological data from deposits of earlier ages, and to other quantitative palaeoecological variables, such as sediment chemistry, plant macrofossils, diatoms, cladocera, mollusca, foraminifers, ostracods, and dinoflagellates.

H. J. B. B. wishes to record his very great debt to the late J. B. Birks, E. J. Cushing, I. C. Prentice, R. A. Reyment, and T. Webb III for stimulating and encouraging his interests in quantitative methods of analysing Quaternary palynological data and for freely sharing ideas, insights, and experiences.

Numerous other friends and colleagues, including P. Adam, S. Th. Andersen, C. W. Barnosky, P. W. Beales, K. D. Bennett, B. E. Berglund, H. H. Birks, S. Björck, A. P. Bonny, R. H. W. Bradshaw, M. E. Edwards, K. Faegri, F. S. Gilbert, B. Huntley, C. R. Janssen, P. D. Kerslake, H. F. Lamb, J. M. Line, L. J. Maher, Jr., P. D. Moore, R. W. Parsons, S. M. Peglar, M. Ralska-Jasiewiczowa,

J. C. Ritchie, L. Rymer, J. Turner, W. A. Watts, W. Williams, and H. E. Wright, Jr., have also contributed to this work by either generously providing challenging data sets, often with unexpected analytical problems, or supplying thoughtful and critical discussion, encouragement and interest, patience and tolerance, or healthy constructive scepticism and palynological common-sense. In addition, B. Huntley and J. M. Line have generously helped with programming and data analysis. Jane Allard Grimm, K. D. Bennett, and J. M. Line have given valuable assistance in the preparation of the bibliography. Sylvia Peglar has provided indispensible and meticulous help in data preparation over the last ten years, and has skilfully drafted almost all the figures in this book. Tricia Brown, Irene Donaldson, and Shirley Lees have carefully and patiently typed several drafts of the book.

All or part of the manuscript has been critically read by C. W. Barnosky, H. H. Birks, R. M. Cormack, F. S. Gilbert, I. C. Prentice, R. A. Reyment, and T. Webb III, who have contributed valuable suggestions for its improvement. To all these people and to others who, directly or indirectly, have influenced the contents of this book, we express our sincere thanks.

We should be most grateful to readers who draw our attention to any errors or obscurities in the text, or suggest other improvements.

This book is a contribution to International Geological Correlation Programme Project 158, Subproject B.

June 1984

H. J. B. Birks
A. D. Gordon

CHAPTER **1**

The Nature of Quaternary Pollen Analytical Data

1.1 Pollen Analysis as a Quaternary Palaeoecological Technique

The primary aim of Quaternary palaeoecology is the reconstruction of the past environments and ecosystems of the last 1–2 million years of earth's history. As past environments and ecosystems cannot be observed directly, they must be reconstructed from the fossils and the sediments in which the fossils are found. Sediments of Quaternary age occur commonly in both continental and marine situations, and their palaeoecology attracts considerable scientific attention. This is because the Quaternary period is unique in earth's history as the period when man evolved and when the climate oscillated, in the latitude of Europe and North America, between temperate, so-called interglacial phases of 10,000 to 20,000 years duration and cold phases of 50,000 to 100,000 years duration during which glaciation commonly occurred.

The period of time represented by Quaternary deposits has been studied more intensively than any other time span of comparable magnitude. Quaternary palaeoecology can only consider and be based on those groups of organisms that are found as fossils—foraminifers, molluscs, arthropods, vertebrates, algae, bryophytes, and vascular plants, all of which contain compounds resistant to decay such as calcite, aragonite, chitin, silica, cutin, lignin, and sporopollenin (see Cushing and Wright, 1967; H. J. B. Birks and H. H. Birks, 1980). Pollen grains and spores of vascular plants are by far the most abundant type of fossil preserved in terrestrial Quaternary sediments, with the result that terrestrial Quaternary palaeoecology is largely dominated by the technique of stratigraphical pollen analysis (Cushing and Wright, 1967).

Pollen analysis, the principles of which we discuss in this chapter, provides a means of reconstructing the past flora (the distribution of individual plant taxa in time and space), the past plant populations (the abundance of plant taxa in time and space), and the past vegetation (the distribution of plant assemblages or communities in time and space) (see Cushing, 1963; H. J. B. Birks, 1973b). The reconstruction of past plant communities represents a major step towards the reconstruction of the past ecosystem, as the plant community is the most complex part of any ecosystem. After the community has been reconstructed, inferences can be made about the environment of the past ecosystem, assuming that the ecological requirements and tolerances of the species and the communities are known (see M. B. Davis, 1978; H. J. B. Birks and H. H. Birks, 1980; H. J. B. Birks, 1981c).

Interpretations of Quaternary pollen analytical data are derived almost entirely from the extrapolation of present-day ecological observations backwards in time. Past communities, environments, and ecosystems are reconstructed by analogy with present-day communities and ecosystems and with known ecological preferences of the taxa and communities involved (D. Walker, 1978). Implicit in all Quaternary pollen analytical studies, as in all palaeoecology, is the assumption and philosophical principle of methodological uniformitarianism (*sensu* Gould, 1965), or actualism, which states that the nature of modern processes is the same as in the past and thus that modern laws of nature can be extended backwards in time and used to reconstruct and explain past events (see Rymer, 1978, for a critical review of uniformitarianism in relation to Quaternary pollen analysis). As H. J. B. Birks and H. H. Birks (1980) discuss, there is no way to prove or disprove methodological uniformitarianism ('the present is the key to the past'), as it is the basic logic and methodology of all historical sciences, including Quaternary palaeoecology; all reconstructions of the past require some application and extrapolation in time of modern ecological or geological knowledge.

Quaternary pollen analysis closely parallels descriptive plant ecology in its scientific development. Qualitative, often rather generalised, descriptions of fossil pollen assemblages preserved in peats were prevalent in the 1880s and 1890s (Erdtman, 1943; Manten, 1967; Faegri and Iversen, 1975), just as qualitative, broad-scale descriptions of modern vegetation were common at that time (Whittaker, 1962). Quantitative descriptions of vegetation were first attempted in the 1910s and 1920s, and quantification in descriptive plant ecology developed rapidly (McIntosh, 1974, 1975). In 1916, the potentialities of the quantitative analysis of pollen grains preserved in Quaternary sediments such as peats were first demonstrated by the Swedish geologist Lennart von Post (1918, reprinted in English in 1967). Following the work of Gustaf Lagerheim, von Post counted samples of fossil pollen preserved at different levels in a peat profile and presented his counts as percentages of the sum of pollen grains counted (Manten, 1967). He displayed his results as stratigraphical diagrams with pollen percentages plotted against depth through the stratigraphical column (see Fries, 1967). Von Post demonstrated strong similarities in pollen profiles from sites within the same region and marked differences between sequences from sites in different parts of Sweden. He was thus able to add the fourth dimension, namely, time, to the study of vegetation and to the elucidation of such problems as the nature of past vegetation, the history of major vegetation formations, the patterns of vegetation change over long periods of time, and the timing and magnitude of post-glacial climatic change. As Deevey (1967, p. 65) so aptly comments, 'Von Post's simple idea, that a series of changes in pollen proportions in accumulating peat was a four-dimensional look at vegetation, must rank with the double helix as one of the most productive suggestions of modern times.'

The basic principles of Quaternary pollen analysis or palynology are as follows (Godwin, 1934; Erdtman, 1943; West, 1971; Faegri and Iversen, 1975: H. J. B. Birks and H. H. Birks, 1980):

1. Pollen grains and spores are small (10–100 microns) and are produced in great abundance by vascular plants (flowering plants and ferns), but only a few of these ever fulfil their natural function of fertilisation. Most eventually fall to the ground.
2. Before reaching the ground, pollen and spores are well mixed by turbulence in the atmosphere, resulting in a more or less uniform pollen rain within an area.
3. The organic compounds (sporopollenin, cellulose, pectins, callose, proteins, etc.) that comprise pollen grains and spores rapidly decay unless the processes of biological decomposition are inhibited in some way. This inhibition occurs where there is a lack of oxygen, for example in permanently waterlogged areas such as bogs, fens, lake bottoms, and the ocean floor. The sporopollenin of the outer wall or exine of pollen and spores is well preserved in non-oxidising sediments that accumulate in these areas.
4. The taxonomy of pollen grains and spores is relatively well known, at least in the Northern Hemisphere (see, e.g., McAndrews *et al.*, 1973; Faegri and Iversen, 1975; P. D. Moore and Webb, 1978). The major types are identifiable to various taxonomic levels (family, genus, species) using the transmitted-light microscope.
5. As the composition of the pollen rain depends on the composition of the vegetation that produced it, the pollen rain is a function of the vegetation of the area. A sample of the pollen rain will thus be a reflection of the vegetation, both local or lowland (aquatic and wetland communities) and regional or upland (forest, grassland, heathland communities, etc.), at that point in space and time.
6. If a sample of the pollen rain preserved in peat or mud of known age is examined and the various pollen and spore types preserved are identified and counted, the pollen spectrum is a reflection of the vegetation surrounding the site of deposition at the time the sediment and its contained pollen and spores were deposited. Because pollen grains and spores are small and very abundant (up to 10^6 grains cm^{-3}), only small amounts ($0.5–1\ cm^3$) of

sediment are needed for detailed pollen analysis. This small sample volume contrasts with the requirements for studies of fossils such as seeds and fruits, molluscs, insects, and vertebrates, where relatively large amounts of sediment are needed to provide an adequate sample of the fossil population.

7. If samples of the pollen rain preserved at several levels throughout the sediment are examined, the fossil pollen assemblages provide stratigraphical records of the past vegetation and its development through the time period represented by the sedimentary record.
8. If two or more series of stratigraphical pollen assemblages are obtained from several sites, it is possible to compare the pollen spectra and to detect similarities and differences in vegetation through time at different points in space.

A Quaternary palaeoecological study involving pollen analysis usually proceeds along the following lines. After the aims of the investigation and the hypotheses to be tested have been defined, the site(s) of interest and relevance to the study is visited. The selection of suitable sites for pollen analysis is of critical importance; it will depend on the aims of the study, the spatial and temporal scales of interest, and the availability of suitable sites. Jacobson and Bradshaw (1981) discuss the question of site selection in detail. Cores of sediment are collected from the site, after a series of trial borings has been made to establish the gross sediment stratigraphy and morphometry of the basin. If the sediments are exposed, for example, in gravel pits, eroding peat hags, or sea cliffs, samples can be conveniently collected from the exposed face. The lithology and sediment characteristics of the samples, either from the cores or from the exposures, are then described (see H. J. B. Birks and H. H. Birks, 1980).

H. J. B. Birks and H. H. Birks (1980) discuss ways in which investigators select the positions in sediment cores or sedimentary sections at which samples of material are extracted for analysis. In Quaternary pollen analysis, the most commonly used sampling strategy is search sampling (*sensu* Krumbein and Graybill, 1965), in which the core or section is initially sampled fairly sparsely. The results of analysing this first set of samples indicate the positions where subsequent more detailed investigation would be advisable. Gordon (1974) describes sequential sampling strategies for use in the uncommon situation in which one has fairly precise information about some feature of the preserved pollen record (e.g., the rise of *Ambrosia*-type pollen in eastern North America some 80–150 years ago) and one wishes to identify rapidly the position in the core at which this occurs.

Samples extracted from the core are prepared for pollen analysis using a standard laboratory procedure, which aims to concentrate the pollen and to remove as much of the sediment matrix as possible. Details of field and laboratory techniques are given by Faegri and Iversen (1975), West (1977), and H. J. B. Birks and H. H. Birks (1980).

After the sample has been prepared for pollen analysis, the residue is mounted on a microscope slide and pollen identification and counting can begin. Identification of pollen and spores can reliably be done only by careful comparison between fossil material and modern reference material of pollen and spores collected from plants of known identity and prepared in the same way as the fossil material (see Hansen and Cushing, 1973, and H. J. B. Birks and H. H. Birks, 1980, for a discussion of identification procedures in Quaternary pollen analysis). Problems of deteriorated pollen and of unknown grains can arise in some samples (e.g., Cushing, 1967a; H. J. B. Birks, 1973b, 1981a).

Pollen counting is carried out along regularly spaced traverses of the microscope slide at a magnification of ×300–400. The number of grains counted depends on the problem being investigated. A sufficient number of grains should be counted to obtain reliable estimates, that is, to achieve broadly constant percentages of the pollen types of interest when their counts are expressed as percentages of the pollen sum. In general, a total of 300 to 500 grains is usually adequate (Bowman, 1931; Crabtree, 1968; Maher, 1972b), but in certain studies counts of 1000 or more grains per sample are essential to obtain the precision required to answer particular questions. Traverses should be positioned evenly over the whole slide (and not concentrated near the edge) to avoid any effects of non-random distribution of pollen and spores on the slide (Brookes and Thomas, 1967). Questions of the statistical reliability of pollen counts and of confidence intervals for pollen percentage data are discussed in Section 2.2.

Having obtained a stratigraphical sequence of pollen counts, the next stage in a palynological investigation is to present the results in the form of a graph or other diagram, prior to the interpretation of the data. Pollen analytical data are invariably complex and are most effectively presented in the form of a pollen diagram (see Section 1.3). This is a series of graphs of the values for different pollen and spore types plotted against their stratigraphical depth or, more rarely, against their age. Nearly all pollen analytical data are relative proportions of the different pollen and spore types (the proportions being with respect to some specified pollen sum). In some studies, estimates are made of the 'absolute' numbers of pollen grains per unit volume or unit weight of sediment, or per unit area of sediment deposited per unit time. In such studies the counts of the different pollen and spore types are independent of each other (which is not the case with relative percentage data) and, in the estimation of number of grains per unit area per unit time, are also independent of changes in the sediment-accumulation rate within the sequence (see Section 1.2).

For percentage data, a critical decision is the specification of the taxa which will be included in the pollen sum. Cushing (1963), H. J. B. Birks (1973b), and H. J. B. Birks and H. H. Birks (1980) argue that the choice should be based on the principle that all members of the 'universe' of interest and under study should be included. In general, the main interest is usually centered on regional vegetational history, and in this case all pollen and spores which could have originated from the upland vegetation should be included in the pollen sum. Pollen and spores of plants confined to the lowland local aquatic and mire vegetation (e.g., obligate aquatic plants and bog and fen species) should be excluded from the pollen sum because they are locally produced from a different vegetation from that with which the investigation is primarily concerned. Special pollen sums can be used for selected taxa only, and for specific palaeoecological problems (see Wright and Patten, 1963; H. H. Birks, 1972; Janssen, 1981b; Cwynar, 1982). The question of pollen sums is discussed further in Section 2.5.

Pollen diagrams (e.g., Figures 1.2 and 1.3) are often complicated and difficult to comprehend rapidly because they present a large amount of data in a graphical form. Prior to interpretation of the data, it is often useful to subdivide each pollen stratigraphical sequence into smaller units, so-called pollen zones, for ease in description, discussion, comparison, interpretation, and correlation. A variety of numerical methods can be used for zoning single pollen sequences, and these methods are described in Chapter 3. Numerical methods can also be used for comparing two or more pollen sequences, and these techniques are discussed in Chapter 4. Comparisons permit the detection of similarities among pollen sequences and hence the delimitation of regional pollen zones. Comparisons can also emphasise differences between profiles, and such differences may be of considerable palaeoecological interest, particularly at a local scale.

Once the pollen sequence has been divided into pollen zones and compared with other sequences, the task of interpretation can begin. In general, interpretation should follow the logical sequence of reconstruction of past flora, reconstruction of past populations, reconstruction of past vegetation, and reconstruction of past environments. H. J. B. Birks and H. H. Birks (1980; see also H. J. B. Birks, 1973b) present six basic questions of interpretation which can be asked of a pollen diagram.

1. What taxa were present in the past flora?
2. What were the relative abundances of the taxa present in the past?
3. What plant communities were present in the past?
4. What space did each community occupy in the past?
5. At what time did each community occur in the past?
6. What was the past environment of the plant communities at that time and space?

The answer to the first question can never be complete because the fossil pollen and spore record is never complete. Some plants produce pollen that is rarely, if ever, preserved (e.g., *Juncus, Luzula, Naias flexilis*) or produce pollen in such small amounts (e.g., *Linum*) that the probabilities of its pollen ever becoming incorporated in sediments and being detected by a pollen analyst are extremely low. In addition, some plants, such as grasses and sedges, produce abundant pollen, but the pollen is not specifically or generically distinct from pollen of other members of the same family. Clearly, the degree of floristic information that can be extracted from the pollen stratigraphical record depends on the quality and reliability of the pollen identifications and on the taxonomic level to which the identifications are carried out. Numerical methods can assist the pollen analyst in separating mixtures of pollen grains of similar morphology but which differ in size and/or shape (see Section 2.4).

An additional problem in reconstructing the past flora from pollen stratigraphical data is that some pollen may have been transported long distances by air masses or other currents. There is a non-zero probability of finding a pollen grain that originates from a plant anywhere in the world (Cushing, 1963). Some pollen types (e.g., *Pinus* and *Ephedra*) are notorious for their long-distance transport, and allowance should be made for this in any floristic reconstruction. H. J. B. Birks (1981b) has applied the theory of point processes (Cox and Lewis, 1966; Reyment, 1969b, 1976b, 1980b) to the occurrences of 27 pollen and spores of presumed long-distance transport in sediments of late-Wisconsin age in Minnesota and has demonstrated that the occurrences of some types (e.g., *Carya* and *Platanus occidentalis*) show significant trends in the rate of occurrences between 8400 and 20,500 years B.P. (before present), whereas such trends were not evident for other species (e.g., *Acer negundo* and *Sarcobatus vermiculatus*).

The second question involves interpretation of the numerical data of the pollen spectra in terms of the past abundances or population sizes of the plants present in the past. Unlike many fossil groups (e.g., trilobites, brachiopods, belemnites, and diatoms) where one fossil reflects one living individual, in pollen analysis, one individual plant may produce many millions of pollen grains. Before the fossil pollen values can be interpreted in terms of past abundances of plants, the pollen production and representation of modern plants must be studied. Modern pollen-representation factors (e.g., R values *sensu* M. B. Davis, 1963) can be derived and applied to the fossil pollen counts to derive estimates of past plant abundance.

Modern pollen-representation factors for taxon k can be estimated by collecting surface samples, such as the top 1–2 cm of mud accumulating in a lake, and by comparing p_{0k}, the observed proportions of pollen type k in the surface sample 0, with v_k, the observed proportions of taxon k in the surrounding modern vegetation (see Figure 1.1). (In Section 5.4 we discuss an appropriate manner of defining v_k). The modern representation factor for taxon k is estimated by

$$\hat{R}_k = p_{0k}/v_k$$

The R values vary between taxa by several orders of magnitude (M. B. Davis, 1963; Andersen, 1970) because abundant species in the vegetation may produce little or no pollen whereas rare species may produce abundant pollen.

To estimate the relative abundance of taxon k 8000 years ago (see Figure 1.1), we examine the core of sediment from the same place as our surface sample, sample the core at the appropriate depth, and estimate u_{ik}, the proportion of pollen of taxon k at the time period represented by sample i, by preparing the sample i for pollen analysis and counting an appropriate number of pollen grains m_{ik} on the microscope slide. The proportion of pollen type k in sample i is p_{ik}, and by assuming that our modern pollen-representation factor is invariant in time and space, we can estimate f_k, the proportion of taxon k in the past vegetation, by

$$\hat{f}_k \propto p_{ik}/\hat{R}_k,$$

where the constant of proportionality can be evaluated by noting that the past vegetation proportions must sum to 1. A fuller discussion of the assumptions

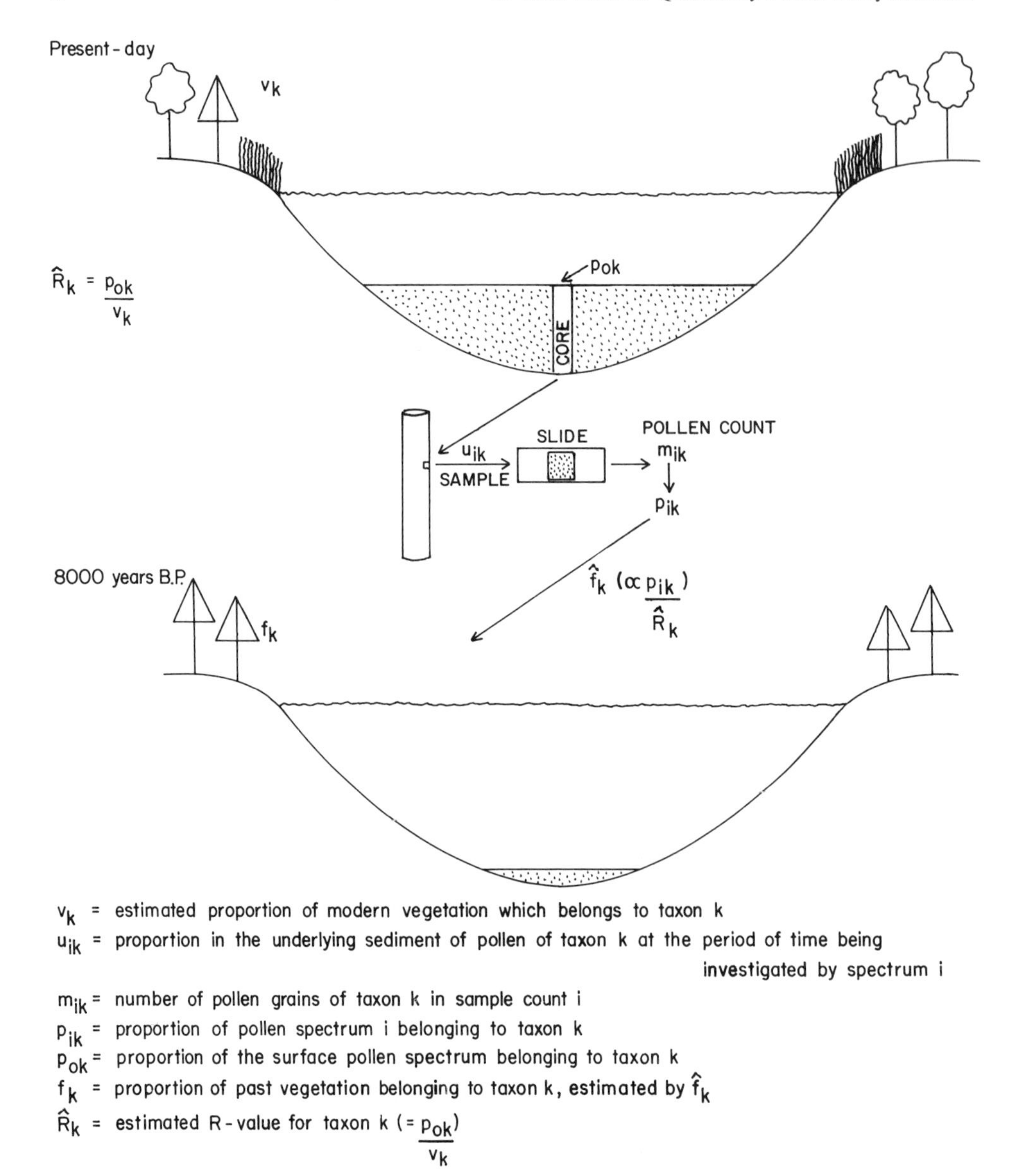

Figure 1.1 Stages in the quantitative reconstruction of the relative abundance of taxon k in the vegetation around a lake 8000 years ago using estimates of modern pollen-representation factors $\hat{R}_k$ to transform fossil pollen spectra into past abundances f_k of the taxon. See text discussion of variables and equations.

involved in this approach is given in Chapters 5 and 6, which describe mathematical methods for deriving pollen-representation factors from modern pollen counts and the use of such factors in estimating past abundances and population sizes of plants (see Sections 5.4 and 6.3).

The reconstruction of past vegetation and of past plant communities involves comparisons of the fossil pollen spectra with modern pollen spectra from areas of known vegetation (Wright, 1967; H. J. B. Birks and H. H. Birks, 1980). If the modern and fossil spectra are similar, it can be concluded that they were produced by similar vegetation. A modern analogue can thus be suggested for the past vegetation. If, however, no analogue can be found, it may either be concluded that the past vegetation has no modern analogue or that modern spectra should be sought elsewhere. Numerical methods for analysing modern pollen spectra are described in Section 5.3, and the use of quantitative techniques for comparing modern and fossil pollen spectra and thus for reconstructing past vegetation is discussed in Section 6.4.

Other approaches to vegetational reconstruction involve the extension backwards in time of known sociological and ecological preferences of individual taxa, the so-called indicator-species approach (see Janssen, 1967b, 1970, 1981b; H. J. B. Birks, 1973b; D. Walker, 1978; H. J. B. Birks and H. H. Birks, 1980) and the search for groups of fossils that are significantly associated together in a series of samples within and between stratigraphical sequences, so-called recurrent groups (H. J. B. Birks and H. H. Birks, 1980). Recurrent groups can, in some instances, suggest which taxa grew together in the past; the approach is described in Section 6.5.

Answers to the question of where particular plant communities grew in the past are critically dependent on our knowledge and understanding of the complex processes of pollen transport and dispersal and on the choice of site, particularly its size and potential pollen source area (Janssen, 1981b; Jacobson and Bradshaw, 1981). Pollen sequences from several sites within a small geographical area are required to detect patterns of vegetational differentiation related, for example, to altitude (Turner and Hodgson, 1979, 1983), soils (D. Walker, 1966; Brubaker, 1975; Jacobson, 1975, 1979), and climate and topography (McAndrews, 1966, 1967). Alternatively, transects of pollen diagrams across a site either in one (Turner, 1965) or two (Turner, 1970, 1975) dimensions may permit the detection of spatial differentiation in the occurrence of particular pollen assemblages and hence of particular plant communities (see H. J. B. Birks and H. H. Birks, 1980).

Detailed, independent chronologies are required to answer the question of when particular communities occurred in the past. In ideal circumstances, radiocarbon dating of organic sediments can give an exact answer (within the precision of the radiocarbon dating technique). However, large errors can arise in the dating of a range of organic materials (peats, wood, charcoal, lake muds), (Broecker, 1965; Ogden, 1967; Godwin, 1969), in laboratory procedures (Pardi and Marcus, 1977), in systematic bias between laboratories (International Study Group, 1982), in the radiocarbon assay of sediments low in organic content (Shotton, 1967;

Olsson, 1979; Sutherland, 1980), and in the dating of sediments subject to input of ^{14}C-deficient carbon from surrounding limestone rocks or calcareous drift, the so-called hard-water effect (Deevey, Gross, Hutchinson, and Kraybill, 1954; Broecker and Walton, 1959; Donner, Jungner, and Vasari, 1971). Other methods of constructing an independent chronology for Quaternary pollen stratigraphical sequences are discussed by West (1977) and H. J. B. Birks and H. H. Birks (1980).

The final phase in any palaeoecological interpretation is the reconstruction of past environments. There are three broad approaches to palaeoenvironmental reconstructions (H. J. B. Birks and H. H. Birks, 1980): (1) using abiotic evidence of past environments, such as sediment lithology, sediment geochemistry, and stable-isotope (e.g., ^{18}O–^{16}O, ^{13}C–^{12}C, deuterium–hydrogen) composition, (2) analysis of stratigraphical patterns within and between fossil pollen assemblages, and determining and interpreting any temporal gradients in the fossil data, and (3) using modern pollen samples from known environments to compare with fossil pollen assemblages.

This latter approach has recently been quantified by deriving mathematically modern calibration or 'transfer' functions that relate contemporary pollen assemblages to present-day climate, comparable to modern pollen-representation factors that relate contemporary pollen to present-day plant abundances. By assuming that these climatic transfer functions also are invariant in time and space, one can apply them to pollen assemblages to derive quantitative estimates of the past environment (H. S. Cole, 1969; T. Webb, 1971; T. Webb and Bryson, 1972; T. Webb and Clark, 1977). This quantitative approach to palaeoenvironmental reconstruction from Quaternary pollen analytical data is discussed in Section 6.6.

1.2 Types of Quaternary Pollen Analytical Data

We have discussed in Section 1.1 the methodology of Quaternary pollen analysis as a palaeoecological technique. We have seen that interpretation of fossil pollen stratigraphical data frequently requires comparisons, either qualitative or quantitative, with modern pollen spectra collected from areas of known vegetation and/or environment, including climate. In addition, interpretation in terms of past plant populations often requires the transformation of the fossil stratigraphical data into estimates of past plant abundance by means of modern representation factors derived numerically from comparison between contemporary pollen frequencies and modern plant abundances (see discussion accompanying Figure 1.1).

Pollen analytical data can thus be classified into two broad types: stratigraphical and non-stratigraphical. In *non-stratigraphical* data, the pollen spectra represent estimates of the pollen composition in samples of the same age, usually but not invariably the present-day (so-called surface samples) (e.g., H. J. B. Birks, Deacon, and Peglar, 1975; H. J. B. Birks and Saarnisto, 1975; Bernabo and Webb,

1977; Huntley and Birks, 1983), but from different geographical, vegetational, and environmental areas. In *stratigraphical data,* the pollen spectra represent estimates of the pollen composition at different specified depths and, hence, times through a stratigraphical sequence at a single point in space. The unique ordering of the samples with depth and age is of paramount importance to the palaeoecologist because it provides the basis for presenting the data in the form of stratigraphical pollen diagrams (see Section 1.3); for detecting stratigraphical changes; for delimiting pollen zones (see Chapter 3) and recognising temporal trends within the sequence; and for interpreting the data as a time-series record of changes in past populations, communities, and, by inference, environments (see Chapter 6).

In the context of Quaternary palaeoecology, the terrestrial pollen stratigraphical record is the most important source of palaeoecological information. The length of the record is usually between 100 and 15,000 years, but sediments from old, deep basins beyond the limits of glaciation can provide continuous stratigraphical sequences for the last 100,000 years or more (e.g., Wijmstra, 1969; Wijmstra and Smit, 1976; Woillard, 1978; Kershaw, 1978; Singh, Kershaw, and Clark, 1981). The sample resolution is usually between 100 and 250 years, but samples from annually laminated lake sediments can have a resolution as low as 10 years (e.g., Swain, 1973, 1978; Saarnisto, Huttunen, and Tolonen, 1977; Cwynar, 1978, Tolonen, 1978; Saarnisto, 1979a), or even single years (e.g., Peglar, Fritz, Alapieti, Saarnisto, and Birks, 1984). T. Webb (1982) presents a critical assessment of the temporal resolution of Quaternary pollen stratigraphical data.

Quaternary pollen analytical data can also be classified in a different way, depending upon the method of calculation of the pollen counts. The commonest are relative or percentage data, where the counts of a given pollen type (m_{ik} in Figure 1.1 and Tables 1.1 and 1.2) in a sample are expressed as proportions (p_{ik} in Figure 1.1 and Tables 1.3 and 1.4) of some specified pollen sum. This conversion to percentages or proportions (see Section 1.4) removes the effects due to the fact that the samples differ in their total number of pollen grains. It does, however, introduce a constraint into the data which is not present in nature; namely, that when one pollen type increases in relative frequency, some other type(s) must decrease because the sum of all proportions must always equal 1 (Fagerlind, 1952). One result of this interdependence is that the expected correlation between two pollen types in a percentage data set is negative (see Sections 2.4 and 6.5) rather than 0, and this may obscure real patterns among the different pollen types. Percentage representation of pollen analytical data can also lead to serious problems of interpretation (M. B. Davis, 1963; M. B. Davis and Deevey, 1964), particularly when tree pollen (e.g., pine or birch) that is wind dispersed over long distances dominates the pollen rain in areas of very low pollen production, such as the arctic tundra today. When calculated on a percentage basis, the pollen rain is dominated by tree pollen, even though the actual pollen deposition of both long-distance-transported tree pollen and herb and dwarf-shrub pollen from the local tundra is low. Thus, although the *ratios* of pollen types in tundra situations today may be the same as in northern forests, the absolute amounts of pollen deposited

Table 1.1

Pollen Counts m_{ik} for the Abernethy Forest 1974 Test Data Set (Test Data Set A)

Sample designation[a] i	Depth (cm)	Sample no. in complete sequence	Number of grains belonging to each pollen type										Local pollen zone[b]
			Betula ($k = 1$)	*Pinus* ($k = 2$)	*Corylus/ Myrica* ($k = 3$)	*Juniperus* ($k = 4$)	*Empetrum* ($k = 5$)	Gramineae ($k = 6$)	Cyperaceae ($k = 7$)	*Artemisia* ($k = 8$)	*Rumex acetosa*–type ($k = 9$)	Pollen sum	
A1	325	6	59	425	12	0	0	2	3	0	0	501	AFP-6
A2	350	11	175	317	37	2	0	11	3	0	0	545	AFP-6
A3	375	20	365	9	150	0	4	19	5	0	0	552	AFP-5
A4	420	24	390	12	98	8	3	14	5	0	1	531	AFP-5
A5	445	29	394	13	12	64	11	34	13	0	5	546	AFP-4
A6	455	31	311	4	1	132	31	78	12	1	1	571	AFP-4
A7	490	38	18	7	1	1	10	13	18	214	11	293	AFP-3
A8	500	39	23	11	1	0	8	19	14	130	12	218	AFP-3
A9	520	43	183	6	4	13	200	69	47	8	26	556	AFP-2
A10	530	45	100	1	1	0	190	81	36	6	97	512	AFP-2

[a] The 'A' refers to the test data set; the number refers to the position in the test data set.
[b] After H. H. Birks and Mathewes (1978).

Table 1.2

Pollen Counts m_{ik} for the Abernethy Forest 1970 Test Data Set (Test Data Set B)

			Number of grains belonging to each pollen type										
Sample designation i	Depth (cm)	Sample no. in complete sequence	*Betula* ($k = 1$)	*Pinus* ($k = 2$)	*Corylus*/*Myrica* ($k = 3$)	*Juniperus* ($k = 4$)	*Empetrum* ($k = 5$)	Gramineae ($k = 6$)	Cyperaceae ($k = 7$)	*Artemisia* ($k = 8$)	*Rumex acetosa*–type ($k = 9$)	Pollen sum	Local pollen zone[a]
B1	110	7	23	105	34	0	0	1	6	0	0	169	AF-4
B2	270	16	18	130	37	0	0	2	3	0	0	190	AF-4
B3	390	25	116	28	119	1	1	11	2	0	2	280	AF-3
B4	430	29	188	7	98	33	5	19	11	0	3	364	AF-3
B5	460	32	230	20	79	178	38	56	32	0	6	639	AF-2
B6	500	36	56	89	26	0	29	39	26	1	33	299	AF-1
B7	530	39	89	9	7	2	196	76	45	11	73	508	AF-1

[a] After H. H. Birks (1970).

Table 1.3

Sample Pollen Proportions p_{ik} for the Abernethy Forest 1974 Test Data Set (Test Data Set A)

Sample designation i	Depth (cm)	Proportions of grains belonging to each pollen type									Local pollen zone[a]
		Betula ($k = 1$)	*Pinus* ($k = 2$)	*Corylus/ Myrica* ($k = 3$)	*Juniperus* ($k = 4$)	*Empetrum* ($k = 5$)	Gramineae ($k = 6$)	Cyperaceae ($k = 7$)	*Artemisia* ($k = 8$)	*Rumex acetosa*–type ($k = 9$)	
A1	325	0.118	0.848	0.024	0	0	0.004	0.006	0	0	AFP-6
A2	350	0.321	0.582	0.068	0.004	0	0.020	0.006	0	0	AFP-6
A3	375	0.661	0.016	0.272	0	0.007	0.034	0.009	0	0	AFP-5
A4	425	0.734	0.023	0.185	0.015	0.006	0.026	0.009	0	0.002	AFP-5
A5	445	0.722	0.024	0.022	0.117	0.020	0.062	0.024	0	0.009	AFP-4
A6	455	0.545	0.007	0.002	0.231	0.054	0.137	0.021	0.002	0.002	AFP-4
A7	490	0.061	0.024	0.003	0.003	0.034	0.044	0.061	0.730	0.038	AFP-3
A8	500	0.106	0.050	0.005	0	0.037	0.087	0.064	0.596	0.055	AFP-3
A9	520	0.329	0.011	0.007	0.023	0.360	0.124	0.085	0.014	0.047	AFP-2
AJ	530	0.195	0.002	0.002	0	0.371	0.158	0.070	0.012	0.189	AFP-2

[a] After H. H. Birks and Mathewes (1978).

Table 1.4

Sample Pollen Proportions p_{ik} for the Abernethy Forest 1970 Test Data Set (Test Data Set B)

Sample designation i	Depth (cm)	Proportions of grains belonging to each pollen type									Local pollen zone[a]
		Betula ($k = 1$)	*Pinus* ($k = 2$)	*Corylus/Myrica* ($k = 3$)	*Juniperus* ($k = 4$)	*Empetrum* ($k = 5$)	Gramineae ($k = 6$)	Cyperaceae ($k = 7$)	*Artemisia* ($k = 8$)	*Rumex acetosa*–type ($k = 9$)	
B1	110	0.136	0.621	0.201	0	0	0.006	0.036	0	0	AF-4
B2	270	0.095	0.684	0.195	0	0	0.011	0.016	0	0	AF-4
B3	390	0.414	0.100	0.425	0.004	0.004	0.039	0.007	0	0.007	AF-3
B4	430	0.516	0.019	0.269	0.091	0.014	0.052	0.030	0	0.008	AF-3
B5	460	0.360	0.031	0.124	0.279	0.059	0.088	0.050	0	0.009	AF-2
B6	500	0.187	0.298	0.087	0	0.097	0.130	0.087	0.003	0.110	AF-1
B7	530	0.175	0.018	0.014	0.004	0.386	0.150	0.089	0.022	0.144	AF-1

[a] After H. H. Birks (1970).

annually, particularly of trees, are very different (see Aario, 1940; M. B. Davis, 1969b; H. J. B. Birks, 1973a; I. C. Prentice, 1978).

With the advent of radiocarbon dating and other independent chronological tools, pollen analysts have attempted to overcome the problems of relative data by developing a second type of presentation, so-called absolute pollen frequency. This estimates annual deposition of a taxon independently of all other pollen types and independently of changing rates of sediment accumulation. As Colinvaux (1978) has emphasised, the terms *absolute pollen frequencies* and *absolute pollen diagrams* are ambiguous because they are used to refer both to pollen concentrations and pollen-accumulation rates. M. B. Davis (1969a) defines pollen concentrations as 'numbers of grains per unit volume of wet sediment' in grains cm^{-3}. Sometimes pollen concentrations are presented as grains per unit weight of sediment in grains gm^{-1}. Davis defines pollen-accumulation rates as 'net number of grains accumulated per unit area of sediment surface per unit time'. Pollen-accumulation rates are also referred to as pollen-deposition rates or pollen influx in units of grains cm^{-2} $year^{-1}$. Thompson (1980) has discussed the use of the term *influx* in Quaternary pollen analysis and suggests that it is an inappropriate term. He emphasises that what pollen analysts refer to as influx, namely, numbers of grains which have accumulated per unit area per unit time, is flux density. In this book we use the term *pollen-accumulation rate* (M. B. Davis and Deevey, 1964) in place of the term *pollen influx*.

Pollen concentrations are independent of changes in other pollen types but are not independent of changes in sediment-accumulation rates. Pollen-accumulation rates are independent of other pollen types and changes in sediment-accumulation rates.

Pollen concentrations can be estimated by a variety of methods (see H. J. B. Birks and H. H. Birks, 1980). Direct estimation involves counting all pollen and spores in a subsample, or aliquot, of known volume or weight. By knowing the fraction that the subsample represents of the original sample, pollen concentrations can be estimated directly (M. B. Davis, 1965a; Jørgensen, 1967). Indirect estimates can be derived by adding a known amount of an exotic pollen or spore type (a type that is morphologically distinctive from all pollen and spore types indigenous to the sample of which the pollen concentration is to be estimated) to a known volume or weight of sample. The exotic pollen (e.g., *Nyssa, Ailanthus, Eucalyptus*) can be added either as a known volume of a suspension whose exotic pollen concentration has been previously estimated using a haemocytometer or a Coulter counter (Benninghoff, 1962; Matthews, 1969; Maher, 1972a; Bonny, 1972) or as tablets containing a known number of exotic grains (Stockmarr, 1971, 1972; Maher, 1980, 1981). After sample preparation, the indigenous and exotic pollen are counted, and the concentration of the indigenous pollen is estimated by simple proportion (Maher, 1972a; H. J. B. Birks and H. H. Birks, 1980). Both the direct and indirect approaches have experimental errors associated with them (see M. B. Davis, 1965a; Maher, 1972a, 1977, 1980, 1981; Bonny, 1972; Peck, 1974; H. J. B. Birks, 1976b; Regal and Cushing, 1979). These errors are proportionately considerably larger than the inherent errors associated with estimating pollen proportions in the same sample (Maher, 1972a, 1972b, 1977, 1980).

To convert pollen concentrations to estimates of pollen-accumulation rates, it is necessary to estimate sediment deposition-times (M. B. Davis, 1969a, defines this as the 'amount of time per unit thickness of sediment' in years cm^{-1}; see also Waddington, 1969) or its reciprocal, sediment matrix accumulation rate (Davis defines this as 'net thickness of sediment accumulated per unit time, after compaction and diagenesis' in cm $year^{-1}$). Deposition times are derived most commonly by obtaining a series of radiocarbon dates at different depths through the sediment sequence and estimating the deposition times from the radiocarbon age–depth relationship, either by linear interpolation between radiocarbon dates (e.g., H. J. B. Birks, 1976b; H. H. Birks and Mathewes, 1978) or by some curve-fitting procedure such as first- or second-order polynomials (e.g., M. B. Davis, 1967b, 1969a; Maher, 1972a; Brubaker, 1975; Likens and Davis, 1975; Cwynar, 1982). Counts of the number of annual laminations between samples from laminated sediments can also be used to provide estimates of sediment deposition-times (e.g., Craig, 1972; Swain, 1973; Cwynar, 1978; Tolonen, 1978; Huttunen, 1980) and hence of pollen-accumulation rates. There are several errors associated with estimating deposition times, particularly from radiocarbon dates (M. B. Davis, 1969a; Maher, 1972a, 1981; H. J. B. Birks and H. H. Birks, 1980). These errors, coupled with the laboratory and counting errors in estimating pollen concentrations, result in estimates of pollen-accumulation rates having proportionately a considerably higher variance than either pollen-percentage or pollen-concentration data (Maher, 1972a).

Besides the errors associated with estimating pollen-accumulation rates, there is considerable inherent variability in the observed pollen-accumulation rates within a lake due to the complex and poorly understood processes of sedimentation, sediment resuspension and redeposition, and sediment focussing (M. B. Davis, Brubaker, and Webb, 1973; Lehman, 1975; M. B. Davis and Ford, 1982). M. B. Davis *et al.* (1973) suggest from their studies on pollen-accumulation rates estimated from different cores collected from the same lake that there is at least a three-fold inherent variability in pollen-accumulation rates within a lake due solely to variations in sedimentation patterns and processes (see also H. J. B. Birks, 1976b, 1981a; Bennett, 1983b).

Pollen-accumulation rates are rarely estimated for modern, non-stratigraphical samples, because of the very considerable problems in estimating sediment deposition-times for recent surface samples. Some modern surface pollen data have been presented as pollen concentrations (e.g., Traverse and Ginsburg, 1967; McAndrews and Power, 1973; R. B. Davis and Webb, 1975; Prentice, 1978; Lamb, 1984) or as pollen-accumulation rates (e.g., M. B. Davis *et al.,* 1973). In addition, counts of pollen deposited in sediment or air traps over a known period of time can provide measures of annual pollen deposition (e.g., Ritchie and Lichti-Federovich, 1967; M. B. Davis, 1967c; Tauber, 1967, 1977; Peck, 1973; Berglund, 1973; Andersen, 1974; Bonny, 1976, 1978, 1980).

Most stratigraphical and modern, non-stratigraphical data are presented as relative percentages because of the difficulties and errors inherent in the methods for estimating absolute pollen frequencies. Some stratigraphical data are presented as estimates of pollen concentrations or pollen-accumulation rates. Only very rarely are modern data presented as pollen concentrations or pollen-accumulation rates. The emphasis in this book is thus on numerical methods applicable to the analysis of relative pollen analytical data, although many of the methods we discuss can also be used with pollen concentrations or pollen-accumulation rates.

1.3 Methods of Presenting Quaternary Pollen Analytical Data

Pollen analytical data, both stratigraphical and from surface-sample surveys, can be presented in a variety of ways. The basic counts of numbers of grains counted per sample (m_{ik} in Figure 1.1) can be tabulated (as, e.g., in Tables 1.1 and 1.2). Such primary data are essential for any subsequent numerical analysis of the data, such as calculating them as proportions of various pollen sums or subjecting the data to numerical zonation procedures. Primary data are, however, rarely published, but some authors make their data available to interested persons either on microfilm or microfiche from a data depository or as copies of the tabulated data.

Tabulations of relative pollen percentages or proportions, of pollen concentrations, and of pollen-accumulation rates are also rarely published. Such tabulations commonly have up to 100 columns (pollen types) and 100 rows (samples) and are too unwieldy to be of any real use to the reader.

Stratigraphical pollen analytical data are most commonly presented as so-called pollen diagrams (von Post, 1967; Fries, 1967; Faegri and Iversen, 1975) in which the vertical axis represents depth or, more rarely, age as determined by estimating the age of each sample from a series of radiocarbon dates; see M. B. Davis (1969a), Brubaker (1975), and H. H. Birks and Mathewes (1978) for examples of pollen diagrams constructed on an age basis. In some instances where the sediments are annually laminated, pollen data can be presented on an age basis derived directly from counts of the laminations (e.g., Swain, 1973, 1978; McAndrews, 1976; Cwynar, 1978; Huttunen, 1980).

The pollen data, expressed as percentages of some specified pollen sum (e.g., total dry-land pollen, total tree pollen), as pollen concentrations, or as pollen-accumulation rates, are plotted for each depth from which samples have been analysed palynologically. The plot may be in the form of a resolved diagram (*sensu* Faegri and Iversen, 1975) in which each pollen type is represented by an individual, non-overlapping curve (see Figures 1.2 and 1.3). The plotted points may be joined up to give a frequency polygon (as in Figures 1.2 and 1.3), so-called silhouette or saw-edge diagrams, or they may be drawn as bar histograms (e.g.,

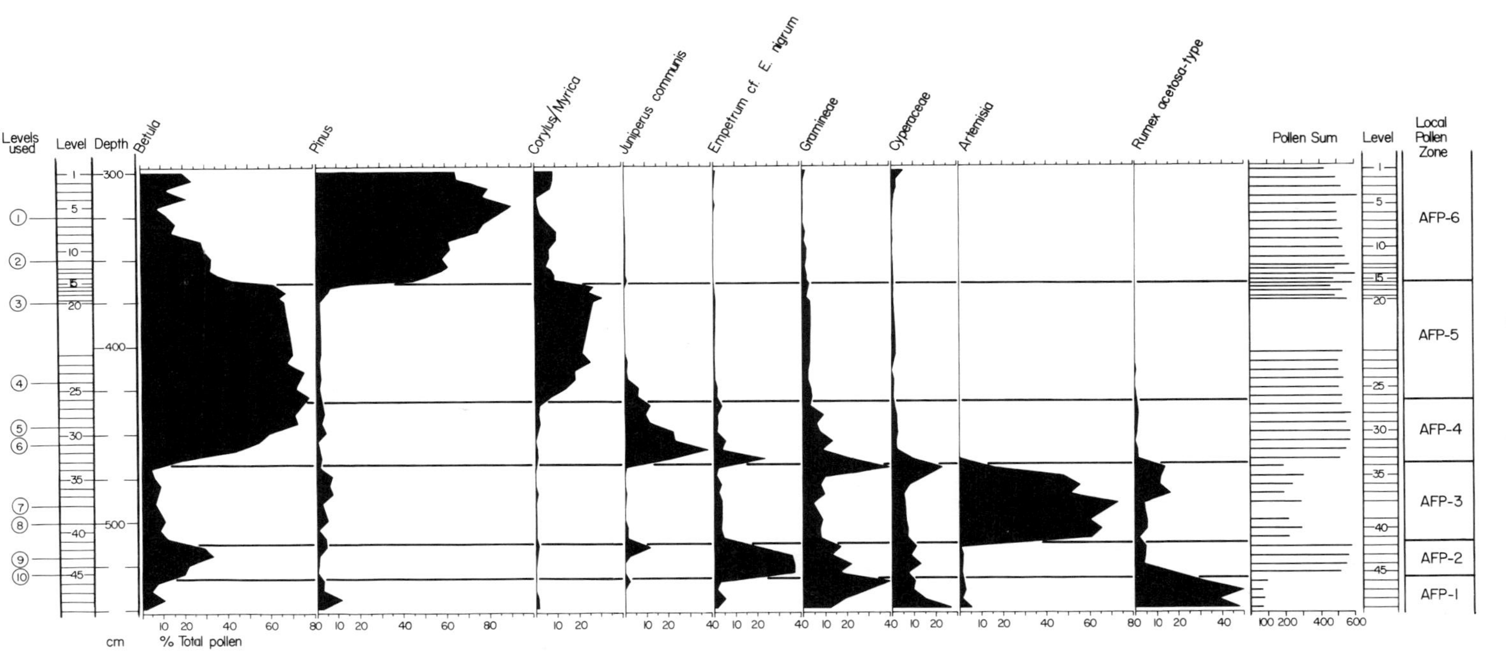

Figure 1.2 Percentage pollen diagram for the Abernethy Forest 1974 data set (nine taxa, 49 samples). The curves represent the amounts of the nine taxa expressed as percentages of the sum of these nine types. The size of the pollen sum, the local pollen zones, and the position of the 10 samples used in the Abernethy Forest 1974 test data set (test data set A) are also shown. (Data from H. H. Birks and Mathewes, 1978.)

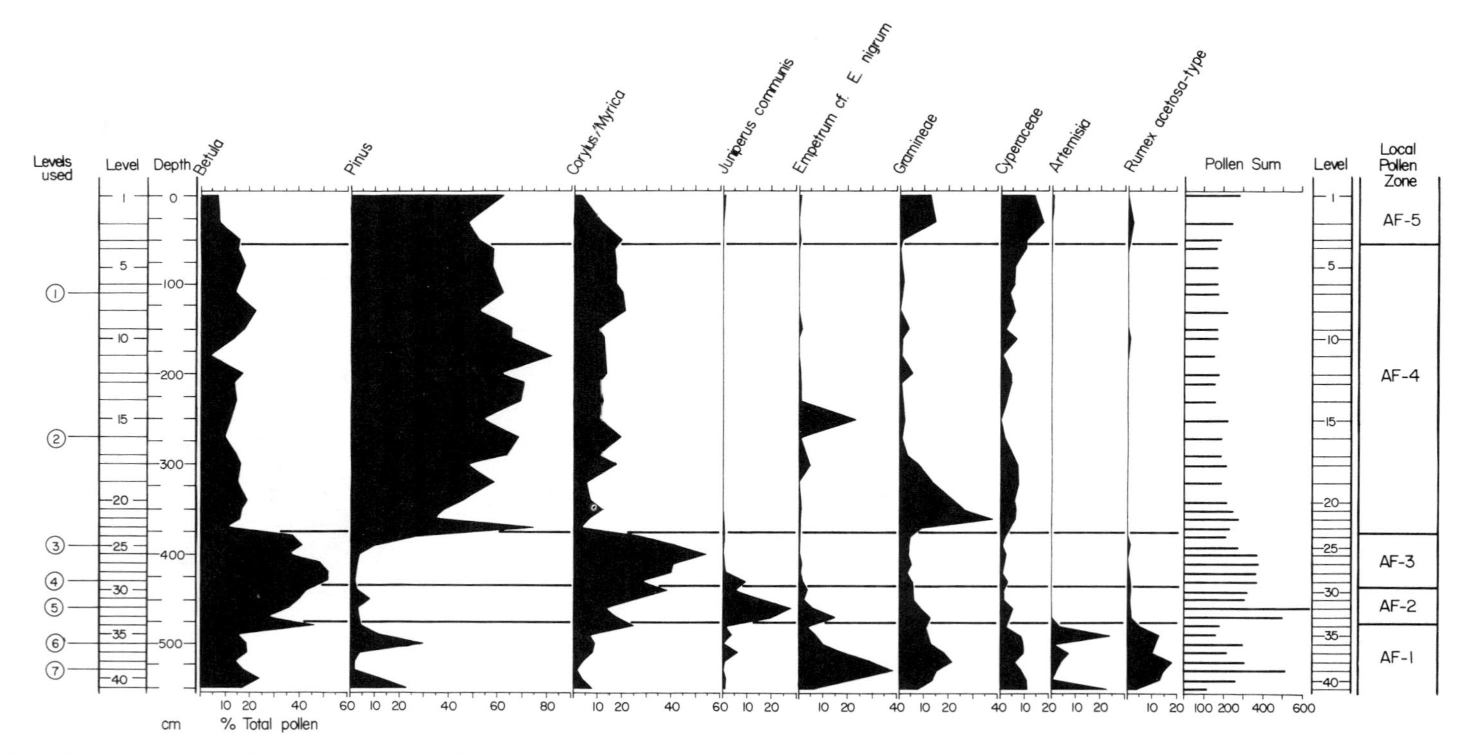

Figure 1.3 Percentage pollen diagram for the Abernethy Forest 1970 data set (nine taxa, 41 samples). The curves represent the amounts of the nine taxa expressed as percentages of the sum of these nine types. The size of the pollen sum, the local pollen zones, and the position of the seven samples used in the Abernethy Forest 1970 test data set (test data set B) are also shown. (Data from H. H. Birks, 1970.)

Turner 1965; Pennington and Bonny 1970; J. A. Webb and Moore, 1982). European pollen analysts often condense their diagrams by plotting overlapping frequency polygons, each with a standard symbol (e.g., X for *Abies,* ● for *Pinus,* ○ for *Betula,* — for Quercus, · · · for *Ulmus;* see Faegri and Iversen, 1975) to form a so-called composite diagram.

Composite diagrams are often difficult to follow, particularly if the curves for many taxa are plotted together. North American and British pollen analysts generally favour resolved diagrams, as these are easy to follow because there is no overlapping or crowding of curves. The one disadvantage of resolved diagrams is that, if many pollen and spore types have been identified and all their frequencies are plotted, the diagram can become very long (e.g., H. J. B. Birks, 1976b). Horizontal guide lines are required in these cases to help the reader trace changes across the diagram. Histogram plots on resolved diagrams provide the most accurate representation of the data, but unless the samples are close together, the diagram can be difficult to follow, particularly for pollen types that only occur sporadically.

Within the two main types of pollen diagrams, composite and resolved, there are many different styles of presentation. The most important feature of any pollen diagram is that the data, which are invariably complex, should be presented in a clear, precise, and unambiguous way. Considerable care is thus required in the design and drafting of pollen diagrams (see Wright, 1974).

The vast majority of stratigraphical pollen diagrams show the pollen percentages, concentrations, or accumulation rates estimated for the levels from which samples have been analysed. For particular palaeoecological problems, pollen curves can be transformed to allow for differential pollen production and representation of different plants by means of R values or other pollen-representation factors (see discussion of Figure 1.1). Examples of pollen diagrams based on transformed data are given by Iversen (1949, 1969), Donner (1972), and Andersen (1973, 1975). Such transformations can, in many instances, provide more valuable estimates of past vegetational composition (f_k in Figure 1.1) than are given by the untransformed pollen data. The derivation and use of such pollen-representation factors are discussed in Sections 5.4 and 6.3.

Pollen analysts occasionally present pollen curves that have been subjected to some form of stratigraphical smoothing or filtering, such as a three-term or five-term moving average (Brenner, 1974; Ritchie, 1982) or a weighted three-term moving average (Usinger, 1978a, 1978b); in this latter, the value plotted at depth i for taxon k is

$$\frac{p_{(i-1),k} + 2p_{ik} + p_{(i+1),k}}{4}$$

Many techniques have been proposed for smoothing stratigraphical data, but they are all moving averages of some sort (see J. C. Davis, 1973, pp. 227–230). The various techniques differ in the relative weight attached to the observed value at depth i and the decreasing weights assigned to depths $i - 1$, $i - 2$, $i + 1$, $i + 2$, and so forth. Smoothing of pollen concentration data or pollen-accumulation rates is often useful (see Ritchie, 1982) because such data have a higher variability than percentage data. The smoothing of stratigraphical data should generally be avoided in the presentation of new, primary data. Borrowing terminology from communication theory, one can describe the usual aim of smoothing as the removal of 'noise' from the data with minimal disturbance of the 'signal'. To do this, one requires *a priori* knowledge of the structure of the data, and this knowledge is rarely available. Smoothing can be very useful, however, for generalised, schematic pollen diagrams (e.g., Iversen, 1973; West 1977; H. J. B. Birks, 1977a) in which selected pollen types are plotted to illustrate particular changes and trends in pollen stratigraphy.

An alternative approach, which also emphasises the variability inherent in pollen data, is to include in the diagram an indication of the precision of estimation (see Section 2.2), for example, by enclosing each datum point within a confidence interval (e.g., Pennington and Bonny, 1970; Maher, 1972a, 1972b; Pennington, 1973; Adam, 1975; Shackleton, 1982; J. A. Webb and Moore, 1982).

Other methods of presenting pollen stratigraphical data include survey diagrams (*sensu* Berglund, 1979b) in which pollen types are grouped into ecological, phytosociological, or phytogeographical categories and composite curves for these groups are plotted stratigraphically to illustrate particular palaeoecological or plant geographical changes. For example, Andersen (1964, 1966, 1969) groups pollen and spore types into plants intolerant of competition and/or shade; plants characteristic of fertile, mull soils; and plants tolerant of acid, mor humus and peats (see also Berglund, 1966). H. J. B. Birks (1973b) groups pollen types into woodland, grassland, tall-herb, shrub–heath, montane–heath, spring, chionophilous, and chionophobous vegetation types.

Stratigraphical plots of mean pollen values for zones can also be constructed (Maher, 1982) and can serve as survey diagrams. The disadvantage of zone mean values is that information about variability within a zone and about stratigraphical trends is lost. All survey diagrams should be clearly designed for specific purposes to show particular features of the data that may otherwise not be readily apparent in the full pollen diagram.

If several pollen diagrams are available from a specific geographical area, and if the investigator is interested not only in temporal changes in pollen composition but also in spatial patterns of variation in pollen composition and, by inference, in past vegetation, circular or 'clock-face' diagrams (e.g., Godwin, 1940; Ritchie, 1976) can be drawn. In these, the proportions of different pollen types in a particular time interval are drawn as segments of a circle, the angle of each segment being proportional to the percentage of the particular pollen type. Unless the composition of the pollen assemblage is very simple, these diagrams can be difficult to comprehend and the geographical patterns difficult to discern. Faegri and Iversen (1975) recommend the use of so-called resolved maps for individual taxa in preference to circular diagrams. In a resolved map, the proportion of the pollen type at the time of interest is shown by the size of circle; the higher the proportion, the larger the circle (e.g., Berglund, 1966).

A second approach to the display of spatial patterns in pollen assemblages is to construct maps for individual pollen types at selected time intervals and to contour the pollen values, so-called isopolls (Szafer, 1935). Such isopollen maps can be constructed on a range of geographical scales, ranging from continents (Huntley and Birks, 1983), to sub-continents (Bernabo and Webb, 1977) to countries (Firbas, 1949; H. J. B. Birks, Deacon, and Peglar, 1975; H. J. B. Birks and Saarnisto, 1975; Ralska-Jasiewiczowa, 1983). Isopollen maps are particularly useful in summarising large amounts of pollen stratigraphical data and in illustrating changes in pollen assemblages in both space and time. Such maps clearly require many well-dated data points and great care is required to ensure that only pollen spectra of comparable age are mapped (H. J. B. Birks and Saarnisto, 1975; Huntley and Birks, 1983). Pollen data can be mapped onto topographical or geographical maps to display patterns of variation in pollen composition through time in relation to landscape features such as topography, elevation, and soils (e.g., T. Webb, Richard, and Mott, 1983). The data can also be analysed numerically to display the major patterns of variation in pollen composition (see Section 5.3) and thus in past vegetation (e.g., H. J. B. Birks, Deacon, and Peglar, 1975; Huntley and Birks, 1983). It can be difficult to retrieve the original data from isopollen maps; indeed, many contouring procedures provide some spatial smoothing (Webb, Laseski, and Bernabo, 1978). Isopollen maps are thus unsuitable for presenting primary pollen stratigraphical data.

Surface-sample data are most commonly presented as resolved bar-histogram diagrams with the pollen spectra grouped together according to the vegetation type (e.g., Bent and Wright 1963; H. J. B. Birks, 1973a, 1973b, 1977b, 1980; O'Sullivan, 1973; Ritchie, 1974; A. M. Davis, 1980) or broad geographical region (e.g., M. B. Davis, 1967a; Wright, McAndrews, and Van Zeist, 1967; Lichti-Federovich and Ritchie, 1968; I. C. Prentice, 1978) from which the surface samples were collected. When the data set is large, mean pollen percentages for surface samples for particular vegetation types or regions can be calculated and plotted (e.g., Ritchie, 1967; H. J. B. Birks, 1973b, 1980; Birks, Webb, and Berti, 1975; Janssen, 1981a). This type of presentation is open to the same criticism as the calculation of means for pollen zones, namely, that the palynological variation within a vegetation type or a region is ignored and the existence of any spatial trends within the data is concealed. Of course, it may be a useful exercise initially to employ such summaries to allow the main features of a data set to be perceived more clearly, and then to undertake a more detailed examination within each group; the general usefulness of this approach will depend on the context of the investigation.

Surface samples collected along transects that pass through different vegetation types can be presented as pollen-transect diagrams (e.g., Janssen, 1966, 1973, 1981a; McAndrews and Wright, 1969; Andersen, 1970; Tinsley and Smith, 1974;

Tamboer-Van den Heuvel and Janssen, 1976; Caseldine and Gordon, 1978; Caseldine, 1981) in which the pollen percentages are plotted on the ordinate and distances along the transect on the abscissa. Variations in vegetation can be superimposed on the diagrams (e.g., Heim, 1962, 1970; Janssen, 1966, 1973). Such diagrams are particularly valuable in studies of local pollen deposition in relation to fine-scale variation in vegetation (see Janssen, 1966, 1973, 1981a; Tamboer-Van den Heuvel and Janssen, 1976; Caseldine, 1981).

Geographical patterns in pollen composition of surface samples collected from lakes or bogs that primarily sample the regional pollen rain can conveniently be presented as isopollen maps (e.g., T. Webb, 1974a, 1974c; H. J. B. Birks, Webb, and Berti, 1975; R. B. Davis and Webb, 1975; T. Webb and McAndrews, 1976; T. Webb, Laseski, and Bernabo, 1978; T. Webb, Yeracaris, and Richard, 1978; T. Webb *et al.,* 1983; I. C. Prentice, 1978, 1983a; Lamb, 1982, 1984; Huntley and Birks, 1983; P. A. Delcourt, H. R. Delcourt, and Davidson, 1983). The isopollen maps can be superimposed onto present-day vegetation maps to illustrate broad-scale pollen–vegetation relationships (e.g., H. J. B. Birks, Webb, and Berti, 1975; R. B. Davis and Webb, 1975; T. Webb and McAndrews, 1976), or onto topographical maps to display pollen–landform relationships (T. Webb *et al.,* 1983). Isopollen maps can also be compared with maps showing the percentages or absolute abundance of plants (generally trees) in the contemporary vegetation (e.g., T. Webb, 1974a; Prentice, 1978; P. A. Delcourt *et al.,* 1983). As with pollen stratigraphical data, surface-sample data can be analysed numerically, and a variety of suitable methods is discussed in Sections 5.3 and 6.4.

Faegri and Iversen (1975) and Berglund (1979b) discuss in detail various standard conventions in the design, construction, and drafting of pollen diagrams for both pollen stratigraphical and surface-sample data.

An alternative way of representing pollen analytical data is to consider the samples, either fossil samples from a core or modern surface samples, as a set of points in a multidimensional space in which each axis represents a single pollen type. If there are n samples and t pollen types, the data can be represented as a set of n points in a t-dimensional space. Within this space, the structure of the data is represented by the patterns of the sample points. Samples which are similar in pollen composition will be positioned close together in the space, whereas samples which are dissimilar will be positioned far apart. Spatial relationships between samples will be lost unless specified by labels on the points. Stratigraphical relationships will also be lost unless the points are labelled or joined up in stratigraphical order.

A multidimensional space is impossible to visualise if there are more than three dimensions, but there exists a variety of numerical methods of analysis which represent the samples by a set of points in a low-dimensional space (usually two, less commonly three, dimensions are used), which preserves the main features of the structure in the pollen data. Such methods are discussed further in Section 2.5 and are used extensively throughout this book.

In the past 10 years, with the development of high-speed electronic computers and the increased availability of fast axillary graph plotters, several computer programs have been written to assist in the calculation of pollen analytical data and the drawing of pollen diagrams (e.g., Squires, 1970; Squires and Holder, 1970; Damblon and Schumacker, 1971; Dodson, 1972; Voorrips, 1973, 1974; King, 1976; H. J. B. Birks and Huntley, 1978; Eccles, Hickey, and Nichols, 1979; Cushing, 1979; Veldkamp, Hogen, and Van der Woude, 1981; Vuorinen and Huttunen, 1981). The aim of all these programs is to read in raw pollen counts, express them as proportions of some user-specified pollen sum, write out these values, and draw a pollen diagram of the results.

A versatile FORTRAN IV program for the basic handling of pollen analytical data is POLLDATA.MK5 (H. J. B. Birks and Huntley, 1978). The program reads in the basic pollen counts, sorts the samples into stratigraphical or any other specified order, sorts the pollen and spore types into categories and into any order specified by the user, deletes any samples or pollen types if required by the user, calculates pollen proportions on the basis of a specified pollen sum, draws preliminary pollen diagrams on a line printer, and plots pollen diagrams of various sizes on a CALCOMP or BENSON graph-plotter with either a depth or age axis. The program also includes options to calculate 95% confidence intervals for percentage data (see Section 2.2), to calculate pollen concentration values and/or pollen-accumulation rates, to transform the pollen counts by means of modern representation factors, and to smooth concentration data or pollen-accumulation rates by moving averages, the number of terms in which is specified by the user. In addition there is a variety of plotting options (joined-up silhouette or saw-edge curves, bar histograms, etc.). The program also has facilities for handling plant-macrofossil data.

The program POLDATA written by Cushing (1979) has many sophisticated options for the plotting of pollen curves and for the drawing of sediment lithologies using the system of sediment description and symbols defined by Troels-Smith (1955). The program uses a Varian Statos 31 electrostatic plotter.

Programs such as POLLDATA and POLDATA are inevitably installation-dependent, as the plotting sub-routines use local plotter procedures that are often exclusive to particular computer centres. Because POLLDATA.MK5 is heavily dependent on local plotter procedures, the program is not listed here, but listings are available on request to H. J. B. Birks.

1.4 The Data Used

Throughout Chapters 3 and 4, and in parts of Chapters 2 and 6, we use two standard sets of pollen stratigraphical data to assist in the exposition of various numerical methods and to illustrate the application of these techniques to actual data. These two test data sets are subsets of two large sets of pollen stratigraphical data that we also use to illustrate particular numerical analyses with more complex data.

The large data sets are from stratigraphical pollen analyses of cores of peat, lake muds, silts, and sands from a channel bog within Abernethy Forest, Inverness-shire, on the northern side of the Cairngorms in the pine forest region of the eastern Scottish Highlands. Hilary H. Birks (1970) prepared a long pollen stratigraphical sequence covering the post-glacial (Flandrian or Holocene) and the Devensian (Weichselian) late-glacial. The data from this study are here called the *Abernethy Forest 1970 data set*. In 1974 H. H. Birks recored the site, and she and R. W. Mathewes studied the new cores in greater detail than in her initial study. In 1978 they published the results of their new investigations involving pollen percentages, radiocarbon dating, pollen-accumulation rates, and plant-macrofossil analysis (H. H. Birks and Mathewes, 1978). The percentage pollen stratigraphical data from their study form the *Abernethy Forest 1974 data set*. The 1970 and 1974 data comprise 41 and 49 samples and 65 and 88 pollen and spore taxa, respectively. These data are too large to present here, but they are available on request. The percentages of the nine quantitatively most important pollen taxa included by H. H. Birks and Mathewes (1978) in their pollen sum are plotted stratigraphically for the 1974 and 1970 data in Figures 1.2 and 1.3. In addition, the size of the pollen counts for each level is shown on Figures 1.2 and 1.3, along with the local pollen zones proposed by H. H. Birks and Mathewes (1978) and H. H. Birks (1970). Details of the ecological setting and sediment stratigraphy of the site, of the field and laboratory methods used, and of the palaeoecological interpretation of the two profiles are presented in full by H. H. Birks and Mathewes (1978) and H. H. Birks (1970).

The two subsets of these data which were selected are called the *Abernethy Forest 1974 test data set* (or *test data set A*) and the *Abernethy Forest 1970 test data set* (or *test data set B*). Set A comprises 10 samples of the 1974 data set, and set B comprises 7 samples of the 1970 data set. The positions of the samples within the two full stratigraphical sequences are shown at the extreme left of Figures 1.2 and 1.3.

The test data sets are presented in Tables 1.1 and 1.2 as raw pollen counts m_{ik}, where m_{ik} denotes the number of grains in sample i which were identified as belonging to pollen type k. For example, $m_{A1,2} = 425$, since 425 of the grains in sample A1 belong to type 2, *Pinus* (see Table 1.1). The stratigraphical details of the samples in the test data sets are also given, as is the identity of the local pollen zone from which the individual samples were derived. The samples in the two test data sets were deliberately chosen to illustrate the range of pollen stratigraphical variation within the two complete sequences (see Figures 1.2 and 1.3); it was not the intention to select random subsets of the 1970 and 1974 data.

In Tables 1.3 and 1.4, the test data are presented as proportions; p_{ik} denotes the proportion of grains in sample i which belong to pollen type k, only the nine pollen types specified in Tables 1.1–1.4 being included in the pollen sum. For example, there is a total of 501 pollen grains of these nine types in sample A1, and so the proportion of sample A1 which is composed of *Pinus* pollen is $p_{A1,2} = 425/501 = 0.848$, to three decimal places (see Table 1.3).

CHAPTER 2

Basic Statistical Concepts

2.1 The Rôle of Statistics in Pollen Analysis

In recent years there has been an increasing use of quantitative methods in the biological and geological sciences. Quaternary pollen analysis has always had a quantitative basis (von Post, 1918, 1967) in that the primary data are invariably recorded in numerical form as the numbers of different types of pollen grains and spores counted in a sample. There has recently been both a growing awareness of the possibility that statistical methodology can contribute to Quaternary palynological investigations, and, at the same time, some scepticism about the size of that contribution (e.g., Faegri and Iversen, 1975; Ogden, 1977a, 1977b; P. D. Moore and Webb, 1978).

We welcome both views; we firmly believe that statistics has a contribution to make to both Quaternary palynology and palaeoecology, but we are aware that excessive claims for, and inappropriate or unnecessary use of, statistical techniques have in some cases retarded a fuller acceptance of the rôle of statistics in some biological and geological disciplines. This is regrettable; a solution is for research workers to maintain an inquisitive, but also constructively critical, attitude to the methods proposed.

Some statistical methods of analysis which we believe can be of assistance in certain situations are described in the remainder of this book. Some specific advantages and disadvantages that we see in these methods are outlined in the text, for example, in the final section of Chapter 3, but it might be useful to make some general comments here on the types of methods of analysis incorporated within the portmanteau term *statistics*. In discussing the types of approach in mathematical ecology, Pielou (1977) finds it helpful to consider three broad headings: (1) ecological model building, (2) the statistical study of populations and communities, and (3) quantitative descriptive ecology.

Palynology has seen comparatively little work in the first category, although some of the methods described in Chapters 5 and 6 can be regarded in this light. A. M. Solomon and Harrington (1979) present a review of model building in Quaternary palynology. A. M. Solomon, Delcourt, West, and Blasing (1980) have used pollen stratigraphical data as a means of testing an ecological simulation model of forest growth over a period of 16,000 years. A. M. Solomon, West, and J. A. Solomon (1981) extended their use of forest simulation models to study the separate and combined effects of climatic change and of species immigration upon forest history. They used pollen stratigraphical data to estimate past climatic conditions, and, not surprisingly, they found that the simulation that most closely resembled the fossil pollen sequence was the simulation in which climatic conditions were varied (cf. I. C. Prentice, 1983b). A. M. Solomon *et al.* (1981) did not, however, describe a potentially critical simulation experiment in which the vegetational effects are modelled of a single major climatic change at about 12,000 years ago and of subsequent lags in immigration of trees following this climatic change (cf. Hare, 1976; M. B. Davis, 1976, 1978).

D. G. Green (1983) used time series procedures (see Section 6.2) to investigate the consequences of sediment mixing, sampling intensity, and assumptions about the variability in annual amounts of pollen deposited, under several models of vegetational change. R. B. Davis (1974) investigated the effect that tubificids can have in mixing the uppermost layers of sediment, and obtained the expected age distribution of pollen grains at various depths under a specified model of vertical mixing of sediment.

Outside palynology, the only attempt at mathematical model-building in the context of Quaternary palaeoecology of which we are aware is the work by P. S. Martin (1973) and Mosimann and Martin (1975) on simulating the spread of prehistoric man in North America and the possible effects of early man on the American megafauna. The application of mathematical models to a variety of pre-Quaternary palaeoecological problems is discussed by Reyment (1967), Harbaugh and Bonham-Carter (1970), and Tipper (1980), and to a range of modern ecological problems by Jeffers (1978).

Some valuable general comments on mathematical modelling are made by Kac (1969), Hedgpeth (1977), and D. L. Solomon (1979). Kac (1969, p. 699) argues that the main rôle of models is 'to polarize thinking and to pose sharp questions'. In this context, computer simulations such as those by A. M. Solomon *et al.* (1980, 1981) can be helpful in presenting one with the logical consequences of ideas and assumptions which have had to be clearly specified. However, as Mosimann and Martin (1975, p. 313) stress, 'simulations can prove nothing about prehistory'; they can only indicate which models are unlikely to be adequate descriptions of conditions and processes in the past.

D. L. Solomon (1979) distinguishes two different motivations for modelling: (1) explanatory, in which one seeks to understand causal mechanisms, and (2) empirical, in which one might not have much biological understanding of the relationships between component parts of the model, but would still be able to make predictions about features being modelled. Ideally, one would like to have an explanatory model but most models contain empirical parts.

At present there would seem to be something of a reaction against ecological modelling using large-scale computer simulations, doubtless caused in part by excessive and inappropriate recourse to them in the recent past (McIntosh, 1980). Such methods of analysis are more likely to prove helpful when a facility with mathematical models and computing is combined with ecological insight, and we would recommend the discussion in Section 6.3 of Pielou (1977) as essential reading for all prospective palynological and palaeoecological model builders. More recently, Pielou (1981) has presented a very thoughtful discussion of the potential and limitations of ecological model building.

Pielou's (1977) third and second categories correspond roughly and inexactly to the distinction made by Tukey (1977) between, respectively, exploratory and confirmatory data analysis. In exploratory data analysis, one carries out an initial investigation of a data set without any explicit ideas about the structure of the data; the aim is to use a variety of summarising and manipulating techniques to discover what, if any, structure exists within the data. This initial phase of investigation can be useful in suggesting various hypotheses about the underlying process by which the data arose. Confirmatory data analysis is more concerned with testing such hypotheses (on new data sets), and with more structured investigations of data sets; it concentrates on specific properties. In that one can expect a set of data to possess many different characteristics, it is advisable not to attempt to fit it into the hypothesis-testing straitjacket too early in the analysis.

Sections 2.2 and 2.3 describe statistical distributions which are of use in palynology, and some examples of more formal statistical studies are outlined in Section 2.4. The bulk of the work described in this book may be described under the heading 'exploratory data analysis', but little attempt is made to formulate specific hypotheses which can be tested on further data; as will be seen, the data are sufficiently complex to suggest that it may be difficult to formulate realistic and useful null hypotheses which could lead to analytically tractable test procedures. Some general remarks on exploratory data analysis, and in particular on the body of methods known under the general heading of 'classification', are made in the last two sections of this chapter.

2.2 Description of the Binomial and Multinomial Distributions

This section introduces two statistical distributions which are fundamental to many quantitative palynological investigations. We consider an examination of the pollen grains contained in a specified volume of sediment. This volume may correspond to our estimate of the amount of sediment accumulation cm^{-2} in, say,

20 years at a particular epoch of time in the past; or it may be less precisely defined. M. B. Davis (1969b) has noted that it is not uncommon to find 500,000 pollen grains cm^{-3} of sediment, and higher estimates have been obtained for some sediments (e.g., H. J. B. Birks, 1976b). Clearly, it would be very laborious to count all the pollen grains in even such a small volume of sediment, and pollen analysts invariably take smaller samples. From their results, they wish to make statements not only about the grains in the counted sample, but also about the situation existing in the larger body of unexamined sediment, and this is where statistics enters.

Let us assume that the volume of sediment being considered contains a proportion u of arboreal pollen (AP) grains, and that it contains a proportion $(1 - u)$ of non-arboreal pollen (NAP) grains. The pollen analyst prepares a sample from this sediment in the manner outlined in Section 1.1, and starts counting the various pollen grains, identifying them as either AP or NAP. Let us suppose that the analyst continues counting until he or she has identified a total of n grains, this number being determined by criteria not depending on the results of the count; for example, n could be fixed beforehand at 1000, or as the total number of grains contained on a slide. (This is sometimes referred to as counting within the pollen sum.) Let the number of AP grains counted be denoted by M. The value of M can be anywhere between 0 and n, although some values will be more likely than others; we say that M is a random variable. If certain conditions (to be discussed in the next section) are fulfilled, namely, if the binomial model is valid, we can make the following statement about the probability that M takes the value m, or $\text{pr}(M = m)$:

$$\text{pr}(M = m) = \frac{n!}{m!(n - m)!} u^m(1 - u)^{n-m}, \qquad (2.2.1)$$

where this formula holds for m taking any integer value between 0 and n inclusive, and $n! \equiv \prod_{i=1}^{n} i = n(n - 1)(n - 2), \ldots, 3 \cdot 2 \cdot 1$.
For example, if $n = 5$ and $u = 0.8$,

$$\text{pr}(M = 3) = \frac{5 \cdot 4 \cdot 3 \cdot 2 \cdot 1}{(3 \cdot 2 \cdot 1)(2 \cdot 1)} (0.8)^3(0.2)^2 = 0.2048,$$

and the probabilities of each possible value of m in this case are:

m	0	1	2	3	4	5
$\text{pr}(M = m)$	0.00032	0.00640	0.05120	0.20480	0.40960	0.32768

The estimate of u obtained from the sample is given by the proportion of AP grains in the sample, m/n, and will be denoted by p (see Figure 1.1). It can be seen that the most likely value of m in the above example is 4, and so the most likely estimate is $p = \frac{4}{5} = 0.8$, the true value. However, with probability $(1 - 0.4096)$ this value will not be obtained; if we imagine repeating this experiment of identifying five grains many times, we would expect to get an estimate other than 0.8 over half the time.

The solution, of course, is to specify a larger value for n, and no pollen analyst would base inferences on a sample as small as five grains; several hundred or even thousands of grains are commonly counted. As the value of n increases, so the distribution of values for p becomes more closely concentrated about the true value u. However, the example does illustrate the point that in addition to giving our 'best' or 'point' estimate, p, of u, it is helpful to give an indication of how reliable that estimate is. This can be done by quoting the standard error associated with the estimate, or by specifying a confidence interval for u. The larger that an estimate is relative to its standard error, the more reliable the estimate. The standard error associated with the binomial estimate p is $[u(1 - u)/n]^{1/2}$; since u is not known, this is sometimes approximated by $[p(1 - p)/n]^{1/2}$ when n is large.

A more precise summary of reliability is given by a confidence interval. A (classical) confidence interval is a random interval which with specified probability will contain the unknown true value u. Approximate confidence intervals can readily be obtained by approximating the binomial distribution by a normal distribution (Mosimann, 1965; for a more complete description of confidence intervals, see Mood, Graybill, and Boes, 1974, Chapter VIII; for other approximations see N. L. Johnson and Kotz, 1969, Section 3.7). For example, an approximate 95% confidence interval for u is given by

$$\left(\frac{p + (1.92/n) - 1.96[p(1 - p)/n + (0.96/n^2)]^{1/2}}{1 + (3.84/n)}, \frac{p + (1.92/n) + 1.96[p(1 - p)/n + (0.96/n^2)]^{1/2}}{1 + (3.84/n)}\right) \quad (2.2.2)$$

Maher (1972b) gives nomograms for calculating confidence intervals from this relation.

The concept of a classical confidence interval can be elusive; we are not making a probability statement about u, which is a fixed, albeit unknown, parameter. If we imagine repeatedly taking a sample of n pollen grains and evaluating p, the proportion of grains in the sample which are AP, then each value of p will define an interval by substitution into the expression (2.2.2). We would expect 95% of the intervals constructed in this manner to contain u. In reality, we only take one sample of size n; the interval constructed from the value of p obtained may or may not contain u (we do not know whether or not it does), but we have 95% confidence, in the above sense, that the interval does contain u. The length of the interval gives an idea of the precision with which we have estimated u; it can be seen from expression (2.2.2) that for a fixed pollen proportion, the length of the confidence interval is essentially proportional to $n^{-1/2}$; in order to halve the length of the confidence interval, it is necessary to count about four times as many pollen grains. Maher (1980) gives an interesting discussion of the balance between increased precision and labour of extra counting.

As an illustrative example, we can consider sample A3 from the Abernethy Forest 1974 test data set (see Table 1.1). A total of $n = 552$ pollen grains were counted; of these, 365 were identified as *Betula* (and 187 as belonging to other taxa). The underlying proportion of *Betula* pollen in the sediment, u, is not known, but an estimate of it is $p = 365/552 = 0.661$, and, from expression (2.2.2), a 95% confidence interval for it is (0.621, 0.699); we have 95% confidence that u lies somewhere between 0.621 and 0.699.

Throughout the above description, we have considered only two different categories of pollen grain: AP or NAP; *Betula* or not-*Betula*. Pollen analysts are invariably interested in more than two taxa, and a generalisation of the binomial model is appropriate. We can consider the situation where there are t taxa, and the proportion of pollen grains in the volume of sediment which belong to the kth taxon is $u_k (k = 1, 2, \ldots, t)$. A total of n pollen grains is counted, n being specified as before (so that again counting is 'within the pollen sum'); let M_k denote the number of these grains which belong to the kth taxon ($k = 1, 2, \ldots, t$; clearly $\sum_{k=1}^{t} M_k = n$). Under the multinomial model, we can make the following probability statement about the random variables $M_1, M_2, \ldots, M_t$:

$$\text{pr}(M_1 = m_1, M_2 = m_2, \ldots, M_t = m_t) = n! \prod_{k=1}^{t} [u_k^{m_k}/(m_k!)], \qquad (2.2.3)$$

where $\sum_{k=1}^{t} m_k = n$.

It can be seen that the case $t = 2$ is just the binomial description given in Equation (2.2.1); indeed, if the t categories are combined into two categories in any way (as was done in the example considered above, in which the nine original taxa were summarised as *Betula*/not-*Betula*), then a binomial model holds.

The point estimate of u_k is m_k/n, denoted by p_k ($k = 1, 2, \ldots, t$). Ways of obtaining simultaneous approximate confidence intervals for the set of unknown parameters $\mathbf{u} \equiv (u_1, u_2, \ldots, u_t)$ are described by N. L. Johnson and Kotz (1969, Section 11.2), and some specific tests of hypotheses about $\mathbf{u}$ are given by Mosimann (1965). Earlier work on comparing two binomial parameters using an exact test is described in Westenberg (1947a, 1947b; see also Westenberg 1964, 1967).

The extent to which the assumptions of the binomial and multinomial models hold in palynology is considered in the next section. This section is concluded, however, with a brief description of two distributions related to the binomial and multinomial.

In the binomial and multinomial models, the total number of grains counted, n, was fixed using criteria not dependent on the results of the count. An alternative method would be to continue counting until a fixed number of grains of a specified taxon (labelled taxon number t, say) had been counted, recording the number of grains belonging to each of the other $(t - 1)$ taxa which had been identified when counting stopped. This is sometimes referred to as counting outside the pollen sum; relevant distributions for modelling the number of grains belonging to taxa other than the tth taxon are the negative binomial (if $t = 2$) and the negative multinomial (if $t > 2$); for further details, see Mosimann (1963, 1965).

2.3 The Binomial and Multinomial Distributions in Palynology

The previous section introduced the binomial and multinomial distributions. This section examines the extent to which palynological data can be adequately described by these models. An abstract representation of the multinomial model (and hence also the binomial model, as the special case when $t = 2$) would be as follows:

We carry out a sequence of n independent experiments. Each experiment has one of t mutually exclusive outcomes. For each experiment, the probability that the kth outcome occurs is u_k. After n experiments (where n does not depend on the outcome of the experiments), we note the number of times, M_k, that the kth outcome occurred ($k = 1, 2, \ldots, t$). The variables $M_1, M_2, \ldots, M_t$ are multinomial random variables.

In a palynological example, an 'experiment' corresponds to the identification of a single pollen grain on the slide, the 'kth outcome' to identifying that grain as belong to the kth taxon, and u_k denotes the probability that a grain will be identified as belonging to the kth taxon. Suppose that u_k represents the proportion of grains belonging to the kth taxon in some larger population of grains from which the sample is counted; the extent of this population will be discussed below. If $u_1, \ldots, u_t$ are taken to refer to the proportions of grains in some volume of underlying sediment, the assumption has to be made that the chemical preparation of the samples does not alter these proportions, for example, by differential destruction of some taxa. If $u_1, \ldots, u_t$ are taken to refer to the proportions of grains at the time when the pollen rain was being deposited, similar assumptions have also to be made for the time period between deposition and extraction of the core.

If the counting strategy involves examining only part of a slide, the assumption has to be made that different types of grain and spore are not concentrated unevenly in different parts of the medium. Brookes and Thomas (1967) noted that if an insufficient amount of mounting medium were used, larger grains tended to be proportionately over-represented at the centre of the slide, being less able to move freely in the medium, but if due care is taken in preparation of slides, various workers have noted the absence of uneven concentrations of pollen of different species (Eyster-Smith, 1977).

The multinomial model requires the assumption that each grain has the same probability u_k of belonging to the kth taxon; it is of interest to know the largest volume of sediment in which this assumption may be regarded as valid, that is, the extent of the 'larger population' of pollen grains from which the sample is taken. For example, does the same multinomial model hold

1. for new preparations taken from the same location?
2. for samples taken over synchronous surfaces?
3. for samples taken at neighbouring stratigraphical levels?

Early investigations in this area were carried out by Barkley (1934) and Woodhead and Hodgson (1935). Studies by Faegri and Ottestad (1948) and Deevey and Potzger (1951) indicate that the hypothesis that the model is the same cannot be rejected in the first situation nor usually for samples taken reasonably close together in the second situation. The observations of M. B. Davis (1968), M. B. Davis, Brubaker, and Beiswenger (1971), M. B. Davis (1973), and M. B. Davis and Brubaker (1973) on the effects of seasonal water circulation on pollen sedimentation provide a caution against assuming that similar conditions prevail over the whole extent of a synchronous layer in a lake sediment (see also R. B. Davis, Brewster, and Sutherland, 1969; Bonny, 1978). Sudia (1952) describes a set of data taken along a transect from a single stratum in a bog in which the samples could not be regarded as coming from the same multinomial population. Methods of testing these hypotheses are given by Mosimann (1965).

In the third situation, consider examining samples from neighbouring levels down a core which are sufficiently similar that pollen analysts would regard them as constituting a pollen zone reflecting stable vegetational conditions. Examination of such sets of data by P. S. Martin and Mosimann (1965) and Gordon and Birks (1974) (see also Faegri and Ottestad, 1948) indicates that the spectra comprising the zone could not be regarded as samples from the same multinomial population. In other words, when pollen analysts speak of homogeneous conditions (e.g., H. J. B. Birks, 1973b), they would appear to have in mind a less restrictive concept than statistical homogeneity. This feature has implications for the work discussed in Chapter 4, where methods are described for comparing two pollen zones from different cores; it would be inappropriate to test the hypothesis that the spectra from both zones are samples from the same multinomial distribution, for reasons quite separate from ecological considerations such as regional parallelism or vegetational differentiation. Although one could relax this strict definition of homogeneity (Gordon and Birks, 1974, p. 228), we believe that in this situation a hypothesis-testing approach is likely to be less useful than the more informal methods of comparison described in Chapter 4.

2.4 Further Statistical Examples in Quaternary Palynology

The last two sections have investigated the relevance of the binomial and multinomial distributions to palynology, and subsequent sections of this chapter concentrate on exploratory methods of data analysis. This section outlines selected Quaternary palynological applications of more formal statistical methodology, describable by the term *confirmatory data analysis;* no attempt is made to give an exhaustive account, but we hope to illustrate the range of palynological problems which can be assisted by statistical methodology. Some further examples are referenced by Frederiksen (1974).

Greig-Smith (1964, p. 212) sums up the advantages of using a quantitative approach with judgements based on the methods of statistical analysis as follows:

(1) The quantitative approach allows the detection and appreciation of smaller differences.
(2) By using suitable statistical tests the quantitative approach provides a sounder basis of judgement of the significance of differences observed.

Section 2.2 has stressed the value of having some idea of the reliability of a statistical estimate, for example, by evaluating a confidence interval for the proportion of the pollen grains in the sediment which belong to a particular taxon. It is also of interest to estimate the mean number of pollen grains per unit weight or volume of sediment (so-called pollen concentrations), for example, by adding a measured volume of a suspension of exotic grains to a measured weight or volume of sediment containing fossil grains (see Section 1.2). Methods for obtaining confidence intervals for the mean number of pollen grains per unit weight or volume of sediment are described by Bonny (1972) (whose intervals err on the side of caution when specifying the approximate confidence level), Regal and Cushing (1979), and Maher (1981); see also M. B. Davis (1965a), Stockmarr (1971), Maher (1972a), and Peck (1974) for a discussion of errors likely to be involved in methods for determining pollen concentrations. Clark (1982) illustrates point and interval estimation of the amount of charcoal contained in a sample.

In an influential paper on R values (see Sections 5.4 and 6.3 and Figure 1.1), M. B. Davis (1963) described how the application of pollen–vegetation representation factors could lead to a re-interpretation of a pollen assemblage dominated by pine pollen from Brownington Pond, Vermont, as implying that at a particular time in the past about 42% of the surrounding vegetation was composed of larch trees. M. B. Davis (1969a, 1969b) has since retracted this interpretation, but it is interesting to note that her original 1963 results can be criticised from a purely statistical viewpoint. If the R-value model is valid, then an estimate of the proportion of larch in the vegetation surrounding the lake is indeed 0.42. However, Parsons, Gordon, and Prentice (1983) have shown that the standard error associated with this estimate is of the order of 0.5; in other words, the estimate is statistically very unreliable. This is an example of the value of requiring not only point estimates, but also some idea of how reliable these estimates are (see Section 6.3 for a further discussion of this approach).

Faegri and Iversen (1975, p. 192) have stressed that 'no mathematical statistics, however complicated, can replace botanical common sense'. This is perfectly true, but the relevant verb describing the relationship between the two disciplines should surely be *assist* rather than *replace*. Another example where such assistance could be useful is in the separation of mixture of morphologically similar pollen taxa. Some circumboreal tree genera such as *Betula, Picea,* and *Pinus* comprise a number of species which produce morphologically similar pollen. Although there are slight differences among the pollen grains produced by each species, the degree of overlap is sufficiently great to prevent unambiguous identification of many grains. Reliable estimates of the proportions in the underlying sediment of each of the species can increase the amount of relevant palaeoecological information available to an investigator, and in recent years increasing attention has been focussed on methods of obtaining such estimates (e.g., Eneroth, 1951; A. R. H. Martin, 1959; Járai-Komlódi, 1970; Hansen and Cushing, 1973; Usinger, 1975, 1978a, 1978b; Gordon and Prentice, 1977; Andersen, 1980a; H. J. B. Birks and Peglar, 1980; I. C. Prentice, 1981; Gordon, 1982d; Hodge, 1983). There are, however, fundamental difficulties associated with methods which seek to estimate the proportions by identifying each individual pollen grain in the mixture (see Gordon and Prentice, 1977; Gordon, 1982d). The statistical method of maximum likelihood (Mood *et al.*, 1974, Chapter VII) can be used to model the data as a mixture of multinomial distributions (Gordon and Prentice, 1977) or normal distributions (Gordon, 1982d; Hodge, 1983); in addition to providing precise estimates of the mixing proportions, this method gives associated standard errors, allowing an assessment of the reliability of the estimates.

A further example where statistical testing is important in Quaternary palynology relates to experiments on the effects of different chemical treatments, mounting media, and sedimentary environment on the size of pollen grains. Although much early work had demonstrated the feasibility of using size as a means of distinguishing morphologically similar pollen grains (e.g., Cain, 1940, 1948; Cain and Cain, 1948), the demonstration that pollen grain size could vary with different chemical treatments, mounting media, slide thickness, and sediment type led Christensen (1946) to propose that pollen size should be measured relative to the size of pollen of a standard taxon, such as *Corylus avellana*. Faegri and Iversen (1975, p. 45) demonstrated that differences in pollen size of several taxa resulting from different treatments are proportional to differences in the size of *Corylus* pollen. Andersen (1960, 1978b) has described a large number of experiments to study the effects of preparation procedure, chemical treatment, mounting media, and sediment type on the size of *Corylus* pollen. The results obtained were analysed statistically using a variety of significance tests. In these instances the adoption of suitable statistical tests provided 'a sounder basis of judgement of the significance of differences observed' and a means for the 'detection and appreciation of smaller differences' (Greig-Smith, 1964, p. 212). Other numerical studies of the variation in pollen grain size include Faegri and Deuse (1960), Whitehead (1964), Whitehead and Langham (1965), Reitsma (1969), and Whitehead and Sheehan (1971).

The final example in this section is concerned with investigating the relationship between two taxa over a period of time, as recorded in the pollen stratigraphical record. For example, did the proportions of *Pinus* and *Abies* pollen vary together in the same direction over time (i.e., did they tend to increase together or decrease together during the period of time under investigation?), or did the proportions vary in opposite directions? The relationship between two taxa can be measured by the correlation coefficient, which takes values between $+1$ and -1, positive values indicating variation in the same direction, and the absolute magnitude of the coefficient (i.e., its size, disregarding the sign) registering the strength of the relationship. Mosimann (1962) gives an example of 73 levels from a core, with the correlation between *Pinus* and *Abies* pollen taking the value −0.3. At first sight, this would appear to suggest a negative relationship between these two taxa, but some care is needed in the interpretation of this figure; we would expect some inbuilt negative correlation because of the constraint that for each spectrum the proportions must sum to 1. The rationale is similar to that considered in the discussion of confidence intervals and standard errors in the earlier examples: is the observed value significant? Mosimann (1962, 1970; see also P. S. Martin and Mosimann, 1965, and Chayes and Kruskal, 1966) constructs null models representing the situation in which there is no real relationship between the two taxa, and investigates the expected properties of the correlation coefficient in the null situation, discovering that a value of −0.3 is quite likely to occur under the null model. This approach is discussed in more detail in Section 6.5.

The above selected examples are designed to show that the 'confirmatory' approach of parameter estimation and hypothesis-testing does have a rôle to play in Quaternary palynology in certain specific and well-defined situations. The more involved mathematics associated with the final example does, however, illustrate some limitations of the confirmatory approach, at least in the present stage of development: there are still gaps in the methodology of multivariate statistics for handling data in the form that Quaternary stratigraphical palynological data are recorded. If the concepts of significance and formal models are set aside, methods of exploratory data analysis can make further useful contributions to the investigation of palynological data.

2.5 Exploratory Data Analysis and Classification

The body of statistical methods referred to by the term *confirmatory data analysis* concentrate on specific investigations: if the *R*-value model is adequate, what is the estimate for the abundance of larch in the vegetation, and how reliable is it? Is the observed correlation over time between the proportions of two taxa significant? Data can, however, possess many different properties and an excessive concentration on particular aspects of the data can cause one to ignore other features of interest. It can be valuable to investigate data by a variety of different techniques designed to allow particular types of structure and relationships to emerge more clearly. This is particularly true when the data set is large and complex, as many palynological data sets are.

A range of multivariate techniques is now available for exploratory data analysis (see, e.g., Seal, 1964; Blackith and Reyment, 1971; W. T. Williams, 1976; Gnanadesikan, 1977; Everitt, 1978), and several of these techniques are illustrated on palynological data in subsequent chapters. However, the body of methods referred to as *classification* is used repeatedly, and so the present chapter is concluded by some general comments on this topic. An influential review of the subject was presented by Cormack (1971), and a critical introduction is given by Gordon (1981).

Classification studies involve the examination of a collection of 'objects', where each object is described by a set of 'variables'. In palynology, an object could correspond to a pollen spectrum, and a variable to a taxon which contributed pollen grains to the spectrum. Thus, for example, A1 (325 cm) in Table 1.1 would be regarded as an object described by nine variables.

The aim of classification studies is to establish whether there is any group structure in the collection of objects under investigation; for example, do the objects fall naturally into a number of distinct groups such that objects within a group are in some sense 'similar' to one another, and 'dissimilar' to objects belonging to other groups? The notion of similarity will be left vague for the time being, but we should note that the emphasis has moved away from testing whether a specified group of spectra may be regarded as, for example, samples from the same multinomial distribution; the groups, if any exist, remain to be discovered.

There are two main approaches in many classification studies:

1. The first approach seeks a representation of the objects as points in some low-dimensional space, so that 'similar' objects are represented by points which are close together in the configuration. Two- or three-dimensional configurations are then assessed by eye. Such geometrical methods are referred to as *scaling* methods, or, in the ecological literature, *ordination* methods.
2. The second approach partitions the objects into a pre-specified number of groups (g, say) so as to optimise some mathematical criterion. Often, a nested set of such partitions is sought (for $g = 2, 3, 4, \ldots$) and the resulting hierarchy is displayed in a dendrogram or tree diagram, such as that shown in Figure 4.3. We refer to this body of methods as *partitioning* or *cluster analysis* methods, although there is no general agreement on terminology, and some research workers would use the term *cluster analysis* to refer to methods in both categories 1 and 2.

There are drawbacks to both classification approaches described in categories 1 and 2 above. The assessment of a low-dimensional configuration of points can be very subjective. Forcing the data into a partition of disjoint groups of objects can give misleading results; some examples are given by Everitt (1980, Chapter 5). A hierarchical representation has often been found useful in summarising the relationships defined by the set of all partitions, but this can be even more misleading than a single partition. Sneath and Sokal (1973, p. 209) note that several studies have indicated that ordination methods tend to give more reliable representations of the relationships between major clusters than the later amalgamations in a dendrogram, whereas a dendrogram will tend to be more reliable in depicting the relationships between very similar objects.

There is much to recommend analysing multivariate data by both clustering and geometrical methods (see Gordon, 1981, 1982c), as the strengths and weaknesses of the two approaches are often complementary (Davenport and Studdert-Kennedy, 1972). Although the interpretation of results can be subjective, geometrical methods allow one to ascertain whether any group structure exists within the data; and if the same structure is indicated by the clustering methods, one has more confidence in the reality of the group structure. A combination of both approaches often allows the dangers and drawbacks associated with just one approach to be avoided. Such double studies have been commonly undertaken in the exploratory analysis of Quaternary palynological data (e.g., H. J. B. Birks, Webb, and Berti, 1975; Caseldine and Gordon, 1978).

However, even if it has been decided to follow this course of action, there are still several questions which have to be considered. These include:

1. How should the 'objects' to be studied be selected?
2. Which taxa should be included in the description of each object?
3. How should one measure the 'similarity' between objects?
4. Which particular method of analysis (within each general approach) should be used?

These questions are now considered in turn:

1. In making general recommendations for carrying out a numerical taxonomic study, Sneath and Sokal (1973, p. 303) recommend that the objects to be investigated 'should be chosen to represent appropriately the variation to be studied'. Palaeoecologists and palynologists would seem to have to face fewer problems than workers in other disciplines in specifying the objects to be studied; certainly in pollen stratigraphical studies, once the location for the core has been selected, a decision on what constitutes the set of objects is relatively straightforward and unambiguous. The decisions about the location of coring sites and of surface samples are, however, more complex, depending on various factors (see Section 1.1). Considerable care and thought must go into the selection of sites (see Jacobson and Bradshaw, 1981) simply because pollen analysts cannot obtain a clear idea of the stratigraphical record present at a site until a large amount of time-consuming pollen counting has been completed.

2. Investigators in taxonomic studies often experience difficulties in deciding on relevant variables to be used for the description of objects. Palynologists are again more fortunate in that the correspondence can generally be made between variables and the pollen and spore types present in the pollen spectrum. Decisions still have to be made about the taxonomic level to which the identifications should be made (e.g., Plantaginaceae, *Plantago,* or *Plantago lanceolata*) and about whether some pollen and spore types are relevant to the study and should be included within the pollen sum. Many investigators, when presenting stratigraphical data from lake sediments, do not include pollen or spores from obligate aquatic taxa in the basic pollen sum (see H. J. B. Birks, 1973b, pp. 222–223). In modern, ecologically based pollen analysis, all identifications should generally be made to

the lowest possible taxonomic level, within the limits set by the inherent pollen-morphological variability of the taxa concerned and the extent and range of the modern reference material available (see Hansen and Cushing, 1973; H. J. B. Birks, 1978; and H. J. B. Birks and H. H. Birks, 1980, for a discussion of pollen identification problems). The question of which taxa to include within the basic pollen sum is more complex. In general, the choice of what to include should be on the principle of including all taxa of the 'universe' of interest and study (Cushing, 1963; Wright and Patten, 1963). In some instances, such as the reconstruction of local bog and fen development, a pollen sum based solely on local taxa is appropriate (e.g., H. H. Birks, 1972, 1975); in other instances, such as in the reconstruction of upland vegetation, a pollen sum of all non-local taxa should be used. Janssen (1979, 1981b) provides a thoughtful discussion of pollen sums in relation to the palaeoecological problems under investigation.

Numerical methods of analysing percentage pollen data tend to be affected little by small amounts of rare taxa, and it is common to consider only the most abundant pollen types in the initial numerical investigation, for example, to restrict attention to taxa which occur with percentages of more than 5% in at least one of the spectra under study. This action is not to be taken as implying that rare taxa are unimportant; they may be of considerable interpretative and ecological value at a later stage of the analysis, namely, in the reconstruction of past floras, vegetation, and environment (see H. J. B. Birks and Berglund, 1979, for discussion of this point).

3. Even a cursory examination of the research literature gives an impression of the large number of different ways in which the similarity between objects may be measured. Further discussion is postponed to the next section, in which a measure is evaluated for the spectra constituting the two test data sets from Abernethy Forest used in this book.

4. A major problem in classification studies is that each method of analysis is particularly well suited to uncovering and displaying particular kinds of structure in data. For example, a criterion which partitioned the data into groups so as to minimise the total within-group sum of squared deviations is found to favour 'spherical-shaped' clusters, whereas the single link criterion will detect clusters of any shape provided these clusters are well separated from each other and not connected to one another by chains of intermediate objects (for a fuller description of these criteria, see Gordon, 1981). Similarly, geometrical methods can differ in the emphases they place on various aspects of the data (I. C. Prentice, 1977). In other words, each method (particularly each clustering method) is pre-disposed to find or display clusters of a particular type, and may be regarded as distorting the original data to a greater or lesser extent. Experience has indicated that if the structure in the data is pronounced then it is usually detected by most methods with only minor differences of detail, but different analyses of data can lead to differing conclusions. It is possible to overstate this effect, but it is a useful exercise to analyse data by several different clustering methods. If the results agree, one has more confidence in the reality of any group structure indicated: it is

less likely to be purely an artifact of the method employed. Some further discussion of this point, and an approach to synthesising the results of different clustering methods are given by Gordon (1980a). The general approach of examining several different analyses has implications for any claims for the objectivity of classification methods, which are discussed further in Section 3.8.

This still leaves open the question of which method, amongst the many available, should be used. We are reluctant to make firm recommendations because we do not believe that there is one all-purpose method which can be guaranteed to solve all problems. Some bad methods have been proposed, and many methods are almost equivalent in many situations. It is interesting to note the collective opinion of people using cluster analysis methods, as recorded by Blashfield (1976). In the applied research literature, three methods accounted for nearly 75% of the published uses of cluster analysis in 1973, namely single link, (group) average link, and Ward's sum-of-squares method (for definitions, see Gordon, 1981, Chapter 3).

Comparisons of some ordination methods are given by A. J. B. Anderson (1971b), Rohlf (1972), and I. C. Prentice (1977). Studies suggest (e.g., Rowell, McBride, and Palmer, 1973; H. J. B. Birks, Webb, and Berti, 1975) that broadly similar conclusions would often be drawn from the analysis of data by different ordination methods, though there could be differences in emphases; further comparative studies would be valuable. A fuller description of some ordination methods is given in Chapters 4 and 5.

2.6 The Measurement of Dissimilarity

Many classification methods are based, explicitly or implicitly, on a measure of the similarity or dissimilarity between each pair of objects. The account in this section will be in terms of dissimilarities, but only minor modifications are needed for the study to be conducted in terms of pair-wise similarities.

It is assumed that we can summarise the differences between two objects, labelled i and j, say, by a measure $d(i, j)$. There are certain minimum conditions which a measure of dissimilarity should satisfy:

(i) $d(i, j) \geq 0$;
(ii) $d(i, i) = 0$;
(iii) $d(i, j) = d(j, i)$

for all pairs of objects i and j. The larger $d(i, j)$ is, the less similar objects i and j are regarded as being.

The assumption that one can represent the relationship between each pair of spectra by a single value is a considerable one, and is bound to introduce some distortion of the true situation. The hope is that the major features of the data set are preserved by this procedure.

It is quite likely that two studies of the same data set which had different aims would wish to emphasise different features of the data, and so might be based on different measures of dissimilarity.

The conditions (i)–(iii) are satisfied by a wide range of functions, and a relevant question is that posed in the previous section: how should the dissimilarity be measured? Some discussion of further desirable properties of any measure of dissimilarity used in palynology is given by I. C. Prentice (1980). L. A. S. Johnson (1970) lists some decisions which have to be made in constructing a general measure of dissimilarity, and while the problems would appear less acute in palynological studies than in general taxonomic investigations, some subjectivity would appear unavoidable. In describing two measures which will be used later, we recognise that other choices might have been equally valid. A criterion of choice would be to seek to specify, as exactly as one can, the features which one believes to be important in the study being undertaken. After considering the variety of dissimilarity coefficients which have been used in taxonomic studies, Sneath and Sokal (1973, Section 4.7) say

> But when all is said and done, the validation of a similarity measure by the scientists working in a given field has so far been primarily empirical, a type of intuitive assessment of similarity based on complex phenomena of human sensory physiology. ... Perhaps the only recommendation that we would care to make at this stage in the development of the field is that, of each type of coefficient considered, the simplest one should be chosen out of consideration for ease of interpretation.

It would be pleasant if a consensus were to emerge on the definition of dissimilarity, but we remain to be convinced that this is likely. The results of robustness studies—such as the analysis of the same data set by principal components analysis and principal coordinates analysis (which essentially carry out the same analysis using different measures of dissimilarity) reported by H. J. B. Birks, Webb, and Berti (1975)—often show that the main features in the data are not markedly affected by the precise choices made, but on occasion such choices can be critical (Prentice, 1982b).

A measure of the dissimilarity between two pollen spectra will be based on differences in the types and proportions of pollen grains recorded in each spectrum. A similar approach could be used for summarising data recorded in other forms, for example, if reliable data were available on the number of grains of each species cm^{-2} per year, but the following account will concentrate on percentage data. Thus, suppose that $p_{i1}, p_{i2}, \ldots, p_{it}$ denote the proportions of the ith spectrum belonging to each of the t species of pollen recorded. For example, if i refers to A1, then the proportions are given in the relevant row of Table 1.3; thus $t = 9$ and $p_{A1,1} = 0.118$, $p_{A1,2} = 0.848$, $\ldots$, $p_{A1,9} = 0.000$. We seek to summarise the differences between each pair of rows in Table 1.3 by a single value, the dissimilarity between the corresponding spectra.

A measure of dissimilarity used later in this book is the city block, or Manhattan, metric, defined by

$$d(i, j) = \sum_{k=1}^{t} |p_{ik} - p_{jk}|;$$

in words, the dissimilarity between the ith and jth spectra is defined as the sum of the absolute differences between their proportions of each species; the symbols "| |" denote that the enclosed expression is to be regarded as positive. For example,

$$\begin{aligned} d(\mathrm{A1}, \mathrm{A2}) &= |0.118 - 0.321| + |0.848 - 0.582| \\ &\quad + |0.024 - 0.068| + \cdots + |0.000 - 0.000| \\ &= +0.203 + 0.266 + 0.044 + \cdots + 0.000 \\ &= 0.533 \end{aligned}$$

The dissimilarities between each pair of spectra in the 1974 test data set are given in Table 2.1. Because $d(i, j) = d(j, i)$, only the terms below the diagonal are shown, the terms above the diagonal being given by symmetry. The dissimilarities between each object in set A and each object in set B are given in the central box in Table 2.2; the interpretation of the values around the perimeter of the box is deferred until Chapter 4. These tables are used to illustrate some of the methods of analysis to be introduced later in the book.

Several points emerge from this example. Firstly, the identity of a taxon does not influence this measure of dissimilarity. For example, if the labels *Betula* and *Pinus* at the heads of the first and second columns of Table 1.3 were interchanged, this would not alter the measure of dissimilarity. Some authors (e.g., P. D. Moore and Webb, 1978, p. 120) argue that extra weight should be given to taxa which are believed to be ecologically important. This approach could be accommodated within the general framework of exploratory data analysis by use of weighting coefficients (see Section 3.4), but there are dangers of circular arguments. We prefer to regard the methods purely as investigatory aids without importing such ecological considerations at this stage.

Secondly, the major contribution to the measure of dissimilarity will usually be made by differences in the proportions of the more abundant taxa. An alternative

Table 2.1

Matrix of Dissimilarities for the Abernethy Forest 1974 Test Data Set (Set A)

Sample designation	Sample designation									
	A1	A2	A3	A4	A5	A6	A7	A8	A9	A10
A1	—									
A2	0.533	—								
A3	1.663	1.138	—							
A4	1.650	1.117	0.193	—						
A5	1.652	1.207	0.501	0.350	—					
A6	1.727	1.282	0.792	0.777	0.449	—				
A7	1.802	1.765	1.737	1.732	1.622	1.647	—			
A8	1.658	1.627	1.645	1.646	1.500	1.473	0.276	—		
A9	1.708	1.263	1.205	1.190	1.030	0.873	1.460	1.258	—	
A10	1.735	1.550	1.500	1.515	1.371	1.170	1.489	1.269	0.375	—
Depth (cm)	325	350	375	420	445	455	490	500	520	530

Table 2.2

Matrix of Dissimilarities between Spectra from the Two Abernethy Forest Test Data Sets, Set A (1974) and Set B (1970)

Sample designation	Depth (cm)	Sample designation											
		(A0)	A1	A2	A3	A4	A5	A6	A7	A8	A9	A10	(A11)
(B0)		0.454	0.454	0.405	1.263	1.282	1.576	1.657	1.738	1.594	1.608	1.635	1.635
B1	110	0.454	0.454	0.405	1.263	1.282	1.576	1.657	1.738	1.594	1.608	1.635	1.635
B2	270	0.375	0.375	0.478	1.348	1.355	1.665	1.740	1.769	1.647	1.721	1.748	1.748
B3	390	1.496	1.496	0.963	0.505	0.670	0.958	1.043	1.702	1.564	1.184	1.487	1.487
B4	430	1.657	1.657	1.124	0.296	0.443	0.507	0.590	1.633	1.531	1.051	1.392	1.392
B5	460	1.634	1.634	1.101	0.899	0.870	0.724	0.471	1.542	1.350	0.848	1.189	1.189
B6	500	1.099	1.099	0.842	1.318	1.319	1.303	1.190	1.461	1.185	0.877	0.796	0.796
B7	530	1.682	1.682	1.529	1.491	1.494	1.350	1.195	1.432	1.214	0.348	0.149	0.149
(B8)		1.682	1.682	1.529	1.491	1.494	1.350	1.195	1.432	1.214	0.348	0.149	0.149
Depth (cm)			325	350	375	420	445	455	490	500	520	530	

measure of dissimilarity which is implicit in some of the methods to be described in Chapters 3, 4, 5, and 6, is the squared Euclidean distance, defined by

$$d_2(i, j) = \sum_{k=1}^{t} (p_{ik} - p_{jk})^2.$$

This measure is likely to emphasise the effect of the more abundant taxa even more strongly, through the square power.

For a fixed value of the sample size n, the nearer the underlying proportion u is to $\frac{1}{2}$, the higher will be the standard error associated with its estimate p, and it could be argued that this implies that some standardisation of the data should be carried out before any analysis is undertaken. One approach would be to replace each p_{ik} by $\sin^{-1}(p_{ik})$, this having the effect of tending to equalise the standard errors. Another approach would be to scale each taxon so that it had unit variance over the data set, thereby increasing the contribution of minor taxa. This latter standardisation is automatically carried out when data are examined by the method of principal components analysis of the correlation matrix. The results described later in the book are not markedly affected by this type of standardisation, but we can conceive of situations in which considerable changes could be brought about by standardisation (see I. C. Prentice, 1982b). A fuller discussion of this topic is given by Sneath and Sokal (1973, pp. 154–157) and by I. C. Prentice (1980).

Data transformations are even more important for pollen concentration data and pollen-accumulation rates, because of their greater inherent variability. Gordon (1982c) presents theory to justify a variance-stabilising transformation for pollen concentrations, which has also been used by Hunt and Birks (1982), but such data are analysed numerically less often than percentage data.

We conclude by noting that the structure in many data sets appears to be detectable by a range of methods. This robustness is reassuring, but should not be presumed always to exist. We would recommend that several different analyses always be undertaken on any data set so as to guard against unwarranted conclusions being drawn on the basis of one particular method of analysis, or standardisation, or measure of dissimilarity.

CHAPTER **3**

The Analysis of Pollen Stratigraphical Data: Zonation

3.1 The Concept of the Pollen Zone

In Quaternary palaeoecology, as in other descriptive historical sciences, there is a necessity for the accurate and unambiguous presentation, analysis, and interpretation of all basic stratigraphical and palaeoecological data. There is also an equally important need for the critical evaluation of such data in relation to current paradigms, concepts, and hypotheses about Quaternary stratigraphy, vegetation dynamics, and environmental history. It is to these ends that the application of the *International Guide to Stratigraphic Classification, Terminology, and Usage* (Hedberg, 1972a, 1972b, 1976) to Quaternary pollen stratigraphy is most valuable, as emphasised by Andersen (1961), Cushing (1963, 1964, 1967b), West (1970), H. J. B. Birks (1973b), and Mangerud, Andersen, Berglund, and Donner (1974). These authors stress the need, in all Quaternary stratigraphy, of applying, wherever possible, the concepts, rules, and terminology of geological stratigraphy. In the field of Quaternary stratigraphical pollen analysis this has been done, for example, by Cushing (1963, 1967b) in Minnesota, by H. H. Birks (1970, 1972, 1975) in Scotland, by H. J. B. Birks (1973b) on the Isle of Skye, Scotland, by Craig (1978) in Ireland, by Mangerud (1970) in Norway, by Donner (1971) and Hyvärinen (1972) in Finland, and by Göransson (1977), Digerfeldt (1977), H. J. B. Birks and Berglund (1979), and Björck (1979, 1981) in Sweden.

Pollen analytical and other stratigraphical data are often so complex in the number of fossil taxa represented, in the number of stratigraphical levels analysed, and the number of individual fossils counted at each level that it is generally necessary to subdivide each stratigraphical sequence into smaller units for ease in (1) the description; (2) the discussion, comparison, and interpretation; and (3) the correlation in both time and space of the sequences. The most useful unit of subdivision of the vertical or time dimension of a pollen diagram is the *pollen zone*.

Although pollen analysts have been delimiting and, in some cases, defining pollen zones for over 50 years, there are very few definitions or guidelines about what a pollen zone is or should be. Faegri and Iversen (1975, p. 198) simply state that 'in order to deal with it (a pollen diagram) more easily the investigator has to subdivide it into parts, usually referred to as zones. Since further work with the diagram depends on these zones, the division must be carried out carefully'. Cushing (1964, p. 1) explicitly defines a pollen zone as 'a body of sediment distinguished from adjacent sediment bodies by differences in kind and amount of its contained fossil pollen grains and spores, which were derived from plants existing at the time of deposition of the sediment'.

Cushing (1963, 1964) has discussed in detail the concept of the pollen zone. He has emphasised that a pollen zone, as defined above, is a biostratigraphical, or biostratic, unit or biozone (*sensu* Hedberg, 1972b, pp. 222–227) and has suggested that of the different types of biozone distinguished by stratigraphers a pollen zone corresponds most closely to the category of an *assemblage zone* (defined by Hedberg, 1972b, p. 223 as being 'characterized by a distinctive natural assemblage of forms or of forms of a certain kind'). Cushing (1964) recommends the *pollen assemblage zone* as the basic unit in Quaternary pollen stratigraphy. Other types of pollen zones are, of course, possible and desirable in some instances, as discussed below, and may correspond to concurrent-range zones, acme or peak zones, interval zones, or taxon-range zones recognised by biostratigraphers (see Hedberg, 1972b, p. 223; 1976, pp. 50–61).

There has been considerable interest in the development and application of numerical methods for the subdivision of pollen stratigraphical sequences. This has come about with the realisation (Cushing, 1964) that pollen zones, as generally used in Quaternary stratigraphy, are biostratigraphical assemblage zones that are (or should be!) defined solely on the observed pollen and spore content of the sediments without any preconceptions or assumptions—explicit or implicit—about sediment lithology, inferred climate, past vegetation, or presumed time equivalence (cf. Janssen, 1980). Numerically derived pollen zones are, by necessity, delimited solely on the basis of specified mathematical criteria, without any reference to inferred climate, vegetation, or chronology (although these latter factors could be included in a statistical analysis). They are thus, by definition, strict pollen assemblage zones.

Various numerical methods of dividing stratigraphical sequences into zones will be described later in this chapter. Such numerically derived zones refer only to a single stratigraphical sequence. They are thus *local* or site pollen assemblage zones (*sensu* Cushing, 1967b; H. J. B. Birks, 1973b, p. 273; Watts, 1977) that describe and summarise the basic features of the pollen stratigraphy for that particular sequence. Such local zones are of limited use in discussion and comparison between profiles, and for these purposes it is useful to define *regional* pollen assemblage zones based on several sequences (see Figure 3.1) (Cushing, 1967b; H. J. B. Birks, 1973b, p. 273; Watts, 1977). Numerical methods for comparing pollen stratigraphical sequences and for aiding in the recognition, delimitation, and definition of regional pollen assemblage zones are considered in Chapter 4.

As mentioned above, alternative formulations of the concept of a pollen zone are possible. Peak or acme zones (*sensu* American Commission on Stratigraphical Nomenclature, 1961; Hedberg, 1972b, p. 225; 1976, pp. 59–60) can, for example, be usefully delimited to characterise pollen spectra of narrow stratigraphical extent that are distinguished by exceptionally high values of one or two pollen or spore types (see examples in McAndrews, 1966; Watts, 1977; Craig, 1978; H. C. Prentice, 1982). Such peak zones generally represent a small fraction of the total stratigraphical sequence, but they are particularly valuable in late-glacial pollen stratigraphy where there may be several rapid and apparently short but complex pollen stratigraphical changes (e.g., Pennington, 1977; Watts, 1977; Hunt and Birks, 1982).

D. Walker (1966, 1972) has suggested that for some purposes, particularly ecological reconstructions of past vegetational dynamics and of past processes of vegetational change and differentiation, a zonation scheme is required that distinguishes between two different types of stratigraphical interval detectable in pollen sequences. These are (1) intervals of relatively 'stable' pollen composition, and

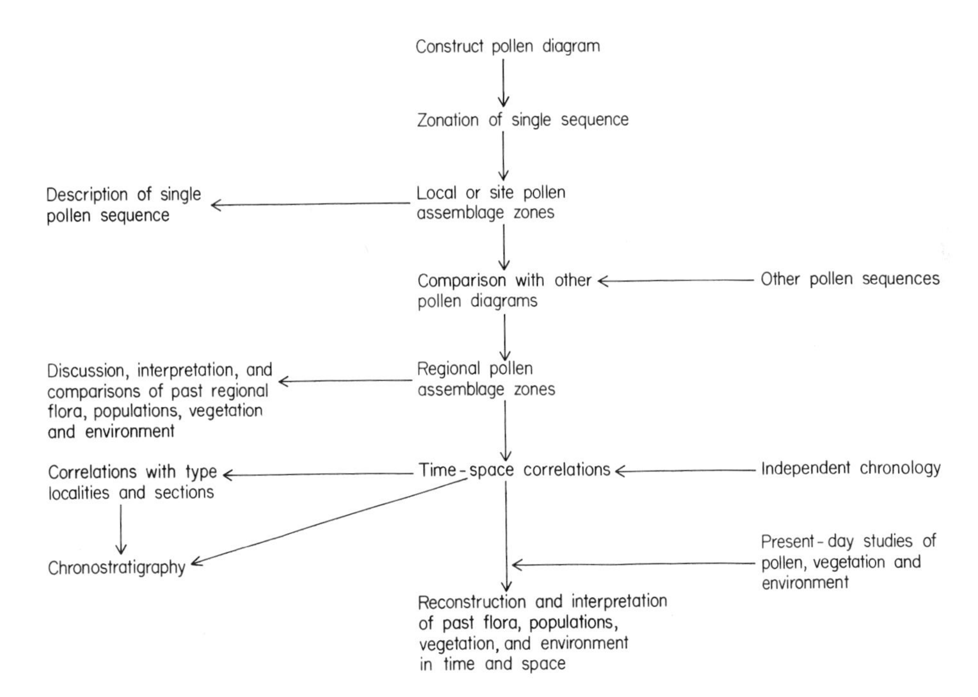

Figure 3.1 A schematic representation of stages in palaeoecological reconstruction and interpretation.

(2) intervals of abrupt or systematic change in pollen composition interspaced between 'stable' intervals (see also Watts, 1973). The first type of interval corresponds to the assemblage zone concept of a pollen zone, with the possible removal of a few transitional levels at the upper and lower edges of the zone. The second type of interval, which is of considerable importance in understanding population and vegetation dynamics, can be used for zonation in particular circumstances (e.g., the Cumbrian zonation of D. Walker, 1966). This approach is explored further in Section 3.6. In general, such an approach tends to increase the number of zones in a profile and thus to decrease their stratigraphical and spatial extent and hence their regional applicability.

It is important to emphasise that the subdivision of any pollen stratigraphical sequence into zones, whether local or regional, peak or assemblage, stable or transitional, should not be viewed as an end in itself but rather as a means to an end. Such an end is most commonly the reconstruction of the past flora, plant populations, vegetation, and environment, and of the dynamics and differentiation, in both time and space, of these reconstructions (see Figure 3.1). The delimitation of pollen zones is merely a step towards that end (H. J. B. Birks and H. H. Birks, 1980).

D. Walker (1972, 1982b), Watts (1973, 1977), D. Walker and Wilson (1978), and D. Walker and Pittelkow (1981) have questioned whether pollen zones should be delimited at all. They argue that pollen sequences are complex stratigraphical records of periods of apparent vegetational stability (cf. D. G. Green, 1982) interspaced by phases of vegetational change. The delimitation of pollen zones and the drawing of zone boundaries assume that pollen sequences consist of distinct, homogeneous units (Watts, 1977), particularly as most pollen zonations ignore the phases of rapid but short-lived (50–150 years) change and the more gradual transitions that occur over relatively long (200–500 years) intervals. Both types of change are of considerable ecological importance (Watts, 1973) in terms of population dynamics and vegetational and environmental history, and ideally they should not be ignored in the basic description and summarisation of pollen stratigraphical data.

Watts (1977) draws the interesting analogy between pollen zonation and one of the fundamental problems in descriptive plant ecology, namely, 'is the classification of plant communities either possible or desirable?' (cf. D. A. Webb, 1954, p. 362). Watts asks whether in plant sociology the underlying nature of vegetation is such that individual stands of vegetation can be viewed as discrete, homogeneous units which can be classified together on the basis of their overall floristic composition, as is the practice in phytosociology, or whether each stand varies continuously in space and time, as Gleason (1939) proposed, in which case stands should logically not be classified together. As the bulk of pollen analytical evidence supports Gleason's individualistic concept of vegetation (see Sections 6.2 and 6.4), it appears at first sight that the classification into zones of pollen spectra that vary continuously in space and time conflicts with an important feature of palynological data (Watts, 1977).

In plant sociology it is frequently convenient and desirable to classify stands of vegetation for the purposes of basic description, vegetation mapping, and nature conservation, even though one's ideas about the underlying nature of vegetation may agree with Gleason's individualistic concept (see van der Maarel, 1975). Gleason (1939, p. 104) himself pointed out that 'a community is uniform, either in space or in time, only to a reasonable degree. This uniformity is sufficient to enable us to recognise the community and to accept it as a unit of vegetation, while its variability, although slight, is sufficient to indicate the impossibility of considering any such area of vegetation as a definitely organized unit'. Similarly in Quaternary palaeoecology, it is often convenient and desirable to define pollen zones for purposes of description, discussion, and interpretation, without implying that vegetation in the past has not varied in an individualistic way (Watts, 1977). Individual pollen spectra and local pollen zones can be viewed as analogous to, respectively, the actual quadrats of vegetation described in the field by a plant sociologist, and the preliminary groupings of such quadrats from a single locality. The delimitation of regional pollen assemblage zones, based on a synthesis of several sequences, is analogous in phytosociology to the recognition of plant associations—which are also abstract, synthetic concepts (*sensu* van der Maarel, 1975)—from the available quadrat data (Watts, 1977). As Gilmour and Walters (1963) have emphasised, the value of any classification should be judged in relation to the purposes for which it is required. For the vast majority of purposes in Quaternary pollen analysis, there is a real need to present and to summarise the large amounts of complex multivariate data that are so characteristic of palynological studies. Local pollen zones, carefully defined on the basis of the available data only (Cushing, 1964, 1967b), and their subsequent grouping into regional pollen zones are probably the most effective and convenient means of presenting succintly the mass of information contained in pollen diagrams (Watts, 1977; Janssen, 1980). If, however, the aims of the investigation are to consider the detailed changes, in time and space, of pollen taxa considered as individuals and as populations, alternative approaches to data analysis, such as those presented by D. Walker and Wilson (1978), are required. These approaches are discussed in Section 6.2.

3.2 Numerical Approaches to Pollen Zonation

There has been much interest recently in the development and application of numerical methods for the subdivision of Quaternary pollen stratigraphical sequences (e.g., Adam, 1970, 1974; Kershaw, 1970; Dale and Walker, 1970; Mosimann and Greenstreet, 1971; Gordon and Birks, 1972, 1974; Yarranton and Ritchie, 1972; Gordon, 1973a; H. J. B. Birks, 1974; Pennington and Sackin, 1975; Mehringer, Arno, and Peterson, 1977; H. H. Birks and Mathewes, 1978; H. J. B. Birks and Berglund, 1979; Hunt and Birks, 1982), as well as for dividing other stratigraphical sequences (e.g., Gill, 1970; Webster, 1973, 1978; Hawkins and Merriam, 1973, 1974, 1975; Hawkins, 1976a; Bement and Waterman, 1977;

Hawkins and ten Krooden, 1979; Shaw and Cubitt, 1979). A brief review of existing methods in pollen analysis is given in this section.

One approach has compared pairs of neighbouring levels in the stratigraphical sequence in an attempt to detect sudden changes in pollen and spore composition. A chi-squared statistic (Kershaw, 1970, 1971; H. J. B. Birks, 1973b), the product-moment coefficient (see Figure 3.2) (Yarranton and Ritchie, 1972; R. B. Davis, Bradstreet, Stuckenrath, and Borns, 1975; Ritchie, 1977; Ritchie and Yarranton, 1978a, 1978b; Tallis and Johnson, 1980), and Spearman's rank correlation coefficient (Maher, 1982) have been used as measures of between-level differences in pollen composition. Such an approach places emphasis on the boundaries between zones, and will be considerably influenced by the practice commonly adopted by many pollen analysts of counting additional levels close to suspected zone boundaries. Yarranton and Ritchie (1972) also considered the behaviour of a 'running mean' of the correlation coefficient for five neighbouring levels (i.e., the correlation coefficients for levels 1 to 5 are averaged, as are those for levels 2–6, 3–7, 4–8, etc.) as a means of distinguishing between fluctuations in pollen composition within a zone and differences in composition between zones. Although the general approach has provided useful and intuitively sensible pollen zones in several instances, it is not wholly satisfactory. In achieving a pollen zonation, more of the data than just the boundaries between possible zones should be considered, as a pollen zone is defined as a 'body of sediment' and not solely by its lower and upper boundaries (see Hedberg, 1972b, p. 223). Further, the product-moment correlation coefficient does not seem to be an appropriate measure of the similarity between two sets of proportions, due to the constraint that proportions must sum to 1.

Standard numerical classification procedures have been applied to pollen stratigraphical data by several workers without taking into account the position of the levels in the stratigraphical sequence. Adam (1970, 1974), Dale and Walker (1970), R. B. Davis *et al.* (1975), Nichols (1975), Andrews, Webber, and Nichols (1979), and Brubaker, Garfinkel, and Edwards (1983) have used different agglomerative clustering procedures, and Mosimann and Greenstreet (1971) have used Friedman and Rubin's (1967) non-hierarchical iterative clustering procedure for zonation purposes. Similar approaches, lacking any stratigraphical constraint, have been widely used in pre-Quaternary biostratigraphy for delimiting fossil assemblage zones (see reviews by Hazel, 1977; Hay and Southam, 1978; and Millendorf, Brower, and Dyman, 1978). All these procedures, by having no stratigraphical constraint, allow the grouping together of levels which occur some distance apart in the stratigraphical sequence. For some purposes it is of palaeoecological interest and interpretative value to note the similarity of levels which are separated by dissimilar levels, for example, in the detection of vegetational reversion. However, the manner in which the data have arisen suggests that, in the absence of independent evidence indicating disturbance of the sedimentary record, one should impose a stratigraphical constraint on the classification, at least for the purposes of zonation. In other words, one should insist that

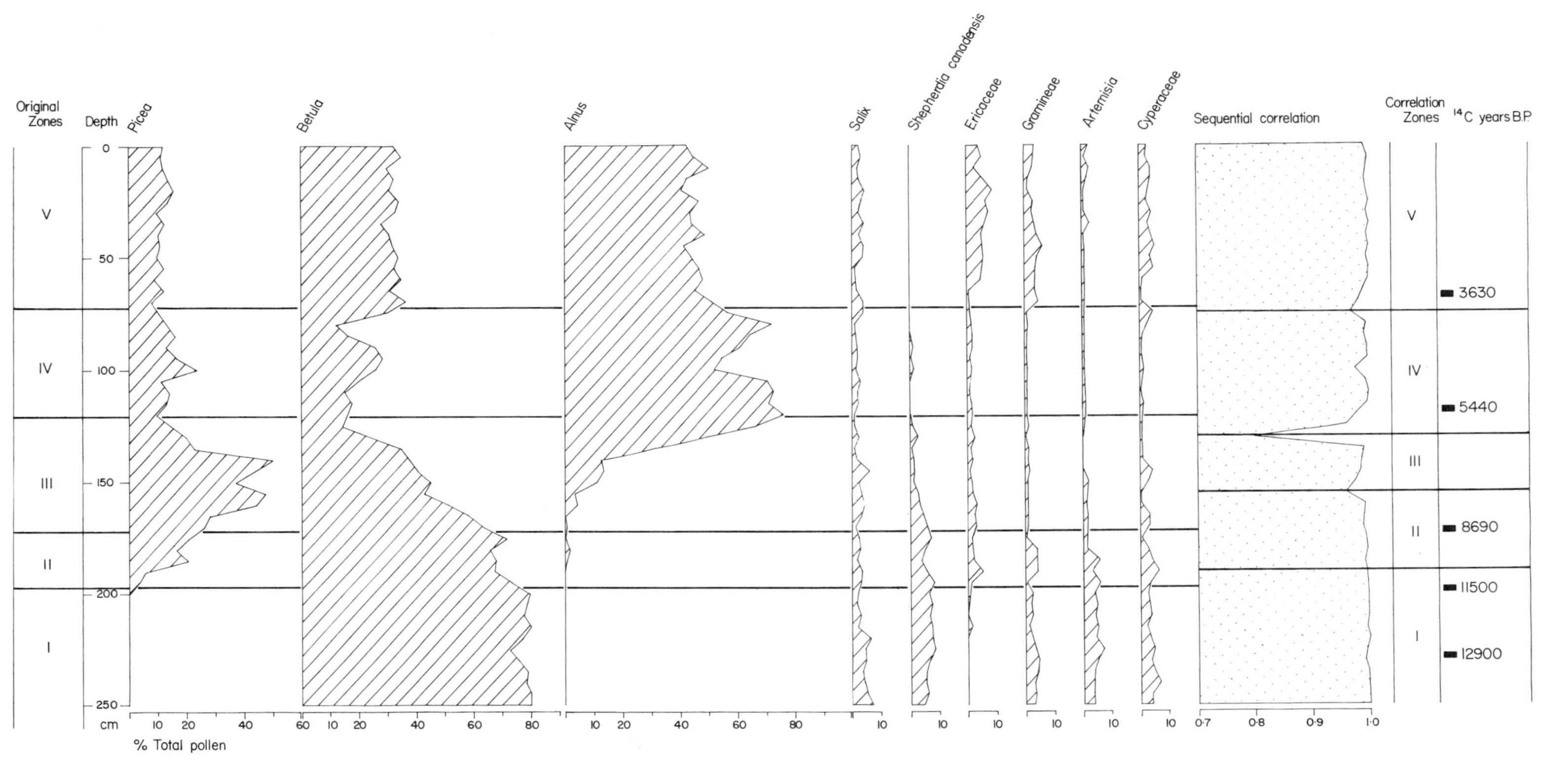

Figure 3.2 Pollen diagram from the Tuktoyaktuk Peninsula, Northwest Territories, Canada (Ritchie and Hare, 1971) illustrating the sequential correlation method of detecting zone boundaries proposed by Yarranton and Ritchie (1972). (After Yarranton and Ritchie, 1972.)

clusters or groups be formed only from sets of neighbouring levels. By imposing a stratigraphical constraint, the number of possible partitions of the data set which have to be considered is considerably reduced.

The introduction of a stratigraphical constraint into numerical classifications for Quaternary pollen zonation purposes was first proposed and implemented by E. J. Cushing (unpublished work cited by D. P. Adam in his discussion of Mosimann and Greenstreet, 1971, and used by Mehringer *et al.*, 1977; and Whitehead, 1979, 1981). Cushing imposed a constraint in Orloci's (1967a) sum-of-squares agglomerative procedure in which only adjacent levels or clusters of levels could be fused together. Dale and Walker (1970) mentioned introducing stratigraphical constraints as an idea to be pursued in future work (see also D. Walker, 1972). Falcon (1976) appears to have applied stratigraphical constraints in her single link cluster analysis of microfossil assemblages in Lower Karroo (Gondwana) sediments in Rhodesia to derive assemblage zones.

An investigation of various methods of constrained classification was carried out on a range of pollen stratigraphical data by Gordon and Birks (1972, 1974) and by Gordon (1973a). Sections 3.3 to 3.6 are devoted to a detailed description of these methods. These methods have been used as aids in the zonation of a wide variety of pollen stratigraphical data by, for example, H. J. B. Birks (1974, 1976b, 1981a); H. H. Birks and Mathewes (1978); H. J. B. Birks and Berglund (1979); Ritchie (1982), and Cwynar (1982). Gordon (1980b) provides a review of currently available methods of constrained classification.

A fourth approach has been to apply standard scaling procedures such as principal components analysis or non-metric multidimensional scaling to pollen stratigraphical data, no use being made in the numerical procedure of the known stratigraphical ordering. The geometrical configuration of points obtained can provide a useful check of whether the zonations suggested by the numerical procedures are a gross misrepresentation of the data. This is particularly relevant for stratigraphical data where there are gradual and often complex transitions from one pollen assemblage to another, for example in late-glacial profiles (Pennington and Sackin, 1975; Björck, 1979, 1981). These methods of analysis can also be of assistance in the comparison of several different pollen diagrams, as discussed in Sections 4.4 and 4.6.

The results of applying a scaling procedure to a stratigraphical data set need not be presented as a geometrical configuration of points; instead, the scores or coordinates of the levels on each axis can be plotted in stratigraphical order (see Adam, 1970, 1974; H. J. B. Birks, 1974; Pennington and Sackin, 1975; Huntley, 1976, 1981; H. J. B. Birks and Berglund, 1979; Björck, 1979, 1981; M. E. Edwards, 1980; Björck and Persson, 1981). This is illustrated in Figure 3.3, in which the scores on the first three principal axes are plotted in stratigraphical order for three post-glacial sequences in southern Sweden. This approach is particularly relevant if the axes of the representation effectively portray trends in the data. In some geometrical methods (such as principal components analysis), the axes have a particular significance, but this is not the case for other methods (such

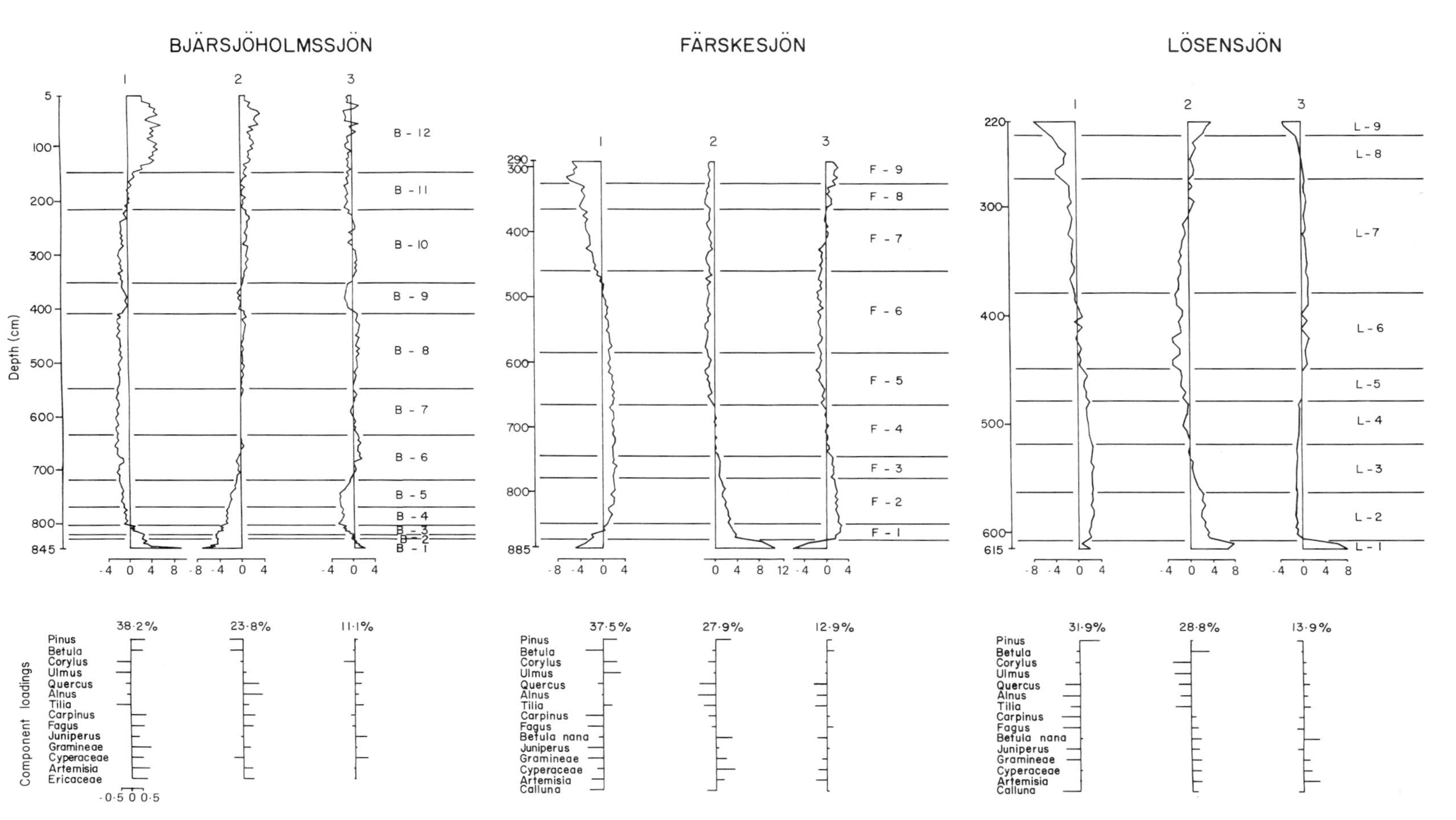

Figure 3.3 Stratigraphical plots of the sample scores on the first three principal components of three individual pollen sequences in southern Sweden. Local pollen zones for the sequences are also shown. The loadings of the pollen types on each of the principal components are given, along with the percentage of the variance accounted for by each component. (After H. J. B. Birks and Berglund, 1979.)

as non-metric multidimensional scaling), and some thought should be given to an appropriate selection of axes. There is the further point that sometimes geometrical configurations display a non-linear trend in the data, which would not be detected fully on any single axis. Nevertheless, such stratigraphical plots can be assessed visually in much the same way as pollen diagrams (the plots are, in a sense, 'composite pollen curves'), and can prove a useful adjunct to the methods described in this chapter, particularly in displaying the nature of transitions between zones and variations within zones.

All the methods described in this chapter are based on the analysis of sample proportions. There are no technical difficulties involved in extending the methods to analyse pollen concentrations or pollen-accumulation rates (see Mosimann and Greenstreet, 1971; Pennington and Sackin, 1975; Gordon, 1982c; Hunt and Birks, 1982). However, as discussed in Section 1.2, such data are subject to large and, at present, unavoidable errors in determination, and inevitably less confidence should be attached to any numerical analyses of them.

3.3 The Constrained Single Link Method

This method of classification is derived from the single link method of cluster analysis, but there is the constraint that at all times classes must comprise sets of stratigraphically neighbouring levels. The method can readily be implemented using an agglomerative algorithm; this is illustrated on the test data set of levels A1–A10 from the Abernethy Forest 1974 core. The dissimilarity matrix to which the method will be applied is given in Table 2.1. For convenience, the suffix 'A' will be dropped for the rest of this chapter, and the levels in the test data set will be referred to by the labels 1–10.

Initially, then, there are 10 separate groups, each containing a single object. Instead of searching the entire matrix of dissimilarities for the smallest dissimilarity to establish when the first amalgamation occurs, we are constrained to examine only the dissimilarities between stratigraphically neighbouring levels. At first, this involves examining only the dissimilarities along the diagonal of Table 2.1. These can be alternatively represented as follows:

```
1      2      3      4      5      6      7      8      9      10    level
   0.533  1.138  0.193*  0.350  0.449  1.647  0.276  1.258  0.375    dissimilarity
```

In this representation, the value on the second line gives the dissimilarity between the pair of levels bracketting it on the line above. Thus, the dissimilarity between levels 3 and 4 is 0.193. This is the smallest value on the second line (this fact being indicated by the asterisk), and so the first amalgamation is between levels 3 and 4, at a height of 0.193.

We now have to specify the dissimilarity between the group of levels (3, 4) and its stratigraphically neighbouring levels, 2 and 5. In the single link method, the dissimilarity between 2 and (3, 4) is defined to be the smaller of the dissimilarity between 2 and 3 ($d(2, 3)$, = 1.138), and the dissimilarity between 2 and 4 ($d(2, 4)$, = 1.117), in this case, the latter. In the same way, the dissimilarity between 5 and (3, 4) is defined to be the smaller of $d(3, 5)$ (= 0.501) and $d(4, 5)$ (= 0.350). None of the other dissimilarities in the representation is affected by the amalgamation of levels 3 and 4. After the first amalgamation, the situation may thus be represented by

1 2 (3, 4) 5 6 7 8 9 10 level
0.533 1.117 0.350 0.449 1.647 0.276* 1.258 0.375 dissimilarity

The next amalgamation is between levels 7 and 8, at a height of 0.276. After recalculating the dissimilarities, the representation becomes

1 2 (3,4) 5 6 (7, 8) 9 10 level
0.533 1.117 0.350* 0.449 1.473 1.258 0.375 dissimilarity

The next amalgamation leads to the group (3, 4, 5). The dissimilarity between 2 and (3, 4, 5) is defined to be the smallest of $d(2, 3)$, $d(2, 4)$ and $d(2, 5)$; that is, $d(2, 4)$ (= 1.117). Similarly, the dissimilarity between 6 and (3, 4, 5) turns out to be $d(5, 6)$ (= 0.449). The representation becomes

1 2 (3–5) 6 (7, 8) 9 10 level
0.533 1.117 0.449 1.473 1.258 0.375* dissimilarity

After the next amalgamation, between levels 9 and 10 at a height of 0.375, we have to evaluate the dissimilarity between the groups (7, 8) and (9, 10). This is defined to be the smallest dissimilarity between any pair of levels, one in each group, that is, the smallest of $d(7, 9)$, $d(7, 10)$, $d(8, 9)$ and $d(8, 10)$. The smallest is $d(8, 9)$ (= 1.258), and so the representation becomes

1 2 (3–5) 6 (7, 8) (9, 10) level
0.533 1.117 0.449* 1.473 1.258 dissimilarity

Subsequent steps may be summarised by the following representations:

1 2 (3–6) (7, 8) (9, 10) level
0.533* 1.117 1.473 1.258 dissimilarity

(1, 2) (3–6) (7, 8) (9, 10) level
1.117* 1.473 1.258 dissimilarity

(1–6) (7, 8) (9, 10) level
1.473 1.258* dissimilarity

(1–6) (7–10) level
0.873 dissimilarity

It should be noted that after the amalgamation of groups (7, 8) and (9, 10) at height 1.258, the dissimilarity between groups (1–6) and (7–10) is given as 0.873. This is

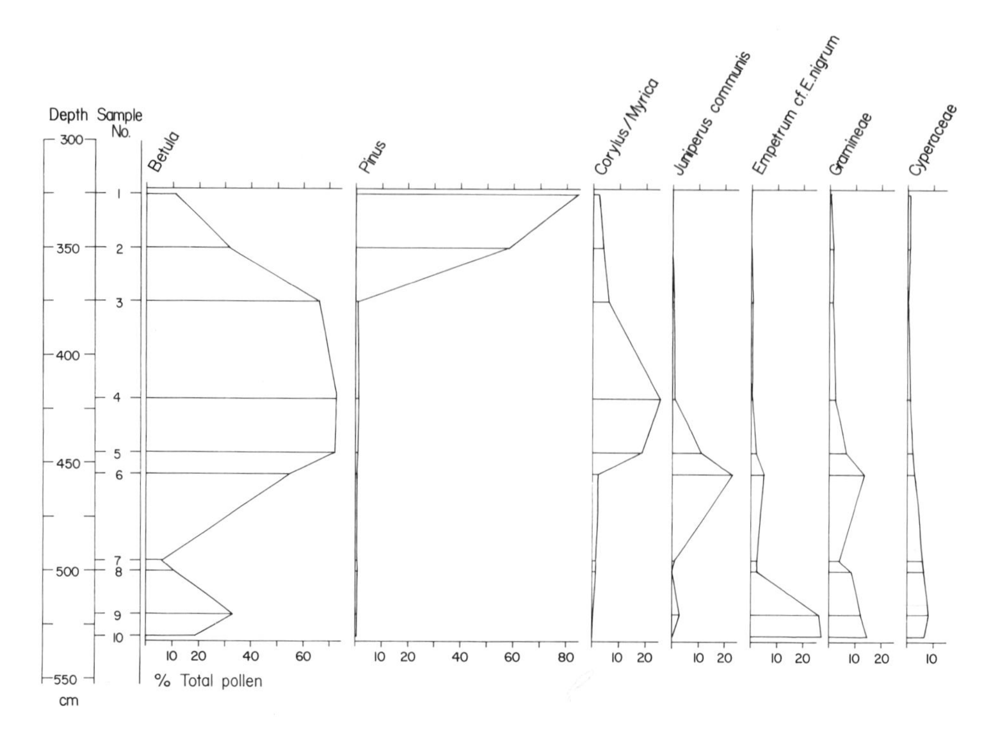

Figure 3.4 Pollen diagram for the Abernethy Forest 1974 test data set, together with the numerical

the dissimilarity between levels 6 and 9, but they were prevented from amalgamating earlier by the stratigraphical constraint; they were not in stratigraphically neighbouring classes until after the amalgamation of groups (7, 8) and (9, 10). The results of the constrained single link analysis will thus differ from the results of an ordinary single link analysis. As the groups (1–6) and (9, 10) cannot amalgamate until after the group (7, 8) has fused with one of them, the height at which the final amalgamation occurs is defined as 1.258.

The algorithm described above is incorporated in the subroutine CSLINK of the FORTRAN IV program ZONATION, listed in the Appendix. The entire set of amalgamations can be represented by the dendrogram, or tree diagram, given on the right-hand side of Figure 3.4. Sectioning the dendrogram at a particular value of the height (1.0, say, as given by the dashed line) will yield the constrained single link groups at that height (the groups at height 1.0 are (1, 2), (3–6), (7, 8) and (9, 10)). The assessment of such groups in an attempt to delimit local pollen zones will be considered in more detail after other methods of analysis have been introduced.

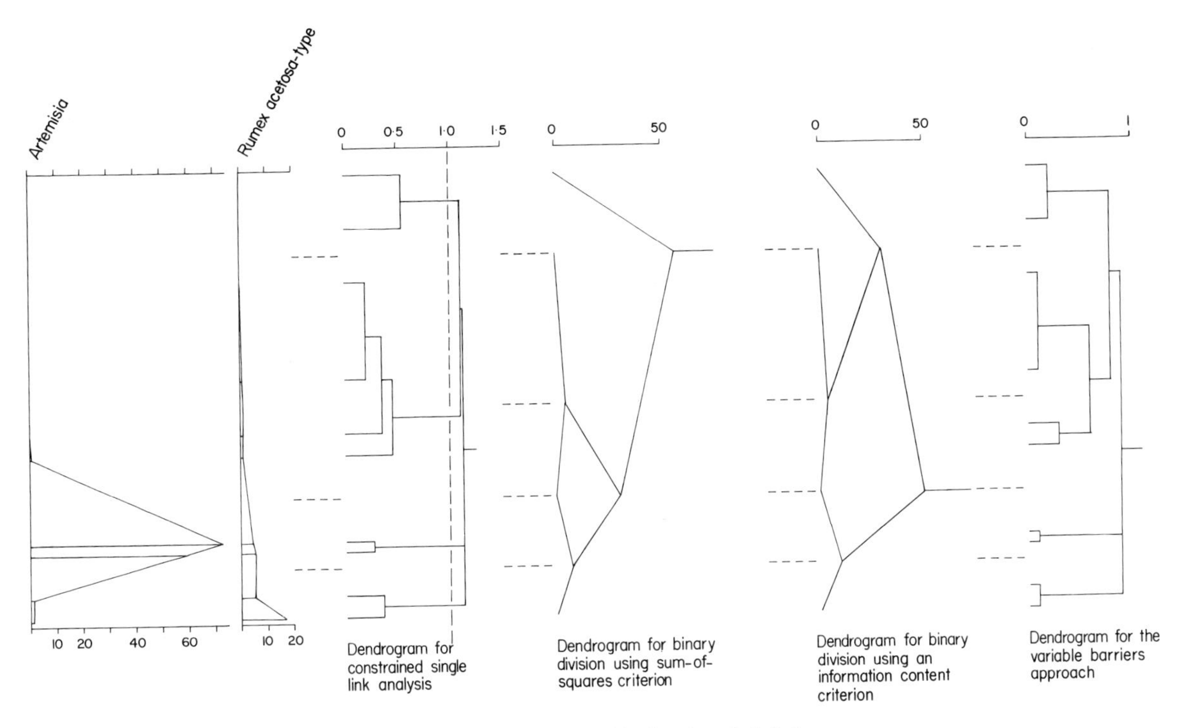

zonations indicated by the methods of analysis described in Sections 3.3–3.5.

3.4 Binary Divisive Procedures

The constrained single link method was found, when applied to large data sets, to have difficulty in discerning changes in parts of pollen sequences where there were numerically small but stratigraphically consistent differences (see Gordon and Birks, 1972). When using the single link criterion, for an object to belong to a group it need be similar to only one other member of the group, and the group as a whole could encompass a wide range of variation. An alternative approach, which is considered in Sections 3.4–3.6, is to define a global measure of variability which will be minimised when the data are partitioned into fairly compact groups of similar objects.

We seek, for example, to partition the objects into a certain number of groups (g, say) so as to minimise the total within-group sum of squared distances about the g centroids. There are, in general, a large number of possible partitions which would all have to be examined if we wished to be sure of identifying the partition with minimum sum-of-squares. However, if we insist that the groups must consist of stratigraphically neighbouring levels, the number of allowable partitions is greatly reduced. We can represent such a constrained partition into g groups by a set of $g - 1$ markers which indicate the boundaries between neighbouring groups. For example, the 1974 test data set of 10 levels can be divided into four groups by specifying three markers. A marker will be labelled by the number of the last level in the stratigraphically higher group. Thus, markers labelled 2, 6, and 8 would divide the sequence into the groups (1, *2*), (3–*6*), (7, *8*), and (9, 10).

The optimal partition of a pollen sequence into g groups for $g = 2, 3, 4, 5, \ldots$ can be found by means of the dynamic programming algorithm described in the next section. However, for many measures of variability the partitions so obtained need not be hierarchically nested (e.g., the optimal partition into three groups need not be obtained from the optimal partition into two groups by dividing one of these two groups; this is illustrated by an example in Section 3.7). It appears that pollen analysts in general find a hierarchical solution most useful, in that it allows an examination of the relationship between zones, sub-zones, and smaller units. An approach which leads to a hierarchical solution is described below.

It is necessary here to recall the symbolic definition of the sample proportions p_{ik} denoting the proportion of the ith level ($i = 1, 2, \ldots, 10$) which is composed of the kth taxon ($k = 1, 2, \ldots, 9$; where $k = 1$ corresponds to *Betula*, $k = 2$ corresponds to *Pinus*, etc.). Thus, for example (from Table 1.3), $p_{1,1} = 0.118$, $p_{1,2} = 0.848$, $p_{1,9} = 0.000$, $p_{2,1} = 0.321$, $p_{2,2} = 0.582$, $p_{2,9} = 0.000$. Consider dividing the test data set into the two groups (1, 2), (3–10), this corresponding to the introduction of a marker labelled 2. The averages of the proportions of levels 1 and 2 are

Q_1	Q_2	Q_3	Q_4	Q_5	Q_6	Q_7	Q_8	Q_9
0.2195	0.715	0.046	0.002	0.000	0.012	0.006	0.000	0.000

(e.g., $Q_1 = \frac{1}{2}(p_{1,1} + p_{2,1}) = \frac{1}{2}(0.118 + 0.321)$). The terms $\{Q_k(k = 1, 2, \ldots, 9)\}$ may be regarded as the sample proportions of a 'dummy' level which summarises the levels in the group (1, 2) by giving the group averages. The sum-of-squares of the group (1, 2) is the sum of the squared distances of each of the levels 1 and 2 from the mean 'dummy' level, that is,

$$(0.118 - 0.2195)^2 + (0.848 - 0.715)^2 + \cdots + (0.000 - 0.000)^2$$
$$+ (0.321 - 0.2195)^2 + (0.582 - 0.715)^2 + \cdots + (0.000 - 0.000)^2 = 0.0570865.$$

If we define $q_{ik} = Q_k$ (for levels $i = 1, 2$; taxa $k = 1, 2, \ldots, 9$), that is, let $\{q_{ik}(k = 1, 2, \ldots, 9)\}$ denote the sample proportions of the mean of the group to which level i belongs, then the sum-of-squares of the group (1, 2) can be represented in symbols as

$$\sum_{i=1}^{2} \sum_{k=1}^{9} (p_{ik} - q_{ik})^2.$$

The more dissimilar the levels in the group are, the larger this expression will become; it gives a measure of the within-group variability of the group (1, 2). We can show, by a similar analysis of the proportions of the other levels given in Table 1.3, that the mean 'dummy' level of the group (3–10) has proportions

Q_1^*	Q_2^*	Q_3^*	Q_4^*	Q_5^*	Q_6^*	Q_7^*	Q_8^*	Q_9^*
0.419125	0.019625	0.06225	0.048625	0.111125	0.084	0.042875	0.16925	0.04275

Defining $q_{jk} = Q_k^*$ (for levels $j = 3, 4, \ldots, 10$; taxa $k = 1, 2, \ldots, 9$), the sum-of-squares of the group (3–10) is (to 7 decimal places)

$$\sum_{j=3}^{10} \sum_{k=1}^{9} (p_{jk} - q_{jk})^2 = (0.661 - 0.419125)^2 + (0.016 - 0.019625)^2 + \cdots + (0.189 - 0.04275)^2 = 1.5638909$$

Hence, the total within-group sum-of-squares for the groups (1, 2), (3–10) is (to 7 decimal places)

$$0.0570865 + 1.5638909 = 1.6209774.$$

This just gives a measure of the variability associated with one possible partition of the data into two groups, namely, the partition associated with having a marker labelled 2. We wish to find the most homogeneous pair of groups, and so wish to examine all possible positions for the marker in order to establish which position gives rise to the smallest total within-group sum-of-squares. Table 3.1 summarises the total within-group sum-of-squares (correct to three decimal places) for each of the nine possible partitions of the test data set into two groups. It is seen that the partition examined in detail above is the optimal one with respect to the sum-of-squares criterion, that is, the most homogeneous pair of groups is the pair (1, 2), (3–10).

This completes the first stage of the analysis. We now wish to find the optimal partition into three groups, where we have the further constraint that this must be obtained by dividing one of the existing groups, (1, 2), (3–10), into two subgroups. This can be formulated as seeking to insert two markers in the sequence, where the first marker must remain fixed at position 2. There are eight possible positions for the second marker. If it were placed in position 6, this would imply

Table 3.1
Partition of the Abernethy Forest 1974 Test Data Set into Two Groups

	Marker								
	1	2	3	4	5	6	7	8	9
Total within-group sum-of-squares	1.898	1.621	1.940	1.961	1.856	1.717	2.143	2.176	2.329

that levels (3–6) have mean proportions q_{ik} (for levels $i = 3, 4, 5, 6$; taxa $k = 1, 2, \ldots, 9$) that equal

(0.6655, 0.0175, 0.12025, 0.09075, 0.02175, 0.06475, 0.01575, 0.0005, 0.00325);

and that levels (7–10) have mean proportions q_{jk} (for levels $j = 7, 8, 9, 10$; taxa $k = 1, 2, \ldots, 9$) that equal

(0.17275, 0.02175, 0.00425, 0.0065, 0.2005, 0.10325, 0.070, 0.338, 0.08225);

the mean proportions of levels (1, 2) are, of course, not changed. It can be shown that this partition, into (1, 2), (3–6), (7–10), is the partition into three groups which has minimum total within-group sum-of-squares among these examined.

The algorithm continues in the same manner, at each stage dividing an existing group into two sub-groups and choosing that division which yields the maximum drop in the total within-group sum-of-squares. The first six divisions are summarised in Table 3.2. The figure on the first line gives the total within-group sum-of-squares when all the levels belong to a single group, and the sum-of-squares at subsequent stages is also expressed as a percentage of this grand total. It can be seen that successive divisions lead to smaller reductions in the sum-of-squares. At some stage, when the introduction of further markers would not lead to a great reduction in the variability, we stop the divisive procedure, and regard the partition into homogeneous groups of stratigraphically neighbouring levels as a suggested zonation. The results can also be represented by the upper part of a 'block' dendrogram, as shown for example in Figure 3.4 for the first four divisions.

The algorithm described above has been implemented in the subroutine SPLSQ of the program ZONATION, listed in the Appendix. Because of the danger, mentioned in Section 2.5, that the clustering criterion might impose unwarranted structure on the data, a similar binary divisive procedure can be carried out, using a different measure of within-group variability. If similar results are suggested by the two criteria, one can be more confident that the groups suggested are not purely an artifact of the clustering criterion used.

Table 3.2
Binary Division of the Abernethy Forest 1974 Test Data Set Using the Sum-of-Squares Criterion

Sum-of-squares	Percentage of total	Markers
2.541		
1.621	63.8	2
0.781	30.7	2 6
0.206	8.1	2 6 8
0.121	4.8	2 4 6 8
0.064	2.5	1 2 4 6 8
0.038	1.5	1 2 4 5 6 8

A second measure of the variability of, say, the levels (3–10) is the information content

$$\sum_{i=3}^{10} \sum_{k=1}^{9} p_{ik} \log(p_{ik}/q_{ik}),$$

where the q_{ik}'s are obtained exactly as before. Unless otherwise stated, in this book *log* is to be interpreted as 'logarithm to base e'; in the context of information theory, as here, it is convenient to use logarithms to base 2. The subroutine SPLINF of the program ZONATION carries out exactly the same binary divisive procedure as described above, using the information content as the measure of variability. When this procedure is applied to the 1974 test data set, the first six divisions are as given in Table 3.3, and the 'block' dendrogram for the first four divisions is shown in Figure 3.4.

It should be noted that these criteria of variability can be modified to allow for the possibility of differential weighting of taxa (or even of levels). Weighting of taxa could, for example, be implemented by defining the measure of within-group variability as a weighted sum-of-squares,

$$\sum_{i} \sum_{k=1}^{9} w_k(p_{ik} - q_{ik})^2.$$

Two possible aims for the weighting factors $\{w_k(k = 1, 2, \ldots, 9)\}$ are

1. to reduce the effect of taxa which are present in large quantities (and which will thus tend to have greater absolute variability) by using some kind of standardisation (see Gordon and Birks, 1972, p. 970); and
2. to give extra weight to taxa which are believed to be of critical ecological importance, though there are clearly dangers of a circular argument in this kind of approach (see Anonymous, 1972; Janssen, 1980).

The essential property of the data which allowed the algorithms described above to be used is not that there is a linear ordering of the objects, but that there

Table 3.3

Binary Division of the Abernethy Forest 1974 Test Data Set Using the Information Content Criterion

Information content	Percentage of total	Markers
10.706		
5.788	54.1	6
3.068	28.7	2 6
1.169	10.9	2 6 8
0.441	4.1	2 4 6 8
0.307	2.9	1 2 4 6 8
0.180	1.7	1 2 4 6 8 9

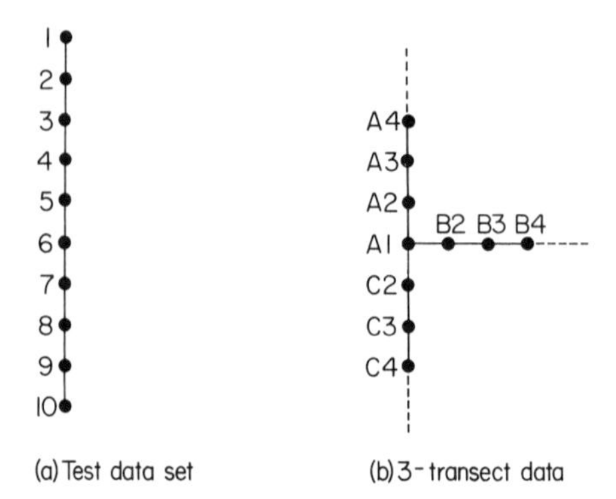

Figure 3.5 Graph-theory representations of two data sets which can be analysed by a constrained classification method: (a) objects have linear ordering imposed, such as the Abernethy Forest 1974 test data set; (b) objects located along one of three transects, starting from a common point. (See Caseldine and Gordon, 1978.)

are contiguity constraints of the following kind. Consider the graph representation in Figure 3.5, where an object is represented by a point, and contiguous pairs of objects have the corresponding points joined by a line. Thus, the test data set can be represented by Figure 3.5a. The necessary condition for the methods of constrained classification to be applicable is that the removal of *any* line will divide the points into two sets of linked points. This condition is also satisfied by the graph given in Figure 3.5b, which represents surface pollen samples taken along three transects (A, B, C) meeting at the centre of a bog. Constrained classifications of these data are given by Caseldine and Gordon (1978) and M. M. Campbell (1978).

3.5 Dynamic Programming Algorithm

When the 1974 test data set was partitioned into three groups by the binary divisive procedures, the groups (1, 2), (3–6), (7–10) were obtained, corresponding to markers in positions 2 and 6. We cannot guarantee that these are the positions for the two markers which give the overall minimum total within-group sum-of-squares, because we fixed the first marker at 2, and only allowed the second marker to find its optimal position subject to this restriction. Examining all possible positions for 2 (then 3, 4, 5, ...) markers soon becomes very time consuming for larger data sets, but the following algorithm greatly reduces the computation required. It is based on the optimality principle of dynamic programming (Bellman and Dreyfus, 1962, p. 15). An early application of this principle to classification was given by W. D. Fisher (1958); the approach has also been used to analyse geological sections by Hawkins and Merriam (1973, 1974, 1975) and a related method (Bement and Waterman, 1977) could be used to identify 'peak' pollen zones.

Let $s(i, j)$ denote the within-group sum-of-squares of the set of levels i to j inclusive; and let $t(g, l)$ denote the total within-group sum-of-squares when the set of levels (1 to l) is optimally divided into g groups. For example, from the previous section we have $s(1, 2) = 0.057$, $s(3, 10) = 1.564$, $t(2, 10) = 1.621$ (all to three decimal places). We wish to find $t(g, 10)$ for $g = 3, 4, 5, \ldots$, together with the location of the optimal $(g - 1)$ markers.

The algorithm proceeds by successively evaluating the elements of the array $t(\ ,\ )$, in the order

$$\begin{array}{r} t(1, 1), t(1, 2), t(1, 3), t(1, 4), \ldots, t(1, 10), \\ t(2, 2), t(2, 3), t(2, 4), \ldots, t(2, 10), \\ t(3, 3), t(3, 4), \ldots, t(3, 10), \\ t(4, 4), \ldots, t(4, 10), \\ \vdots \\ t(10, 10); \end{array}$$

at each stage using previous results in order to make the current evaluation.

For $j = 1, 2, \ldots, 10$, $t(1, j)$ is just equal to $s(1, j)$, the within-group sum-of-squares of the set of levels (1 to j), since the optimal division into one group is just the given group. Successive rows of $t(\ ,\ )$ are evaluated by means of the recursive relation

$$t(g, l) = \underset{g-1 \le i \le l-1}{\text{minimum}} \{t(g - 1, i) + s(i + 1, l)\}$$
$$\text{for } l = g, g + 1, \ldots, 10; \quad g = 2, 3, \ldots, 10.$$

This recursive scheme considers all possible allowable positions for the marker with the highest label (in the above expression, it is labelled i, and can take values between $(g - 1)$ and $(l - 1)$), and chooses that value which minimises the expression within the curly brackets. For example, consider the evaluation of $t(3, 5)$, that is, the sum-of-squares of the optimal partition into three groups of the levels (1–5). At this stage in the analysis, we will already have evaluated all the elements of the array $s(\ ,\ )$, and the first two rows of the array $t(\ ,\ )$. A summary of the information required for the evaluation of $t(3, 5)$ is (all correct to three decimal places):

$$\begin{array}{lll} t(2, 2) = 0.0, & t(2, 3) = 0.057, & t(2, 4) = 0.064, \\ s(3,5) = 0.044, & s(4,5) = 0.019, & s(5, 5) = 0.0 \end{array}$$

Dividing the first five levels into three groups involves inserting two markers into the sequence. The marker with the higher label can be assigned one of the three labels, $i = 2$, 3, or 4.

Setting $i = 2$ means dividing levels (1, 2) into two groups (with sum-of-squares $t(2, 2) = 0.0$), and placing levels (3–5) in the third group (with sum-of-squares $s(3, 5) = 0.044$). Hence, this partition would have a total within-group sum-of-squares of $0.0 + 0.044 = 0.044$.

Setting $i = 3$ would imply dividing levels (1–3) into two groups. The optimal division of the first three levels into two groups has sum-of-squares $t(2, 3) = 0.057$, which would have been evaluated in the previous stage of the algorithm; the third group, levels (4, 5), has sum-of-squares $s(4, 5) = 0.019$, and so this partition would have a total within-group sum-of-squares of $0.057 + 0.019 = 0.076$.

Similarly, the partition associated with $i = 4$ would have a total within-group sum-of-squares of $t(2, 4) + s(5, 5) = 0.064 + 0.0 = 0.064$.

The smallest value of the total within-group sum-of-squares is the value associated with having the higher marker (i^*, say) at 2; $t(3, 5) = \text{minimum}(0.044, 0.076, 0.064) = 0.044$. The position of the lower marker (in this case, clearly at 1) will have been recorded during the evaluation of $t(2, 2)$, and so the optimal partition into three groups of levels (1–5) is (1), (2), (3–5).

For larger values of g and l, the positions of the markers are recovered by a traceback procedure. For example, $t(4, 10) = 0.206$, with $i^* = 8$. This means that the smallest sum-of-squares was obtained when levels (1–8) were divided into three groups and the fourth group comprised levels (9, 10). An examination of $t(3, 8)$ shows that its corresponding i^* is 6, and $t(2, 6)$ has $i^* = 2$. Hence, the markers occur at 2, 6, and 8, and the optimal partition of the 1974 test data set into four groups is (1, 2), (3–6), (7, 8), (9, 10). This is the same partition as was obtained by the binary divisive algorithm, and, in fact, the two algorithms give identical results when applied to the test data set. The binary divisive approximating algorithm has thus in this case produced the overall optimal partition into g groups for all values of g. When the two algorithms are applied to the full data sets, there are some differences (see Section 3.7), although there is a fairly close correspondence of results. For a wide range of data sets which have been analysed, the binary divisive algorithm has been found to give a close correspondence to the overall optimal solution in the great majority of cases.

The dynamic programming algorithm described in this section is implemented in the subroutine DNAMIC of the program ZONATION, listed in the Appendix. Although the program implements the algorithm only for the sum-of-squares criterion, the general approach can be used with any criterion for which the total 'variability' of the partition into groups is simply the sum of the 'variabilities' within each group separately.

Gordon (1982c) has used the dynamic programming algorithm to zone several pollen stratigraphical sequences; it merits more widespread use.

There are two other applications of the dynamic programming algorithm which might be relevant in certain circumstances. Firstly, Hawkins and Merriam (1973) impose further restrictions on the allowable partitions, stipulating that all groups shall contain more than a specified minimum number of objects. Secondly, Bement and Waterman (1977) describe how one can identify segments of sequential data which have maximum variability; this approach could be used to search for 'peak' zones or regions of rapid transition in a pollen diagram.

3.6 THE VARIABLE BARRIERS APPROACH

The algorithms described in the previous two sections might not give an accurate representation of pollen sequences in which there is a large number of transitional levels between regions with homogeneous pollen proportions (the second type of 'interval' described by D. Walker, 1972; see Section 3.1). The algorithms considered so far will tend to place boundaries in the middle of such transitional intervals. An approach which attempts to distinguish such transitional levels is described below.

Consider a generalisation of the situation described in Sections 3.4 and 3.5. Instead of markers between certain pairs of neighbouring levels, indicating the presence of the boundary between two groups, there is a barrier between each pair of neighbouring levels. The height of each barrier lies between 0 and 1. The q_{jk}'s are, as before, an average of the p_{ik}'s for values of i close to j, but there are the following differences:

1. The q_{jk}'s are a *weighted* average of the p_{ik}'s for i lying between certain limits.
2. The limits are found as follows: from j, we can 'reach' any level which involves 'jumping over' barriers whose total height is less than 1; the p_{ik}'s which can be 'reached' from j contribute towards the evaluation of q_{jk}.
3. As before, we seek to minimise the total sum-of-squares,

$$\sum_{i=1}^{10} \sum_{k=1}^{9} (p_{ik} - q_{ik})^2,$$

but there is now the restriction that the sum of the barrier heights should be fixed ($=h$, say).

Let x_m denote the height of the barrier between level m and level $(m + 1)$, for $m = 1, 2, ..., 9$, and consider the set of barriers given in Figure 3.6. By convention, the two end barriers, which are not included in the sum, always have height 1.0; thus, $x_0 = 1.0 = x_{10}$. Let y_{ji} denote the 'weight' which the proportions in the ith level contribute to the value of q_{jk}, that is,

$$q_{jk} = \sum_{i=1}^{10} (y_{ji}\, p_{ik}) \Big/ \sum_{i=1}^{10} y_{ji}$$

for levels $j = 1, 2, ..., 10$, taxa $k = 1, 2, ..., 9$, where y_{ji} is formally defined by

$$y_{ji} = \begin{cases} \text{maximum}\left(0,\ 1 - \sum_{m=j}^{i-1} x_m\right) & \text{if } i > j \\ 1 \quad \text{if } i = j & \\ \text{maximum}\left(0,\ 1 - \sum_{m=i}^{j-1} x_m\right) & \text{if } i < j, \end{cases}$$

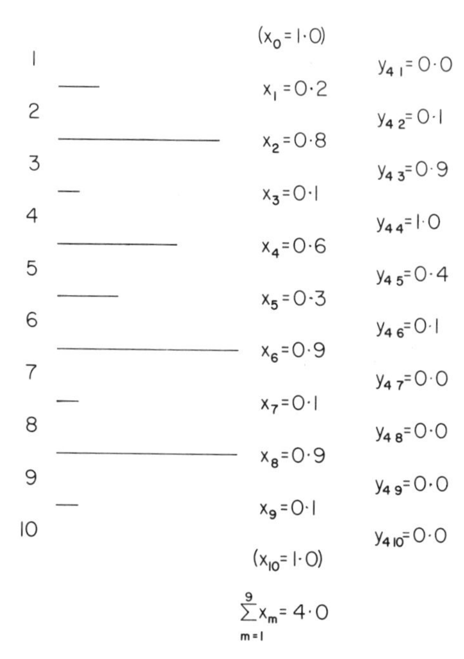

Figure 3.6 The variable barriers approach illustrated by the Abernethy Forest 1974 test data set. A barrier of height x_m is placed between levels m and $m + 1$ (for $m = 1, 2, \ldots, 9$). The weights y_{4j} ($j = 1, \ldots, 10$) are also shown. For an explanation, see the text.

from which it can be seen that $y_{ij} = y_{ji}$. A verbal description of the above formula is as follows: we find y_{ij} by adding up the heights of all the barriers lying between levels i and j, and subtracting the sum of these heights from 1. If this gives a negative answer, substitute the value 0, as we do not wish to have any negative weights. There will thus be a progressively smaller contribution to q_{jk} from levels further away from level j. For example, the contributions y_{4i} ($i = 1, 2, \ldots, 10$) are shown at the side of Figure 3.6. From these, we can evaluate

$$q_{4k} = \frac{0.1p_{2k} + 0.9p_{3k} + p_{4k} + 0.4p_{5k} + 0.1p_{6k}}{0.1 + 0.9 + 1.0 + 0.4 + 0.1};$$

for example,

$$q_{41} = \frac{(0.1)(0.321) + (0.9)(0.661) + 0.734 + (0.4)(0.722) + (0.1)(0.545)}{2.5} = 0.68172.$$

The entire set q_{jk} (for levels $j = 1, 2, \ldots, 10$, taxa $k = 1, 2, \ldots, 9$) is evaluated in this manner, and the corresponding sum-of-squares,

$$S(x) = \sum_{i=1}^{10} \sum_{k=1}^{9} (p_{ik} - q_{ik})^2$$

is obtained. As is emphasised by the notation, $S(x)$ depends on the particular values taken by $x_1, x_2, \ldots, x_9$. We seek to find those values which lead to a minimum of the sum-of-squares, subject always to

$$\text{Condition A:} \quad \sum_{m=1}^{9} x_m = 4 \qquad \text{and} \qquad x_m \geq 0 \ (m = 1, 2, \ldots, 9).$$

In the absence of such a condition, a solution with $S(x) = 0$ would be obtained, with $x_m = 1$ $(m = 1, 2, \ldots, 9)$.

This problem has not been solved analytically. Condition A is satisfied by a change of variables, and a solution is obtained by recourse to an iterative function-minimisation procedure. The solution has values of $x_1, x_2, \ldots, x_9$ close to those shown in Figure 3.6.

It can be seen that restricting the barriers to be all of height either 0 or 1 would reduce the problem to the one considered in the two previous sections: the q_{jk}'s would then be (unweighted) averages of all levels lying between adjacent barriers of height 1 (or markers). For example, setting x_2, x_4, x_6, and x_8 all equal to 1 and all the other barrier heights equal to 0 would divide the sequence into the five groups: (1, 2), (3, 4), (5, 6), (7, 8), (9, 10).

In the variable barriers approach, instead of insisting that each barrier must be of height 0 or 1, we allow the heights of the barriers to take any value between 0 and 1, subject to the constraint that the sum of the heights shall be fixed; this constraint is analogous to specifying the number of groups. Minimising the sum-of-squares subject to this condition will ensure that levels with dissimilar sample proportions will be separated by high barriers, but instead of barriers of height 1 abruptly marking group boundaries, there will be a configuration of lower barriers. Levels belonging to regions of rapid transition will appear between adjacent higher barriers, as will be illustrated in the next section.

An alternative manner of conception is that we can regard the above procedure as defining a measure of dissimilarity, $d^*_{ij} = 1 - y_{ij}$, between each pair of levels. This 'dissimilarity' differs from previous concepts of dissimilarity given in this book in that d^*_{ij} depends not only on the physical difference between levels i and j but also on the relationships between these levels and neighbouring levels. Analysis of the matrix of dissimilarities (d^*_{ij}) using the single link method will yield precisely the same results as the consideration of the barrier heights given above. This is because x_m, the height of the barrier between level m and level $m + 1$, is the 'branch' of the dendrogram associated with level m (for $m = 1, 2, \ldots, 9$; recall also that $x_{10} = 1$. This relationship can perhaps be appreciated most readily by considering the height at which two neighbouring levels come into the same group (e.g., levels 3 and 4 at 0.1; but 0.1 is precisely the height of the barrier separating these levels).

The single link dendrogram can be constructed from the barriers given in Figure 3.6 as follows. Move each barrier up half a 'notch', so that the end of the barrier of height x_m is opposite level m (for $m = 1, 2, \ldots, 10$). From the end of each barrier drop a perpendicular until it hits some barrier below it. This yields the

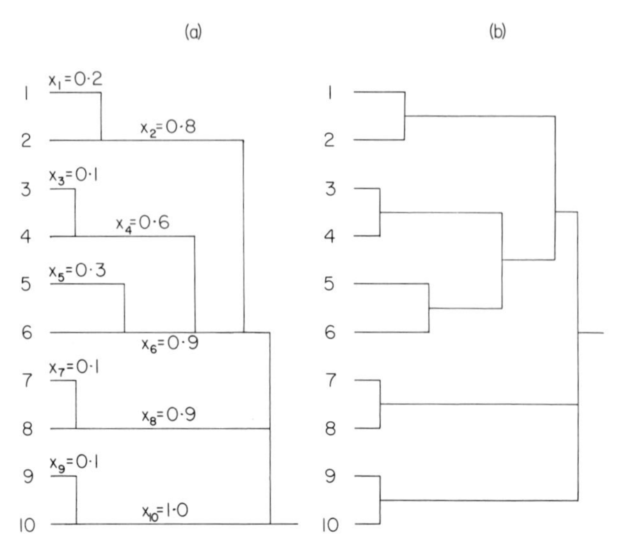

Figure 3.7 The variable barriers approach illustrated by the Abernethy Forest 1974 test data set: (a) barrier of height x_m is identified with branch of dendrogram associated with level m ($m = 1, 2, \ldots, 9$); (b) a conventional BARRIER dendrogram.

dendrogram given in Figure 3.7a. A more conventional dendrogram, in which the branches are realigned after each amalgamation, is shown in Figure 3.7b, and this is the form in which the results for the data set are summarised in Figure 3.4.

It can be seen that transitional levels will now be represented as late additions to groups visualised as growing in an agglomerative fashion. It has been found that the results obtained are not critically dependent on the value of h, the sum of the barrier heights; broadly similar results (with a different scale) are obtained over a range of values for h.

The variable barriers approach to zoning pollen stratigraphical sequences has been used by Gordon and Birks (1974), H. H. Birks and Mathewes (1978), and H. J. B. Birks (1981a). In all instances, the results obtained have been of value in the subsequent interpretation of the stratigraphical data.

3.7 Examples of Numerical Zonations

3.7.1 Abernethy Forest, Inverness-shire

The numerical zonation procedures described in Sections 3.3–3.6 have been applied to the two full data sets from Abernethy Forest, the 1970 and 1974 cores. The optimal sum-of-squares partitions, as given by the dynamic programming algorithm, are listed in Tables 3.4 and 3.5. The six-group solution for the 1974 data and the five-group solution for the 1970 data are shown diagrammatically along

Table 3.4
Optimal Sum-of-Squares Partitions of the Abernethy Forest 1974 Data

Number of groups *g*	Percentage of total sum-of-squares	Markers								
2	59.3	15								
3	28.4	15	32							
4	18.9	15	33	41						
5	14.7	15	33	41	45					
6	10.6	15	32	34	41	45				
7	8.1	15	26	32	34	41	45			
8	5.8	8	15	26	32	34	41	45		
9	4.7	8	15	24	29	32	34	41	45	
10	3.9	8	15	24	29	32	33	34	41	45

with the dendrograms summarising the results of the other analyses on Figures 3.8 and 3.9, respectively. The zones suggested by each of the numerical procedures are also shown next to the dendrograms as broken lines to facilitate ease of comparison with the pollen data and with the original zonation.

The Abernethy Forest 1974 data (Figure 3.8) show considerable consistencies between the five zonations suggested by the numerical analyses, as well as with the original pollen zonation proposed by H. H. Birks and Mathewes (1978). This zonation was itself based on numerical methods but the original analyses used a larger number of pollen types than were used here. All the zonations suggest partitions between levels 15 and 16 (expansion of *Pinus* pollen), 41 and 42 (expansion of *Artemisia* pollen), and 45 and 46 (expansion of *Empetrum* pollen). Most of the five numerical zonations also suggest divisions between levels 32 and 33 and/or levels 34 and 35.

Because the numerical methods make different assumptions about the data, involve different measures and concepts of dissimilarity or variability, and use

Table 3.5
Optimal Sum-of-Squares Partitions of the Abernethy Forest 1970 Data

Number of groups *g*	Percentage of total sum-of-squares	Markers								
2	41.4	23								
3	26.3	23	34							
4	22.3	18	23	34						
5	19.1	18	23	30	34					
6	16.5	19	22	23	30	34				
7	15.2	19	22	23	30	34	40			
8	13.4	19	22	23	30	34	36	40		
9	12.1	2	19	22	23	30	34	36	40	
10	10.8	2	19	22	23	24	30	34	36	40

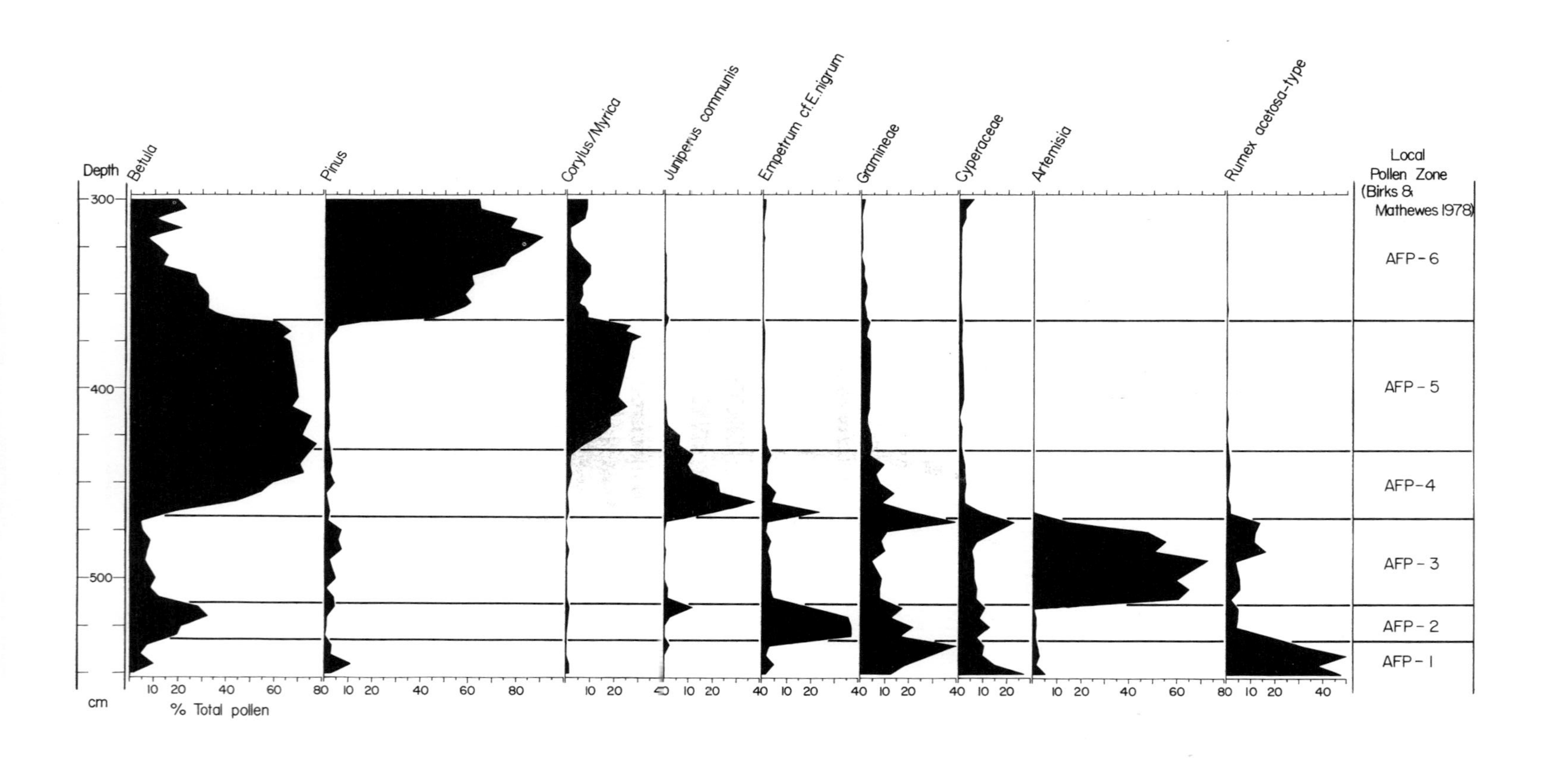

Depth
cm
300
400
500
Betula
Pinus
Corylus/Myrica
Juniperus communis
Empetrum cf.E.nigrum
Gramineae
Cyperaceae
Artemisia
Rumex acetosa-type
% Total pollen
Local Pollen Zone (Birks & Mathewes 1978)
AFP-6
AFP-5
AFP-4
AFP-3
AFP-2
AFP-1

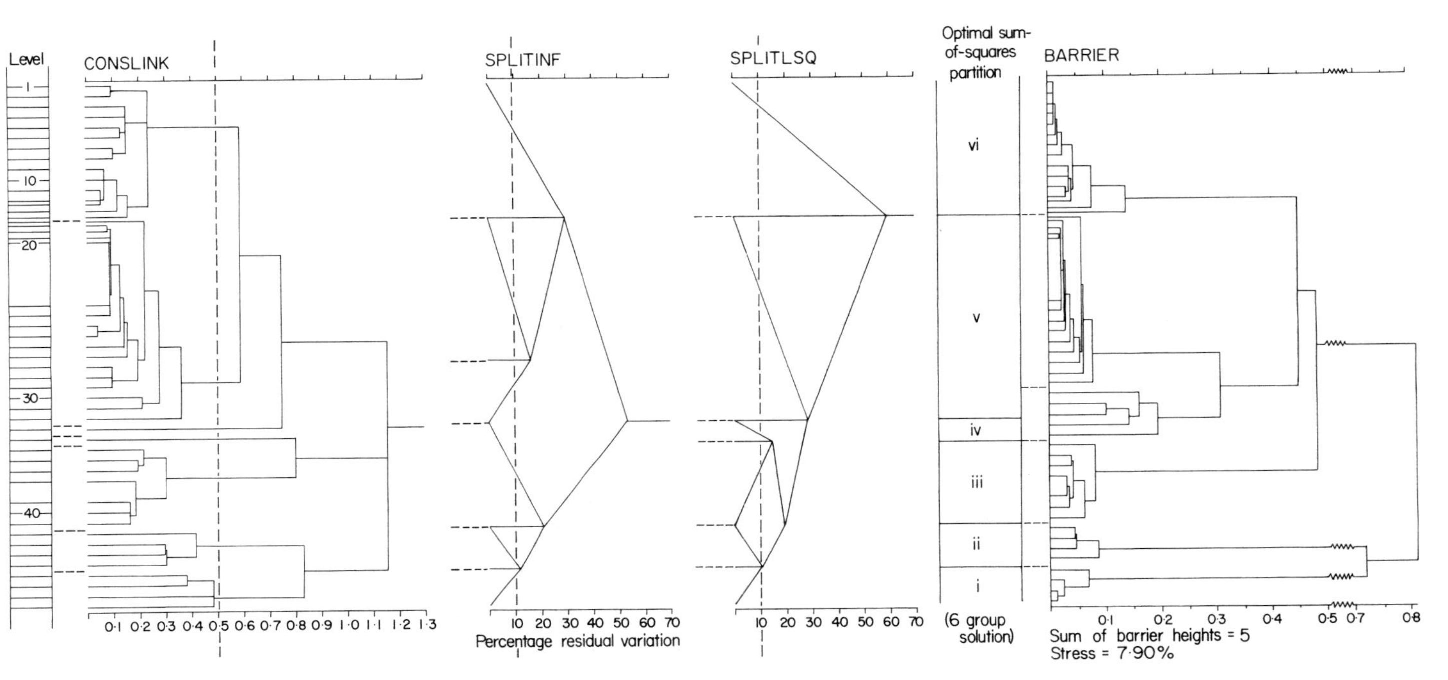

Figure 3.8 Pollen diagram and numerical zonation analyses for the complete Abernethy Forest 1974 data set.

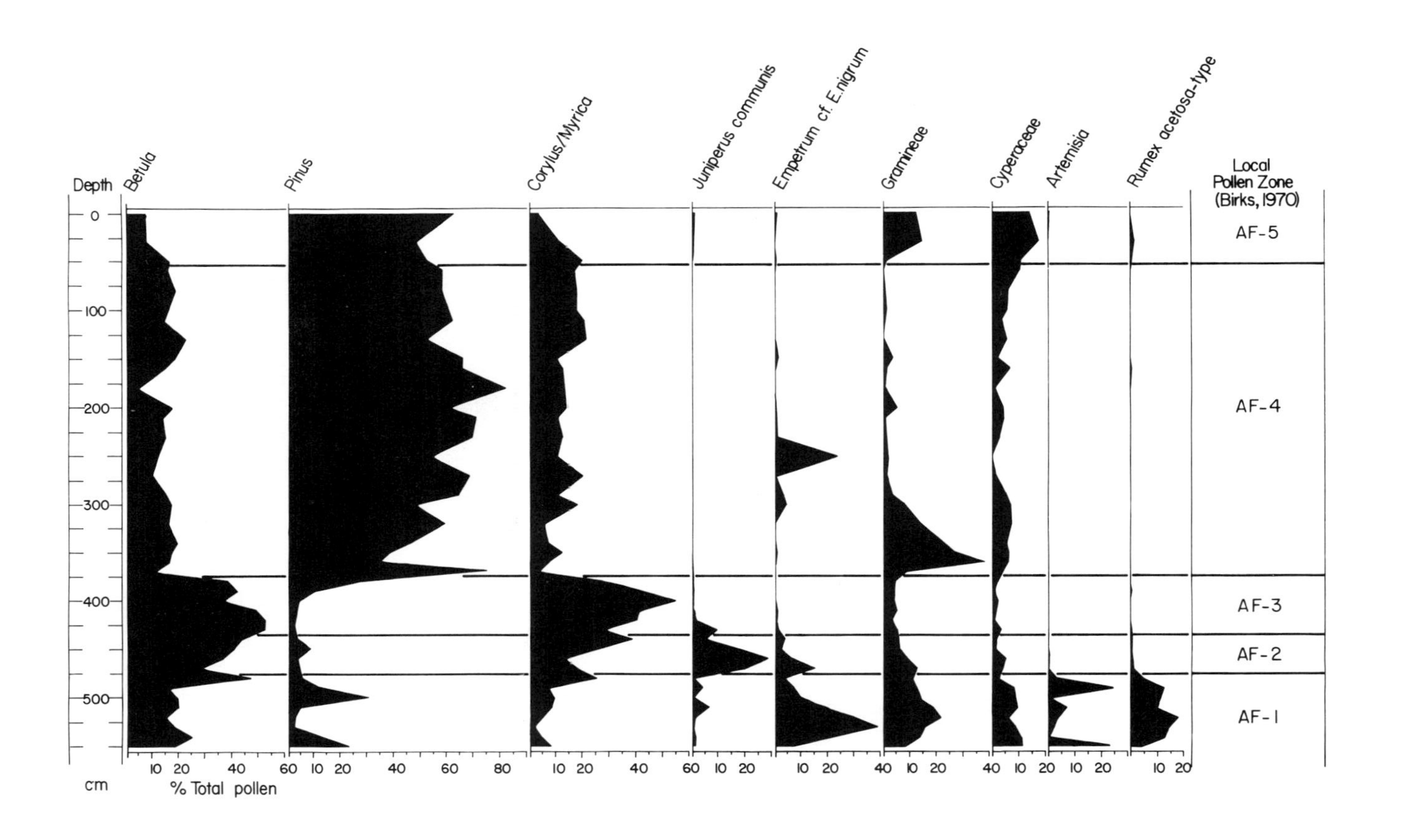
Depth
Betula
Pinus
Corylus/Myrica
Juniperus communis
Empetrum cf. E.nigrum
Gramineae
Cyperaceae
Artemisia
Rumex acetosa-type
Local Pollen Zone (Birks, 1970)
AF-5
AF-4
AF-3
AF-2
AF-1
cm
% Total pollen

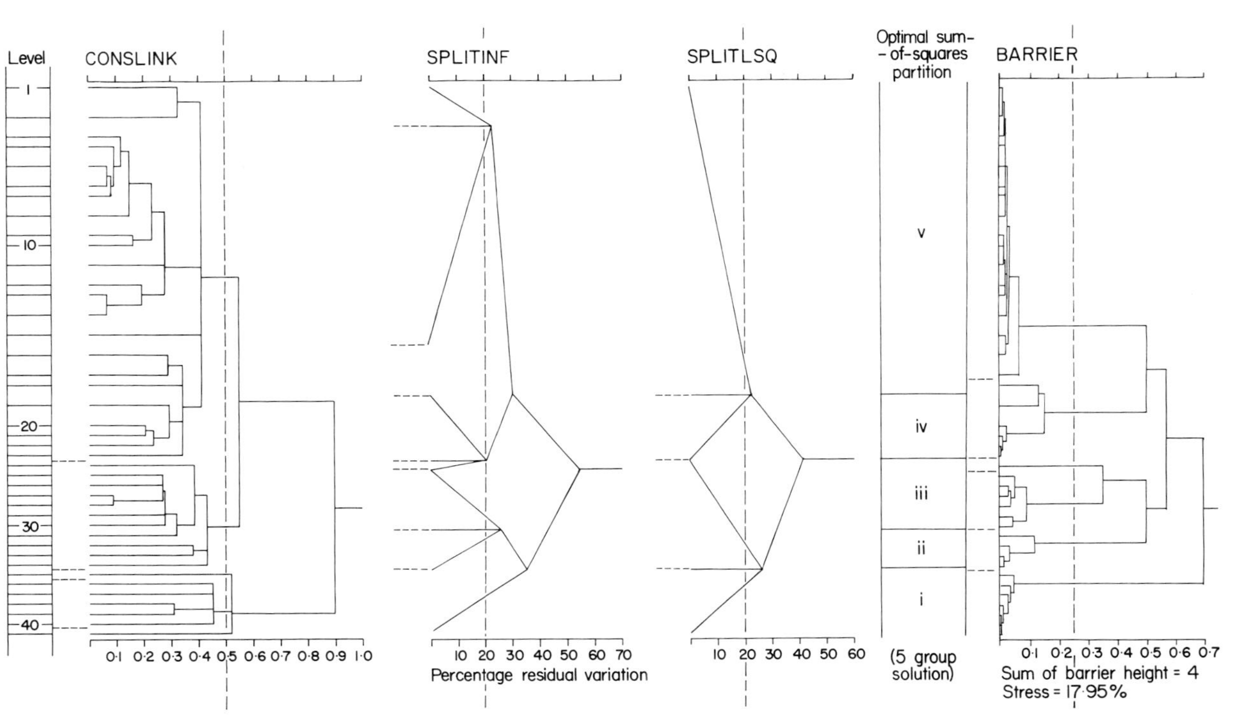

Figure 3.9 Pollen diagram and numerical zonation analyses for the complete Abernethy Forest 1970 data set.

different criteria for partitioning the sequence, the consistency of the results indicates that the proposed zones are not artifacts of any one numerical procedure. The proposed zone boundaries shown in Figure 3.8 may thus represent the most effective discontinuities within the sequence. The numerical analyses have thus provided us with a consistent, unambiguous, and repeatable zonation based solely on the observed pollen stratigraphy and allow the delimitation of a series of local or site pollen assemblage zones.

In comparing the various results obtained, the results of the constrained single link analysis (CONSLINK in Figure 3.8) tend, as is usual (Gordon and Birks, 1972; H. J. B. Birks and Berglund, 1979), to be the most difficult to interpret. Problems can arise in delimiting the extent of the clusters, and hence the stratigraphical extent of the pollen zones (Maher, 1982). A further problem is highlighted by levels 33 and 34, which both amalgamate at a late stage in the classification. Data such as level 33 with very high percentages of *Empetrum* cf. *E. nigrum* pollen and level 34 with a peak of Gramineae pollen could both be single level representatives of pollen zones of narrow stratigraphical extent. Such zones can arise as a result of rapid vegetational change, as a result of a very slow sediment-accumulation rate, or as a combination of both. Using the statistical theory described in Section 2.2, a 95% confidence interval for the underlying proportion of *Empetrum* pollen at level 33 is given as (0.204, 0.278), and a 95% confidence interval for the underlying proportion of Gramineae pollen at level 34 is given as (0.328, 0.467), the greater width of the latter interval being a reflection of the smaller number of pollen grains counted at level 34. It is evident that the difference of these taxa from the corresponding values taken in neighbouring levels is markedly greater than would have been expected from simple statistical fluctuations. The numerical results emphasise the need to re-examine and replicate the counts at these levels and to examine adjacent levels in an attempt to assess the numerical significance of the results in relation to the errors inherent in pollen counting. It should be noted that such pronounced and rapid pollen stratigraphical changes are characteristic of this time interval (Devensian late-glacial/early post-glacial (Flandrian) transition; 10000–9000 B.P.) elsewhere in the British Isles (see Pennington, 1977; Watts, 1977; M. J. C. Walker and Lowe, 1977; Craig, 1978). Such changes are of such narrow stratigraphical extent that in many instances they have been described as peak zones (see Section 3.1).

Binary divisive analysis using a sum-of-squares criterion (SPLITLSQ in Figure 3.8) and optimal sum-of-squares partition both delimit a zone containing levels 33 and 34, whereas the variable barriers approach, with the sum of barrier heights set at 5, distinguishes a rather broad and heterogeneous zone containing levels 30–34. Binary divisive analysis with an information statistic criterion (SPLITINF in Figure 3.8) does not suggest any partition between levels 33 and 41, although levels 33 and 34 are separated from the rest of the group at a later stage in the divisive procedure. If the sum of barrier heights is increased to 6, the resulting dendrogram is virtually identical to that shown in Figure 3.8 except that levels 30–34 now have a higher barrier between levels 32 and 33, suggesting two zones, one containing levels 30, 31, and 32, the other with levels 33 and 34. Which of these various suggested zonations is adopted depends very much on the initial aims of the investigation. For example, if the emphasis of the study is on detailed vegetational dynamics, a fine-scale zonation of these complex stratigraphical intervals would be useful for the purposes of description, discussion, and interpretation. If, however, the principal aim of the investigation is regional pollen stratigraphy and correlation, a rather broad zonation such as is suggested by the optimal sum-of-squares partition would be appropriate.

Although this is not the place to present a detailed interpretation of the pollen stratigraphy of the Abernethy Forest 1974 core, it is useful to comment that the sequence represents both the Devensian late-glacial (zones AFP1–3) and the first half of the post-glacial (Flandrian) (zones AFP4–6). Zone AFP-1 reflects a pioneer vegetation of Gramineae and *Rumex acetosa*-type that is replaced in zone AFP-2 by *Empetrum* heath with some *Betula* (mainly the dwarf-birch *B. nana*). In zone AFP-3, open soils and landscape instability resulted in the expansion of *Artemisia* at the expense of the other pollen types. At the transition to the post-glacial, rapid vegetational changes occurred, initially with the spread of grasses and sedges, followed by *Empetrum,* and then by *Juniperus* and *Betula*. About 8750 B.P., *Corylus/Myrica* pollen expanded, presumably as a result of the spread of *Corylus avellana* into the area. Little change occurred until about 7200 B.P. when *Pinus sylvestris* invaded and formed the Caledonian pine forests characteristic of the Abernethy Forest area today. A detailed account of the vegetational history of the site based not only on pollen percentage data but also on pollen-accumulation rates and on plant macrofossil data is given by H. H. Birks and Mathewes (1978).

The numerical zonations of the Abernethy Forest 1970 data set (Figure 3.9) are, with the exception of the constrained single link analysis, consistent between themselves. However, in contrast to the 1974 data set, the numerical results correspond only in part with the original pollen zonation presented by H. H. Birks (1970). There is consistency between the numerical methods in partitioning the sequence between levels 34 and 35, whereas the original zonation proposed a boundary between levels 33 and 34. Nearly all the numerical results suggest a boundary between levels 30 and 31, in contrast to the initial zonation with a boundary between levels 29 and 30. All the zonations, numerical and non-numerical, propose a zone boundary between levels 23 and 24. Several of the numerical procedures suggest a boundary between levels 18 and 19, although no such division was suggested by H. H. Birks (1970). Such a zone, comprising levels 19–23, is characterised by high Gramineae pollen values (10–35%). H. H. Birks (1970) interpreted these high percentages as reflecting local reed-swamp development at and near the coring site. As her zonation was based primarily on upland pollen types, she disregarded these high grass-pollen values for the purposes of pollen zonation. A further difference between the original zonation and the numerical zonations is the uppermost zone, AF-5. This zone was based primarily on extrinsic criteria, having its type locality elsewhere in eastern Scotland (H. H. Birks, 1970). The zone was defined on the basis of decreased tree pollen values and of correspondingly increased non-tree pollen frequencies. The numerical methods, using only the data of the Abernethy Forest 1970 core, are less emphatic in suggesting a partition of the sequence in this upper 100 cm. (but note the CONSLINK and SPLITINF results in Figure 3.9). Clearly in this case the numerical zonation methods suggest that the original local or site zonation should be re-examined and re-evaluated.

3.7.2 Palaeolimnological Examples

Although the numerical zonation methods discussed in this chapter have been applied here only to pollen stratigraphical data, they are equally applicable in palaeolimnology to the zonation of stratigraphical sequences of other palaeontological variables such as cladocera, mollusca, ostracods, and diatoms. R. B. Davis and Norton (1978) and Norton, Davis, and Brakke (1981) have applied conventional clustering procedures that lack any stratigraphical constraint (see Section 3.2) to zone stratigraphical diatom and cladocera sequences in New England lake sediments. Jatkar, Rushforth, and Brotherson (1979) adopted a similar approach to zone a stratigraphical diatom sequence from Utah. Carney (1982) also used an agglomerative clustering algorithm that lacked a stratigraphical constraint to group diatom percentage counts in 14 samples from different depths in a core from a Michigan lake. Each sample consisted of three replicate counts. Clustering was performed on the 42 counts. The results showed that the replicate counts from each depth were grouped together before clusterings occurred with counts from other depths. Carney supplemented the cluster analysis results with a principal components analysis of the 42 counts. Both procedures gave broadly similar results.

The only application in palaeolimnology of a stratigraphically constrained clustering procedure to zone palaeontological sequences that we know of is Binford's (1982) study of Lake Valencia, Venezuela. He used Orloci's (1967a) sum-of-squares agglomerative procedure with E. J. Cushing's (unpublished work, see p. 54) stratigraphical constraint to derive faunal assemblage zones using ostracod, cladoceran, and dipteran data from three cores. These zones were then characterised by a form of canonical variates analysis (see Subsection 5.3.3) in terms of individual sample scores on the first two canonical variates. Interpretation of the zones was facilitated by means of the canonical loadings for the taxa included in the zonations.

3.7.3 Other Examples

Comparisons of independent zonations of different palaeoecological variables studied in the same sequence can provide valuable insights into the timing and response of stratigraphical changes in different ecological groups and can thus assist the handling and interpretation of diverse data sets, such as are common in palaeolimnology and palaeoecology. In this section, the numerical zonations of two independent palaeoecological variables studied in the same stratigraphical sequence, namely, pollen and plant macrofossils, are discussed (see Gordon and Birks, 1972).

The data discussed come from Kirchner Marsh, Dakota County, Minnesota. A detailed pollen diagram from the site covering the last 13,000 years (Figure 3.10) was published by Wright, Winter, and Patten, (1963). Watts and Winter (1966) presented a plant macrofossil diagram for the section 190–1210 cm (Figure 3.11) of the original core used for pollen analysis. The numerical analyses of the two data sets discussed here (constrained single link analysis, binary divisive analysis, and optimal sum-of-squares partition) are restricted to all fossil types that attain 5% or more in any of the original profiles.

There are some features of interest associated with the numerical zonation of the pollen stratigraphical data (Figure 3.10). As with the Abernethy Forest 1970 data, the numerical procedures consistently delimit a pollen zone at Kirchner Marsh that was not recognized by Wright *et al.* (1963), or, if recognised, was disregarded as being of little importance in terms of regional pollen zonation. All four numerical procedures (Figure 3.10) indicate a pollen zone containing levels 1–7 inclusive at Kirchner March that is characterised by high (25–50%) Cyperaceae pollen values. Wright *et al.* (1963) interpreted this change as reflecting sedge growth in and around the site, and they disregarded it for the definition of their pollen zones.

A further point of interest is that the numerical procedures, particularly the divisive ones, suggest that the lower part of the Kirchner Marsh pollen sequence could be subdivided in a rather different way from that proposed by Wright *et al.* (1963). These authors (p. 1379) defined a lower zone A 'characterised by high and dominant values of *Picea* pollen', and an upper zone B with '*Betula–Alnus* maxima followed by *Pinus* maximum; *Fraxinus* and *Abies* high; *Ulmus* and *Quercus* and other hardwoods rise sharply near top; *Picea* and NAP almost vanish'. Binary divisive analysis using the information statistic suggests a division into levels 58–64 characterised by high *Pinus* and *Ulmus* pollen percentages, levels 65–68 with high *Alnus* and *Betula* values, and levels 69–72 as a transitional unit with increasing *Betula, Alnus,* and *Pinus* pollen (see Figure 3.10). The lower spruce-pollen-dominated zone (zone A) containing levels 73–93 is delimited by both this procedure and the constrained single link analysis. The binary divisive sum-of-squares analysis suggests a slightly different zonation of the sequence, with zones containing levels 58–60, 61–65, 66–69, and 70–93 characterised by high *Pinus* and *Ulmus* pollen percentages, by high *Betula* frequencies, by high *Alnus* and increasing *Betula* values, and by dominant *Picea* pollen, respectively. The optimal sum-of-squares partition into eight zones is very similar. None of the numerical procedures supports the delimitation of a single zone containing levels 57–68. A further difference between the original zonation and the numerically based zonation is that a division directly above or directly below level 43 is suggested by all the divisive algorithms on the basis of a sharp and consistent fall in *Ulmus* pollen percentages, and progressive increases in values of Gramineae, *Artemisia, Ambrosia*-type, and *Chenopodium*-type pollen.

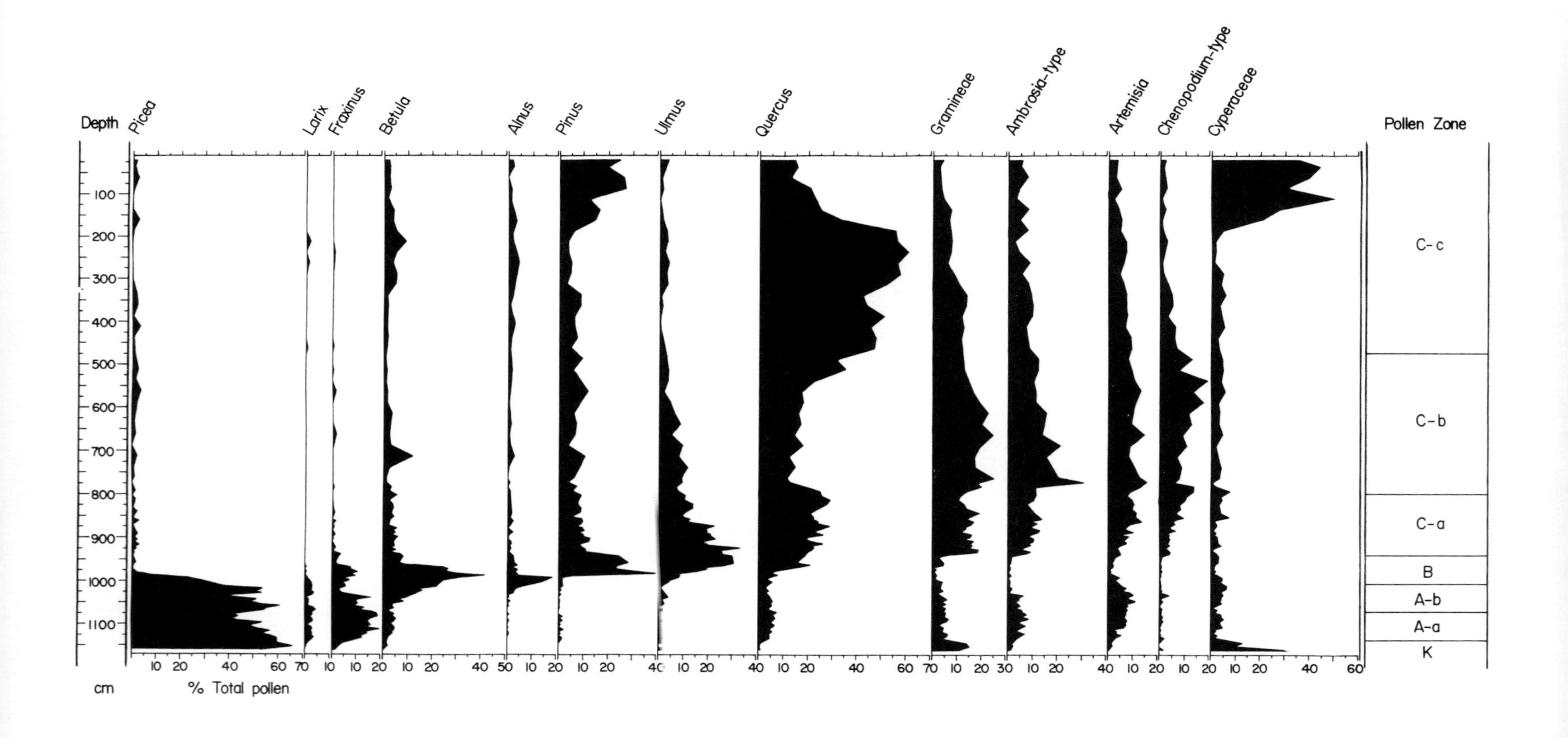
Depth
cm
Picea
Larix
Fraxinus
Betula
Alnus
Pinus
Ulmus
Quercus
Gramineae
Ambrosia-type
Artemisia
Chenopodium-type
Cyperaceae
Pollen Zone
C-c
C-b
C-a
B
A-b
A-a
K
% Total pollen

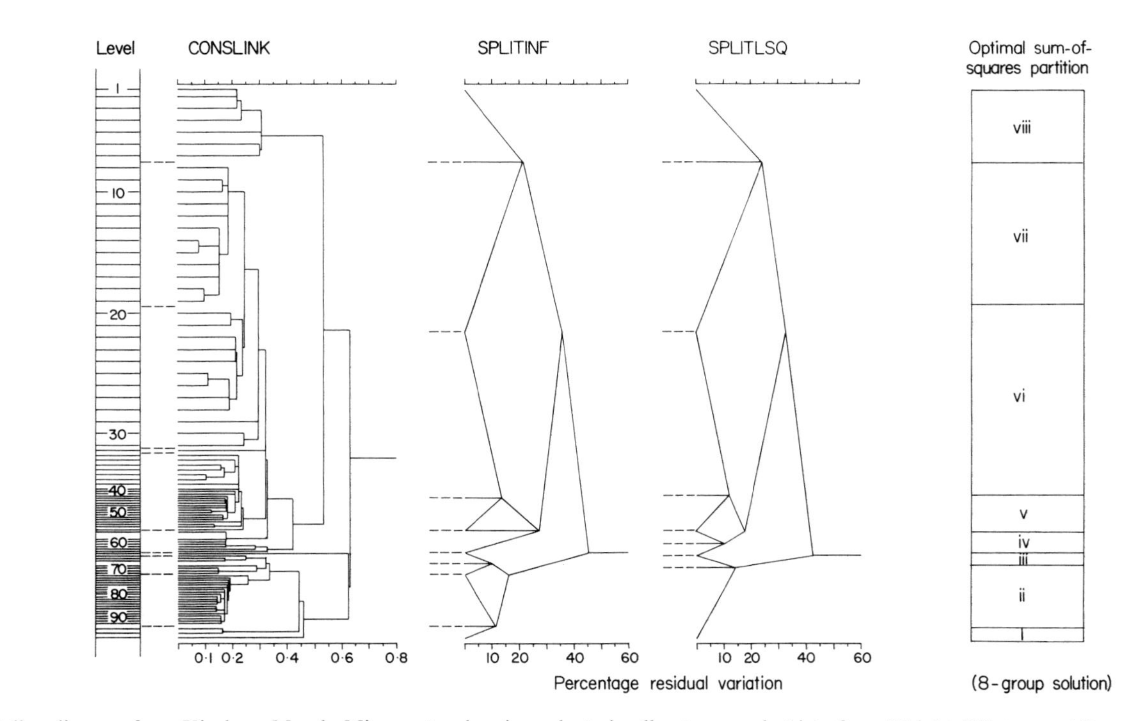

Figure 3.10 Pollen diagram from Kirchner Marsh, Minnesota, showing selected pollen types only (data from Wright, Winter, and Patten, 1963). The original pollen zones and the results of numerical zonations are also shown.

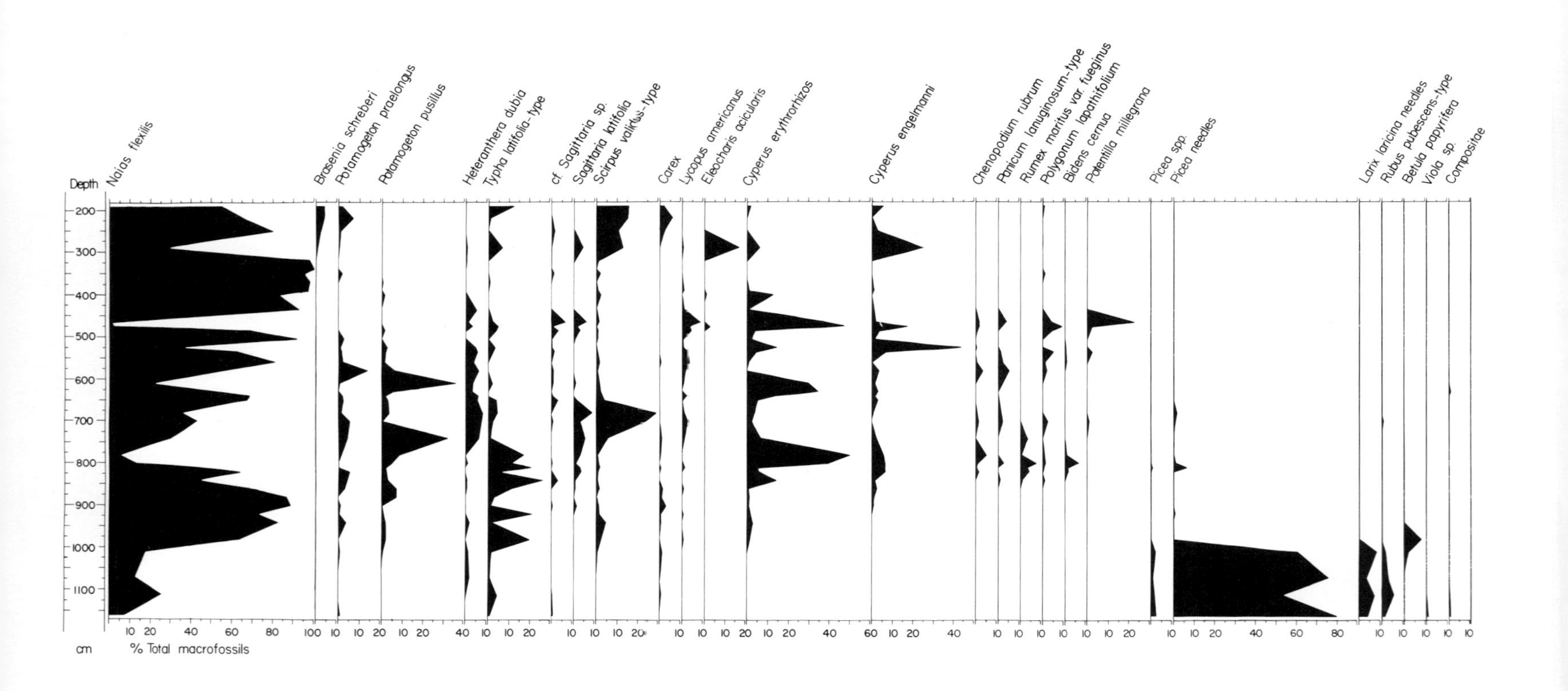
Depth
cm
% Total macrofossils
Naias flexilis
Brasenia schreberi
Potamogeton praelongus
Potamogeton pusillus
Heteranthera dubia
Typha latifolia-type
cf. Sagittaria sp.
Sagittaria latifolia
Scirpus validus-type
Carex
Lycopus americanus
Eleocharis acicularis
Cyperus erythrorhizos
Cyperus engelmanni
Chenopodium rubrum
Panicum lanuginosum-type
Rumex maritus var. fueginus
Polygonum lapathifolium
Bidens cernua
Potentilla millegrana
Picea spp.
Picea needles
Larix laricina needles
Rubus pubescens-type
Betula papyrifera
Viola sp.
Compositae

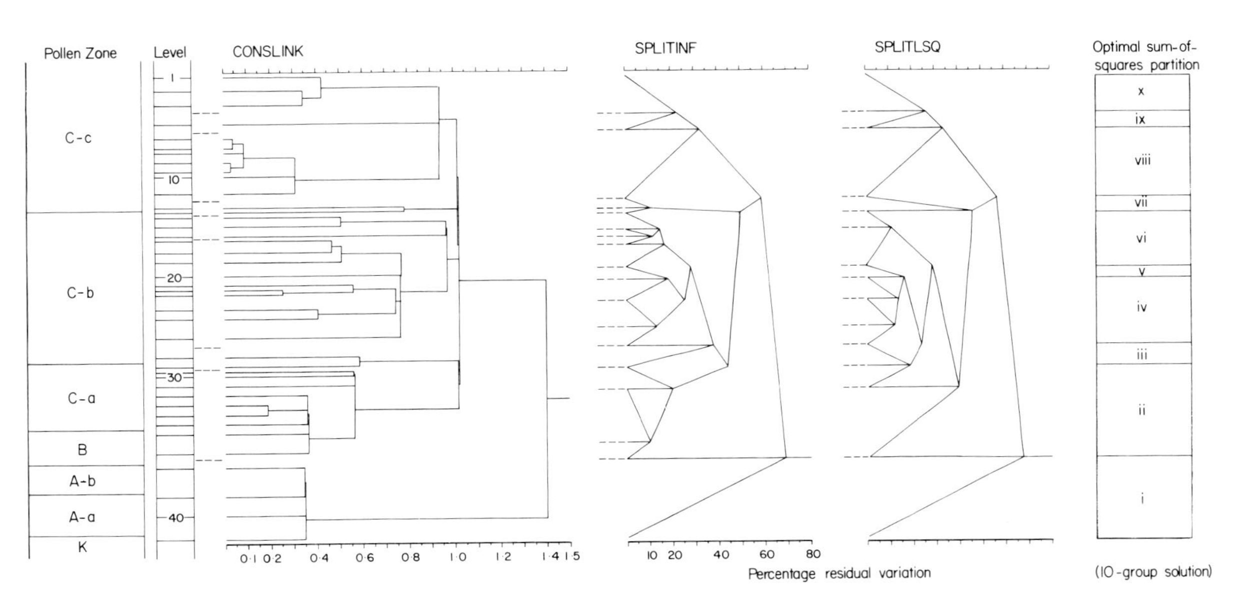

Figure 3.11 Macrofossil diagram from Kirchner Marsh, Minnesota, showing selected macrofossils only (data from Watts and Winter, 1966). The original zones and the results of numerical zonation analyses are also shown.

It is interesting to note here the results of analysing this data set by two different scaling procedures (Gordon, 1982c). In each case, the points corresponding to levels 57–69 followed a general trend in the plot, but were very dispersed compared with other groups of points. The conclusion is that, although stratigraphically neighbouring levels resemble one another, it may be less appropriate to zone this part of the core than to stress the evident trends in the data.

The stratigraphical divisions proposed by the four numerical zonations of the Kirchner Marsh pollen and macrofossil data are summarised in Figure 3.12, along with the original pollen zones of Wright *et al.* (1963) that were used by Watts and Winter (1966) for subdividing the macrofossil data.

There is a broad correspondence between the two sets of zonations. Numerical analyses of both the pollen and the macrofossil stratigraphy suggest a division around 1000-cm depth (zone A–B transition of Wright *et al.*, 1963). This corresponds to a marked decrease in *Picea* pollen values and the disappearance of macrofossils of *Picea, Larix laricina,* and *Rubus pubescens*-type. There is little indication in the macrofossil stratigraphy to suggest any subdivision near the lower boundary of pollen zone C (937.5 cm), whereas the two binary divisive algorithms indicate a division of the macrofossil profile at 850 cm (levels 31–32 in Figure 3.11) that corresponds stratigraphically to a pollen division at 867.5 cm (see Figure 3.10). The division of the macrofossil sequence at 850 cm corresponds to the first increase in the frequencies of *Cyperus erythrorhizos* and *C. engelmanni* fruits and the first appearance of macrofossils of *Bidens cernua, Chenopodium rubrum, Polygonum lapathifolium,* and *Rumex maritimus* var. *fueginus.* At the level of the proposed pollen zone boundary (Figure 3.10), there is a prominent decrease in *Ulmus* pollen values and a rise in *Artemisia, Ambrosia*-type, and *Chenopodium*-type pollen. The numerical methods suggest that the original pollen zone boundary (zone C-a/C-b) may have been misplaced. The striking coincidence of the positioning of both the pollen and macrofossil zones suggests that the stratigraphical changes at these levels are worthy of careful re-examination, as they may indicate an expansion of prairie conditions somewhat earlier than was proposed by Wright *et al.* (1963) and Watts and Winter (1966).

The macrofossil partitions between 450 and 850 cm (see Figure 3.11) reflect the complex phases of high values of *Cyperus* spp., *Polygonum lapathifolium, Potentilla millegrana,* and others, alternating with phases of abundant *Naias flexilis, Potamogeton pusillus, Sagittaria latifolia,* and *Scirpus validus*-type macrofossils. These phases are interpreted by Watts and Winter (1966) as reflecting phases of fluctuating lake levels. Such phases cannot be discerned in the pollen stratigraphy (Figures 3.10 and 3.12).

Zonations of both the pollen and the macrofossil data indicate a partitioning at about 480 cm (Figure 3.12) corresponding to the close of pollen zone C-b. This is marked in the pollen stratigraphy by a rise in *Quercus* pollen values and a corresponding decrease in frequencies of Gramineae, *Chenopodium*-type, and *Ambrosia*-type pollen (Figure 3.10). In the macrofossil sequence (Figure 3.11) there is the disappearance of seeds of many plants of damp ground and ruderal habitats and the onset of an almost total dominance of *Naias flexilis* seeds. A narrow zone

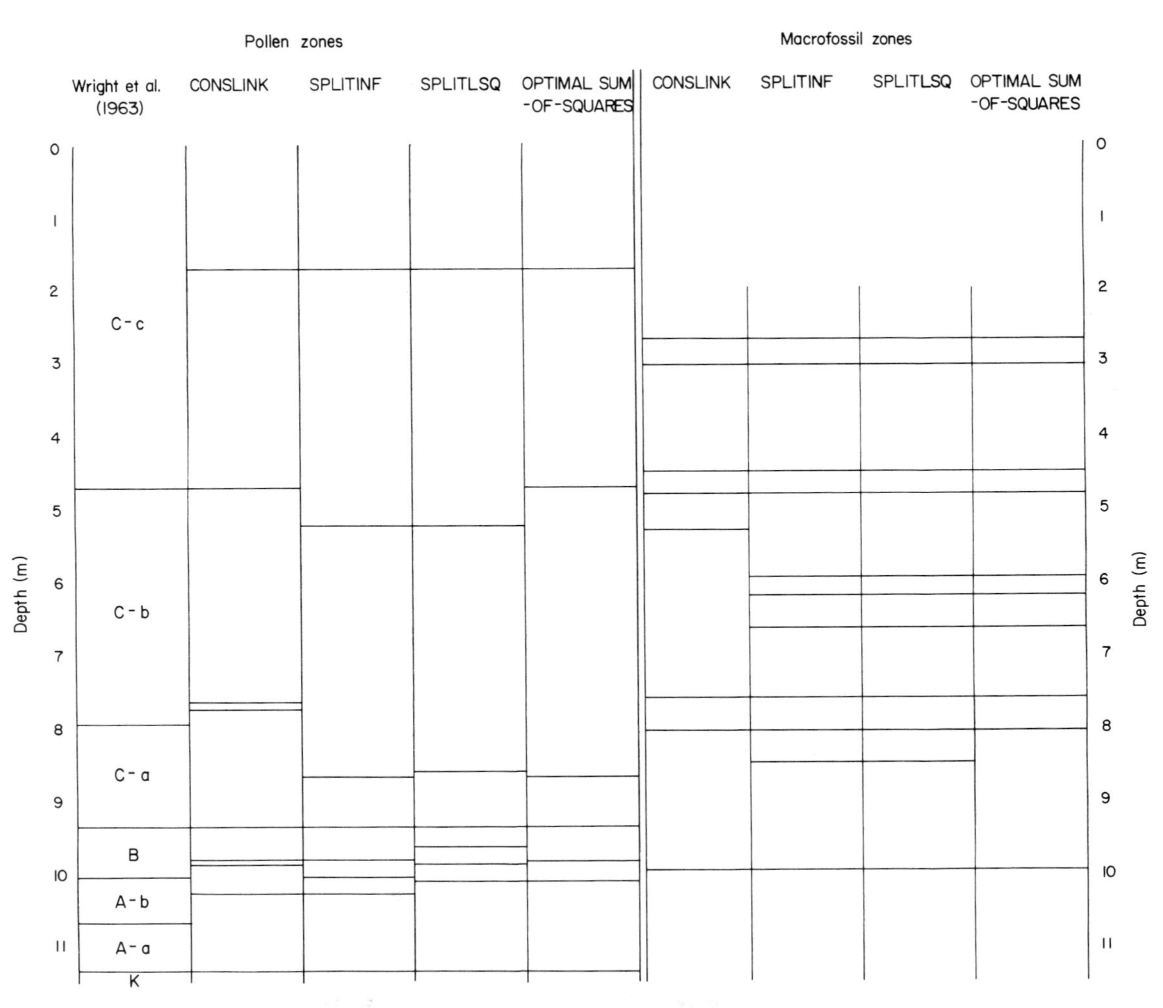

Figure 3.12 Comparison of zonation schemes at Kirchner Marsh on (1) pollen stratigraphy and (2) macrofossil stratigraphy. The zones are derived from Figures 3.10 and 3.11.

between 270 and 305 cm in the macrofossil profile (Figure 3.11) is clearly delimited by all the numerical methods and is characterised by high values of *Eleocharis acicularis* and *Cyperus engelmanni* fruits. There is no corresponding unit in the pollen stratigraphy (Figure 3.12).

It is particularly instructive to subdivide the macrofossil data on the basis of the observed stratigraphy rather than to impose on it a pollen zonation scheme derived from a study of predominantly wind-dispersed palynomorphs. The correspondence (or lack of it) between the pollen and macrofossil stratigraphies provides a basis for considering the floristic and vegetational changes reflected by both data sets. Pollen spectra from sites such as Kirchner Marsh largely reflect the regional upland vegetation, whereas macrofossils tend to record the local lowland vegetation. A comparison of the two zonations can thus provide a background for interpreting the stratigraphical changes in terms of ecological factors that have in some instances influenced both upland trees and locally growing aquatic and damp-ground herbs and in other instances have affected only the upland trees or the local lowland herbs, but not both. Changes in macrofossils that are seemingly independent of the pollen stratigraphy imply local changes only, whereas changes in the pollen stratigraphy that do not correspond to changes in the macrofossils may reflect forest succession, tree migrations, or other non-climatically induced events such as disease, human activity, or soil development. The zonations indicated by the numerical analyses give information more directly relevant to the interpretation of the macrofossil stratigraphy than the use of pollen zones as a means of subdividing the macrofossil sequence. Numerical zonations are particularly valuable as a basis for comparing two or more independent sets of stratigraphical data because they use only the observed data, and are thus not influenced by any preconceived ideas which the investigator might have (see also H. J. B. Birks, 1976b; Ovenden, 1982; Pawlikowski, Ralska-Jasiewiczowa, Schönborn, Stupnicka, and Szeroczynska, 1982).

Other examples of the use of these numerical methods for zoning pollen stratigraphical data are given by Gordon and Birks (1972, 1974), H. J. B. Birks (1974, 1976b, 1981a), Rymer (1974, 1977), Huntley (1976, 1981), W. Williams (1977), Bradshaw (1978), Bostwick (1978), H. J. B. Birks and Berglund (1979), H. J. B. Birks and Madsen (1979), Bernabo (1981), Heide (1981), Turner and Hodgson (1981), Bennett (1982, 1983a), Cwynar (1982), Hunt and Birks (1982), Kerslake (1982), Lamb (1982, 1984), Ovenden (1982), McAndrews (1982), McAndrews, Riley, and Davis (1982), Pawlikowski *et al.* (1982), Ritchie (1982), H. R. Delcourt, P. A. Delcourt, and Spiker (1983), and MacDonald (1983).

3.8 Advantages and Limitations of Numerical Zonations

The advantages of using numerical approaches to the zonation of pollen sequences may be summarised under four broad headings: power, consistency, formulation of criteria, and rapidity of implementation.

Power

The complex nature of palynological data has already been noted in Section 3.1. At some stage of complexity, the human brain starts to experience difficulties in retaining in perspective all of the varying factors which it wishes to influence the final assessment of the data. A suitably programmed computer can more readily hold and balance large amounts of information, and hence help the investigator to gain insights into the structure of the data.

Consistency

There are many advantages in being able to apply the same criteria in the analysis of several different data sets. Such an approach avoids the possibility of bias on the part of an investigator. For example, analysis of the data under investigation could be unconsciously biased by the results of a previous, possibly totally unrelated, study. Alternatively, the investigator may make assumptions, either explicitly or implicitly, about the data based on sediment lithology, inferred vegetation, past climate, or assumed time equivalence (Cushing, 1964). Following the *International Guide to Stratigraphic Classification, Terminology, and Usage,* Quaternary pollen zones—in common with all other biostratigraphical units—should be defined solely on the basis of the observed fossil pollen and spore composition. As discussed in Section 3.1, numerically defined pollen zones are delimited solely on the basis of specified mathematical criteria. No reference is made to inferred climate, vegetation, or chronology, and the zones are thus, by definition, strict assemblage zones, as defined by Hedberg (1972a, 1972b, 1976).

It is also desirable that two scientists presented with the same data reach similar conclusions (Maher, 1982). A computer possesses this consistency, although computing aids have not always been greeted with enthusiasm. In an entertaining diatribe against an ancestor of today's computers, Oliver Wendell Holmes's (1858) Autocrat declaimed

> What a satire, by the way, is that machine on the mere mathematician! A Frankenstein-monster, a thing without brains and without heart, too stupid to make a blunder; that turns out results like a corn-sheller, and never grows any wiser or better, though it grind a thousand bushels of them!

Nevertheless, the Autocrat would seem to have regarded the trait of consistency with some favour (Holmes, 1858); and a suitably programmed computer is one means of achieving such consistency. The ability to derive consistent and repeatable results is a prerequisite for sound stratigraphy and palaeoecology, so that interpretations can be applied in an informative and critical way between sites and from one time to another.

Formulation of Criteria

If the two scientists in the previous paragraph disagreed in their proposed zonations, it would be useful if they could isolate what was causing the disagreement. Because they codify the criteria to be used for zonation, numerical methods require that the investigator define precisely the criteria for analysing the data, and hence state the aims and assumptions of the study. Two scientists who had independently formulated different criteria could then have a more informed and profitable discussion about differences in results and interpretation than could two who were unable to explain clearly how they had arrived at their own subjective assessments. Moreover, such explicit formulation makes it easier for someone not directly acquainted with the primary data to understand and evaluate the criteria used for zonation.

Although some pollen analysts may be sceptical and critical about the use of numerical methods for zonation (e.g., Faegri in Faegri and Iversen, 1975, p. 195), the potential value of applying mathematical approaches is well summarised by D. Walker (1972, p. 101) as follows: 'the more orthodox amongst us should at least reflect that many of the same imperfections are implicit in our own cerebrations and welcome the exposure which numbers bring to the muddle which words may obscure'.

Rapidity of Implementation

The speed with which one can obtain an automated zonation of a pollen diagram can rarely be claimed as a major argument for (or against) the use of numerical methods, since the amount of time which has to be expended in site selection, coring, sample preparation, and pollen counting is considerably greater than the time which would be spent in data analysis and interpretation. However, a weaker claim can be made: the amount of time required to prepare the data in a form suitable for input to a computer program, and the amount of computer time required for analysis are both very small, and should in themselves be no bar to the use of such methods.

Lest it be thought that the results of a numerical analysis should always be accepted without question, we should emphasise the following caveats on the limitations of such methods.

Difficulties of Quantification

It is sometimes difficult to quantify the intuitive feelings of a research worker on how to give prominence to a feature which is believed to be ecologically important (Janssen, 1980). An example is the treatment of *Ambrosia*-type pollen in cores from eastern North America. Many pollen analysts would wish to define a zone at the top of cores, based on an increase in *Ambrosia*-type pollen from low values to about 5 or 10% of the pollen sum (e.g., McAndrews, 1966). This could be achieved in a numerical zonation by a suitable weighting of this taxon, along the lines indicated at the end of Section 3.4. Such a weighting, however, would influence the zonation lower down the core, where *Ambrosia*-type pollen is also present in moderate amounts. The problem is that of seeking to treat a taxon almost as an indicator species in part of the core, but not in this fashion in other parts of the core.

Spurious Objectivity

We would hope that any zonation suggested by one of the methods described in this chapter is not regarded as the definitive zonation against which no objections can be entertained. As indicated in the final sections of Chapter 2, various decisions associated with the classification have to be taken; the inevitable uncertainty about whether the decisions taken are the most appropriate ones suggests that one should not pitch too high any claim for the 'objectivity' of any classification approach (see Janssen, 1980; Maher, 1982).

On the other hand, we have found the methods described in this chapter to be very helpful in the analysis of a large number of data sets. When other investigators' data have been re-analysed by these numerical methods, the results have generally confirmed the main points of the original zonations (see Gordon and Birks, 1972, 1974; H. J. B. Birks, 1974). It might be of interest to conclude this chapter by outlining the kinds of disagreement which have arisen between the original, non-numerical zonations and the numerical zonations.

1. In some instances, the numerical methods suggested pollen zones which were not recognised by the original investigators, or, if recognised, were disregarded as being of little importance in terms of pollen zonation. Examples of this are discussed in Section 3.7 with reference to the Abernethy Forest 1970 data and the Kirchner Marsh pollen data. In both instances, the numerical methods suggested zonations based on high Gramineae or Cyperaceae pollen values which were disregarded in the original zonations on the grounds that these pollen stratigraphical features may result from local grass or sedge swamp development in and around the sites in question.

2. In other instances, the numerical methods failed to detect zones defined by the investigator, for example, the upper *Ambrosia* zone at three of the sites along the Itasca transect in northwestern Minnesota (McAndrews, 1966). As mentioned above, *Ambrosia*-type pollen is being used in the upper parts of these sequences almost as an indicator species, and it is not surprising that these original zones were not detected by the numerical methods.

3. Some of the numerical results have prompted re-interpretation of the original pollen zonation, for example, at Terhell Pond, Minnesota (McAndrews, 1966; see comments in Gordon and Birks, 1974). In a numerical analysis of three long, detailed post-glacial pollen sequences from southern Sweden, H. J. B. Birks and Berglund (1979) found considerable consistency in the results from several numerical methods, but little correspondence between the numerical results and the original pollen zones. Examination of the criteria for the original zonation showed that this lack of correspondence was not, after all, surprising. The original zonations are based largely on the first appearance or major expansion of the main tree pollen types, whereas the numerical zones are based on overall similarities in pollen composition of the spectra considered together. The original pollen zones correspond, in part at least, to *range zones* (*sensu* Hedberg, 1972b) as they are defined by the first appearance of particular fossil pollen types, whereas the numerical zones correspond closely to the concept of an *assemblage zone*. There are no *a priori* reasons why range and assemblage zones should coincide stratigraphi-

cally. This lack of correspondence, although explicable in terms of differing concepts of a pollen zone, focusses attention on the different criteria used for zonation, and encourages the investigators to re-evaluate their criteria for zonation.

In each of these three types of example, the results of the numerical zonation methods have led to a more explicit definition of the zones in the pollen diagrams, and have aided in identifying more clearly which taxa are to be considered in the definition of local pollen zones. Mathematical methods designed specifically to partition quantitative pollen stratigraphical data can, we feel, be viewed as a useful addition to the existing methodology of Quaternary palynology and palaeoecology.

CHAPTER **4**

The Analysis of Pollen Stratigraphical Data: Comparison of Sequences

4.1 Rationale of Comparing Pollen Sequences

The numerical methods described in Chapter 3 provide useful techniques for dividing pollen stratigraphical sequences into a series of pollen zones. These numerically defined pollen zones refer, however, only to single stratigraphical sequences as they are based entirely on the observations from the sequences considered. These zones are thus *local* or site pollen zones (*sensu* Cushing, 1967b; H. J. B. Birks, 1973b; Watts, 1977) (see Figure 3.1) that describe and characterise the basic features of the pollen stratigraphy of the sequences analysed. These local pollen zones are of limited use in the investigation of broad-scale changes in pollen stratigraphy and vegetational history and in the discussion, interpretation, and correlation of the pollen record from several sites within a particular area of study.

Zonation of individual pollen stratigraphical sequences is thus not an end in itself, but rather a preliminary stage in the description and analysis of complex stratigraphical data from several sites. The next step is the comparison of stratigraphical sequences from different sites within the study area (see Figure 3.1). If similar pollen assemblages are found to exist between sequences, *regional* pollen assemblage zones (*sensu* Cushing, 1967b; H. J. B. Birks, 1973b; Watts, 1977) can be delimited and defined. Although comparisons between pollen sequences represent an important stage in many palynological studies (e.g., Godwin, 1940; von Post, 1946; Cushing, 1967b), the criteria used for comparing pollen sequences and for assessing similarities and differences between sequences are rarely, if ever, stated.

This chapter describes three broad numerical approaches to comparing pollen sequences from different sites that can assist in the recognition and delimitation of regional pollen assemblage zones (Sections 4.3, 4.4, and 4.5). Two of the approaches can also be used for comparing surface pollen spectra with either fossil pollen assemblages or modern assemblages from other areas (see Section 6.4). The main emphasis in this chapter is, however, on comparison of stratigraphical sequences as a means of delimiting regional pollen assemblage zones. Once delimited and defined, these regional zones can be ordered stratigraphically and, if an independent means of dating is available (e.g., radiocarbon assay), they can be correlated in time and mapped in space to produce time–space correlation diagrams (see Figure 3.1) (see, e.g., McAndrews, 1966, Plate 1; Cushing, 1967b, Fig. 7; H. J. B. Birks, 1973b, Fig. 27). Such correlations can, if required, be related to a standard chronostratigraphical sequence at a defined type locality and section (see Figure 3.1) (West, 1970). Regional pollen assemblage zones are particularly suitable for describing and comparing broad-scale vegetational changes within different geographical or ecological areas, as regional pollen zones largely reflect the 'regional' pollen rain and hence the regional upland vegetation of the areas of interest (see McAndrews, 1966; Cushing, 1967b).

Besides comparing pollen sequences for the purposes of stratigraphical correlations and the derivation of regional pollen zones, it is often useful to compare them for ecological and interpretative purposes. Numerical methods for comparing several pollen sequences provide useful tools for summarising complex patterns of pollen stratigraphical change and, by inference, of vegetational change within and between areas. By comparing pollen sequences numerically from several sites lying within the same vegetational formation today, it is possible to discover whether the patterns of vegetational change in the past have been the same at all the sites (see Ritchie and Yarranton, 1978a, 1978b). In addition, in some pollen analytical studies, it is valuable to be able to compare pollen sequences in such a way as to be able to emphasise the *differences* between sequences, rather than to seek the similarities, as in regional pollen zonation. In general, similarities in pollen stratigraphy reflect regional vegetation, whereas differences in pollen stratigraphy may result from local vegetational differentiation related to soil type, aspect, and topography (see, e.g., Watts, 1961; D. Walker, 1966; H. J. B. Birks, 1973b; Brubaker, 1975). Jacobson (1975, 1979), Andersen (1978a), Bradshaw (1978, 1981b), and Heide (1981) present methods for displaying quantitatively differences between pollen sequences. Numerical approaches for comparing pollen sequences so as to emphasise patterns of change and differences between sites are discussed in Subsections 4.7.2 and 4.7.3.

4.2 Numerical Approaches to Comparing Stratigraphical Sequences

Several different fields of research are concerned with the comparison of ordered sequences of 'objects'. Two different groups of methods are outlined below, neither of which is completely suitable for analysing pollen stratigraphical sequences. Mann (1979) presents a detailed review of methods used in geology for comparing stratigraphical sequences.

Direct matching between an 'object' in one sequence and an 'object' in another sequence. These methods seek direct matching while retaining the overall ordering within each sequence. Such methods have been used in the comparison of sedimentary rock sequences (e.g., Sackin, Sneath, and Merriam, 1965; Merriam and Sneath, 1967; Read and Sackin, 1971) where an 'object' corresponds to a stratum of rock, and in the comparison of amino-acid sequences in proteins (e.g., Gibbs and MacIntyre, 1970; Needleman and Wunsch, 1970; Sackin, 1971) where an 'object' corresponds to a single amino acid. In pollen stratigraphical sequences, an 'object' would correspond to the pollen spectrum at a single level in the core. What distinguishes pollen data from the other examples, however, is that an 'object' from one pollen sequence cannot be directly matched with an 'object' from another pollen sequence, but merely resembles it to a greater or lesser degree.

Cross-correlation methods of analysing quantitative data. These have been used with considerable success in matching tree-ring sequences (Ferguson, 1970; Fritts, 1971) and varve sequences (R. Y. Anderson and Kirkland, 1966; Dean and Anderson, 1974). However, for such data it is reasonable to expect that a consecutive set of objects in one sequence can be compared with a consecutive set of objects from another sequence with no 'gaps' or 'doubling-up' in the sequences; the 'objects' are produced at regular intervals of time. Similar methods have been applied to the analysis of data for which the investigators expected this assumption to be approximately satisfied, for example isotope, magnetic, or geophysical log sequences (see Rudman and Lankston, 1973; Southam and Hay, 1978; Shaw and Cubitt, 1979; Thompson, 1979; Dearing, 1982). However, in pollen analysis it is extremely rare for one to have available such accurate information about the time intervals associated with each pollen sample. Cross-correlation methods are thus not generally applicable to pollen stratigraphical sequences, although some applications will be described in Section 6.2.

Although superficially a pollen stratigraphical sequence might be considered to have features in common with the types of sequence described above, its properties are sufficiently different to suggest that a different method of analysis might be more appropriate. Section 4.5 describes an approach to comparing two pollen stratigraphical sequences in which one seeks to slot together the two sequences into a single joint sequence with the property that sections of the two initial sequences that have similar pollen composition appear close to one another in the joint sequence. The work described in that section could also be of use in other areas in which one wished to compare data that had an order imposed by, for example, stratigraphy, time, or distance along a transect.

The two other methods of analysis to be described in detail in Sections 4.3 and 4.4 do not directly use the ordering imposed on the pollen samples by the stratigraphy. It is convenient to introduce them by way of describing some quantitative work in the analysis and comparison of soil profiles. A soil profile is conventionally described by its constituent soil horizons. Various workers (e.g., Rayner, 1966; Grigal and Arneman, 1969; A. J. B. Anderson, 1971a, A. W. Moore, Russell, and Ward, 1972) have carried out classifications of sets of soil profiles. As a first step, a single measure of similarity between each pair of profiles is obtained. This is usually constructed by combining the similarities between pairs of horizons, one in each profile, sometimes incorporating some weighting based on the depth of the horizons. In an alternative method, Norris and Dale (1971) and Norris (1971) use a transition-matrix approach, in which consideration is given to pairs of neighbouring samples down a profile. A discussion of these methods is given by Webster (1977, Chapter 7).

While similar methods might be of interest to pollen analysts carrying out very broad-scale investigations, studies which examined the similarities between complete pollen sequences would in most instances be of little interest to a palaeoecologist. There are problems in comparing sequences which do not cover exactly the same period of time. Further, pollen analysts rarely, if ever, expect to find at different sites very similar vegetational histories over extended periods of time. In pollen analysis, one is more concerned with similarities on finer time-scales. The two different scales of investigation comprise the two further methods of analysis to be described in this chapter.

Firstly, one can consider the local or site pollen zones as the basic units for comparison. It is of interest to know whether local zones from different cores are similar (or dissimilar) to one another, and, for the purposes of interpretation, to collections of surface pollen spectra from areas of known vegetation. If so, the area where, and the time when, similar vegetational conditions prevailed can be extended. The local zones could be subjected to various classification procedures (e.g., Ritchie and Yarranton, 1978a, 1978b; see Subsection 4.7.2), but the method described in Section 4.3 is more concerned with what could be termed 'high-level' similarities, or the identification of pairs of zones which are very similar. The work of Ogden (1969) comparing individual fossil pollen spectra with modern spectra using Spearman's rank correlation coefficient can be regarded as a precursor of this approach (see also Ogden, 1977b).

Secondly, one can consider single samples from cores as the basic units for comparison. Work has already been described in Section 3.2 in which researchers sought to establish, using various classification procedures *without applying any stratigraphical constraints* on the classification, whether the pollen samples fell naturally into distinct groups. If one analyses together samples from several different pollen stratigraphical sequences, one can discover whether there are similarities between parts of the several data sets. This approach is described in more detail in Section 4.4.

The next three sections, then, describe three different numerical methods for comparing pollen sequences. These methods are illustrated by applying them in the first instance to the two small test data sets described in Chapter 1 to compare the 7 samples from the Abernethy Forest 1970 data (data set B) with the 10 samples from the Abernethy Forest 1974 data (data set A). The results of applying the various numerical methods to the full Abernethy Forest data sets and to other pollen stratigraphical data are discussed in Sections 4.6 and 4.7.

Although all the methods differ in their basic rationale, a synthesis of the results from the different methods provides, as in the zonation of single sequences discussed in Chapter 3, a fuller picture of the data sets than can be obtained from the results of any single method.

4.3 Comparison of Sequences in the Absence of Stratigraphical Constraints: Zone-by-Zone Comparisons

Suppose that we wished to compare two local pollen zones, labelled zones 1 and 2. The basic idea of this approach is to define a measure of the variability of the spectra constituting each zone, and then to compare this with the variability of the unit composed of all the spectra from both the original zones. An increase in variability which was small would suggest that the original zones were fairly similar. For the reasons discussed in Section 2.3, no attempt is made to obtain significance tests of whether two zones are 'the same'; the approach concentrates on identifying similarities which might repay further study.

One measure of variability which has been used (see Gordon and Birks, 1974) is the information radius (Sibson, 1969). If p_{ik}, as before, denotes the proportion of the ith spectrum, which is composed of the kth taxon ($k = 1, 2, \ldots, 9$), then the information radius of the first zone is defined as

$$H_1 \equiv \frac{1}{n_1} \sum_i \sum_{k=1}^{9} p_{ik} \log(p_{ik}/q_k),$$

where i is summed over all the n_1 spectra in the zone, and

$$q_k \equiv \frac{1}{n_1} \sum_i p_{ik},$$

the average proportion in the zone of the kth taxon. As is usual in an information theory context, log refers to logarithm to base 2; note that 0 log 0 = 0.

As an example, consider a local zone comprising spectra A3 and A4 of the Abernethy Forest 1974 data. The relevant information for evaluating the information radius is summarised in Table 4.1. From this,

$$\begin{aligned} H_1 &= \tfrac{1}{2}[0.661 \log(0.661/0.6975) + 0.734 \log(0.734/0.6975) \\ &\quad + 0.016 \log(0.016/0.0195) + 0.023 \log(0.023/0.0195) \\ &\quad + \ldots \\ &\quad + 0.000 \log(0.000/0.001) + 0.002 \log(0.002/0.001)] \\ &= 0.0168, \end{aligned}$$

one pair of terms arising from each species represented in the two spectra.

A similar calculation shows that the information radius of the local zone comprising spectra B3 and B4 of the Abernethy Forest 1970 data has the value $H_2 =$

Table 4.1

Information for Evaluating the Information Radius of the Zone Comprising Spectra A3 and A4 of the Abernethy Forest 1974 Test Data Set

	Pollen type								
k	1	2	3	4	5	6	7	8	9
p_{3k}	0.661	0.016	0.272	0.000	0.007	0.034	0.009	0.000	0.000
p_{4k}	0.734	0.023	0.185	0.015	0.006	0.026	0.009	0.000	0.002
q_k	0.6975	0.0195	0.2285	0.0075	0.0065	0.030	0.009	0.000	0.001

0.0825, indicating that B3 and B4 differ from one another more than do A3 and A4. The information radius of the combined set of four spectra (A3, A4, B3, B4) is found to take the value $H(1, 2) = 0.0993$. This indicates a slight, though not marked, increase in variability when the two zones are combined, and we may infer that the zones (A3, A4) and (B3, B4) are in fact fairly similar to one another. As is stressed above, the idea of carrying out a significance test of the hypothesis that all four spectra were drawn from the 'same population' seems inappropriate, but it has been found that a useful rule-of-thumb is to define an *H-match* of zones 1 and 2 if $H(1, 2) \leq H_1 + H_2$. The zones examined above just satisfy this requirement, with $H(1, 2) = 0.0993 = H_1 + H_2$. Another example of an H-match occurs when the zone containing spectra (A9, A10) is compared with the zone containing spectra (B6, B7), and there are several cases where this criterion is nearly satisfied.

Several points should be made at this stage. Firstly, if the initial zones for comparison contain a large amount of variability, any resulting H-matches will be less informative than if the initial zones were fairly homogeneous. It is thus advantageous to consider deleting transitional levels from the zones indicated by the methods described in Chapter 3. Secondly, the ordering of spectra within a zone is not taken into account by this method, and thus potentially important information is discarded. However, some simplification of the data can be a useful first step in the identification of similar zones, which can then be subjected to closer scrutiny. Thirdly, although the description of the method has been wholly in terms of local zones from cores, the approach can also be used with collections of modern pollen spectra, and can thus be of assistance in directly comparing modern and fossil pollen spectra. The problems of comparing modern and fossil pollen spectra are discussed more fully in Section 6.4.

4.4 Comparison of Sequences in the Absence of Stratigraphical Constraints: Classification Methods

The method described in the previous section requires specification of the zones or collection of spectra to be compared. Misleading or even invalid conclusions might be drawn if the basic units for comparison were defined incorrectly. In the methods to be described in the next two sections, the basic unit is the pollen spectrum, and hence the results will not be influenced by any prior classification of the data.

This section describes an unconstrained classification study of the two test data sets. The word *unconstrained* means that the data are analysed without the imposition of any constraints based on the stratigraphical ordering of the samples.

As is described in Section 2.5, there are two main approaches in classification studies: (1) obtaining a geometrical representation of the data in a low-dimensional space, and (2) seeking to impose a partition (or set of partitions) on the data. Both approaches are illustrated below.

4.4.1 Geometrical Representation

Each spectrum from the two sequences to be compared may be defined by a point in nine-dimensional space, each of the nine coordinate axes being defined by one of the pollen taxa present in the pollen spectra. Points which are close together in this nine-dimensional space will correspond to spectra which are similar to one another in pollen composition, irrespective of the sequence from which they are derived. This geometrical disposition of the spectra will thus allow the detection of groups of spectra of similar pollen composition, and hence of regional pollen assemblage zones.

Such high-dimensional representations are difficult, however, to assimilate directly, and we seek a lower-dimensional representation which preserves as much as possible of the relevant information about the relationships and distances between the points. Two- and three-dimensional representations are generally the most useful.

There are many different numerical methods which provide low-dimensional representations of multivariate data (see Blackith and Reyment, 1971; Everitt, 1978; Gordon, 1981). Gower (1967a, 1967b), Sibson (1972a), and Gordon (1981, Chapter 5) provide introductions to the ideas and concepts in these geometrical methods of analysis. One such method, which is used extensively in this book, is principal components analysis; I. C. Prentice (1980) reviews the many applications of this method of analysis to Quaternary pollen stratigraphical data.

An informal description of principal components analysis is as follows. We carry out a rigid rotation of the coordinate axes to form new 'principal axes', so that the direction of the first principal axis lies along the direction of the maximum amount of variability within the data. Subsequent principal axes are then selected to account for the largest amount of the residual variability left after the effects of earlier principal axes have been allowed for, but with the constraint that all the axes are orthogonal to each other. If there is marked structure within the data, the first few principal components will usually account for most of the variability, and the higher components can often be disregarded without any serious loss of information. As a result of this rigid rotation of the axes to form principal axes, the coordinates of the spectra or samples on the first few principal component axes provide an efficient summary of the original high-dimensional configuration.

It is worth noting that if the coordinates of the ith point in the original t-dimensional space are specified by the proportions $p_{i1}, \ldots, p_{it}$, the constraint that proportions must sum to 1 reduces to $t - 1$ the dimensionality of the scatter of points representing the spectra. It has been suggested (Aitchison, 1982) that proportions be transformed in some manner before analysis, for example, to

$$z_{ik} \equiv \log(p_{ik}/p_{i*}),$$

where

$$\log p_{i*} \equiv \sum_{k=1}^{t} (\log p_{ik})/t.$$

We are unaware of any applications of this transformation to palynological data; it has the disadvantage that the results can be greatly influenced by very small proportions (which may not have been estimated accurately) and by the strategy adopted to cope with 0 proportions (see Aitchison, 1982).

Seal (1964), Jöreskog, Klovan, and Reyment (1976), Morrison (1976), and Pielou (1977) give detailed mathematical descriptions of principal components analysis, and Gower (1966b, 1967a), A. J. B. Anderson (1971b), Everitt (1978), and Gordon (1981) discuss the method in the general context of geometrical representations of multivariate data. Some comparative comments on principal components analysis and other methods of geometrical representation are postponed to Subsection 5.3.4.

A principal components analysis of the correlation matrix of the nine pollen types was carried out on the set of 17 spectra from the two Abernethy Forest test data sets. The use of a correlation matrix, rather than a covariance matrix, is equivalent, in terms of the above description, to standardising the data to have unit variance along each of the original axes prior to any axis rotation. The results of this principal components analysis are given in Table 4.2, which is to be interpreted as follows. The first principal component is defined by

$$Y_1 = 0.202X_1 + 0.220X_2 + 0.319X_3 + \cdots - 0.118X_8 - 0.443X_9,$$

where X_k denotes the axis in the original space defined by the kth pollen type ($k = 1, \ldots, 9$). This component has high positive loadings for *Corylus*/*Myrica* and high negative loadings for *Empetrum*, Cyperaceae, *Rumex acetosa*-type, and Gramineae pollen (see Table 4.2). Thus, spectra with high positive scores on the first principal axis have relatively high proportions of *Corylus*/*Myrica* pollen and/or relatively low proportions of *Empetrum*, Gramineae, Cyperaceae, and *Rumex acetosa*-type pollen. The second principal component is dominated by high positive loadings for *Pinus* and high negative loadings for *Betula* and *Juniperus* pollen.

The positions of the 17 spectra on the first two principal axes are shown in Figure 4.1, the spectra from the two sequences being joined up in stratigraphical order from the bottom to the top of each sequence. The plot of the first two

Table 4.2

Results of Principal Components Analysis of the Two Abernethy Forest Test Data Sets: Component Loadings for the First Three Principal Components

	Component		
	1	2	3
Betula	0.202	−0.575	0.030
Pinus	0.220	0.488	0.298
Corylus/Myrica	0.319	−0.101	0.299
Juniperus	0.028	−0.502	−0.291
Empetrum	−0.438	−0.066	0.316
Gramineae	−0.442	−0.267	0.057
Cyperaceae	−0.460	0.063	−0.048
Artemisia	−0.118	0.292	−0.757
Rumex acetosa-type	−0.443	0.066	0.240
Eigenvalue	4.031	2.029	1.199
Percentage of total variance	44.79	22.56	13.32
Cumulative percentage of total variance	44.79	67.35	80.67

principal axes accounts for 67% of the original variability, and can thus be regarded as providing a reasonably accurate summary of the information contained in the full nine-dimensional space.

Results such as those presented in Table 4.2 and Figure 4.1 can enable one to assess how large a contribution each taxon makes to the composition of each spectrum. Two geometrical methods of analysis which allow this assessment to be made more directly are the biplot (Gabriel, 1971) and correspondence analysis (Benzécri, 1969, 1973; Hill, 1973, 1974); Gordon (1982c) presents a comparison of these two methods of analysis, illustrating the methodology by applying them to several different palynological data sets.

The basic idea of these methods is that a data matrix describing n pollen spectra in terms of their t constituent taxa is represented by a set of $n + t$ points or vectors in a low-dimensional space, one point being assigned to each spectrum and one point or vector being assigned to each taxon. The geometrical representation has the following properties:

Property 1 The configuration of spectrum points gives an indication of the similarities between the spectra, in a similar manner to principal components analysis.

Property 2 The configuration of the points or vectors representing the taxa gives an indication of the similarities between the behaviour of the taxa over the period of time covered by the observations.

Property 3 The relationships between members of the two sets of points or vectors gives an indication of the importance of each taxon in the composition of each spectrum.

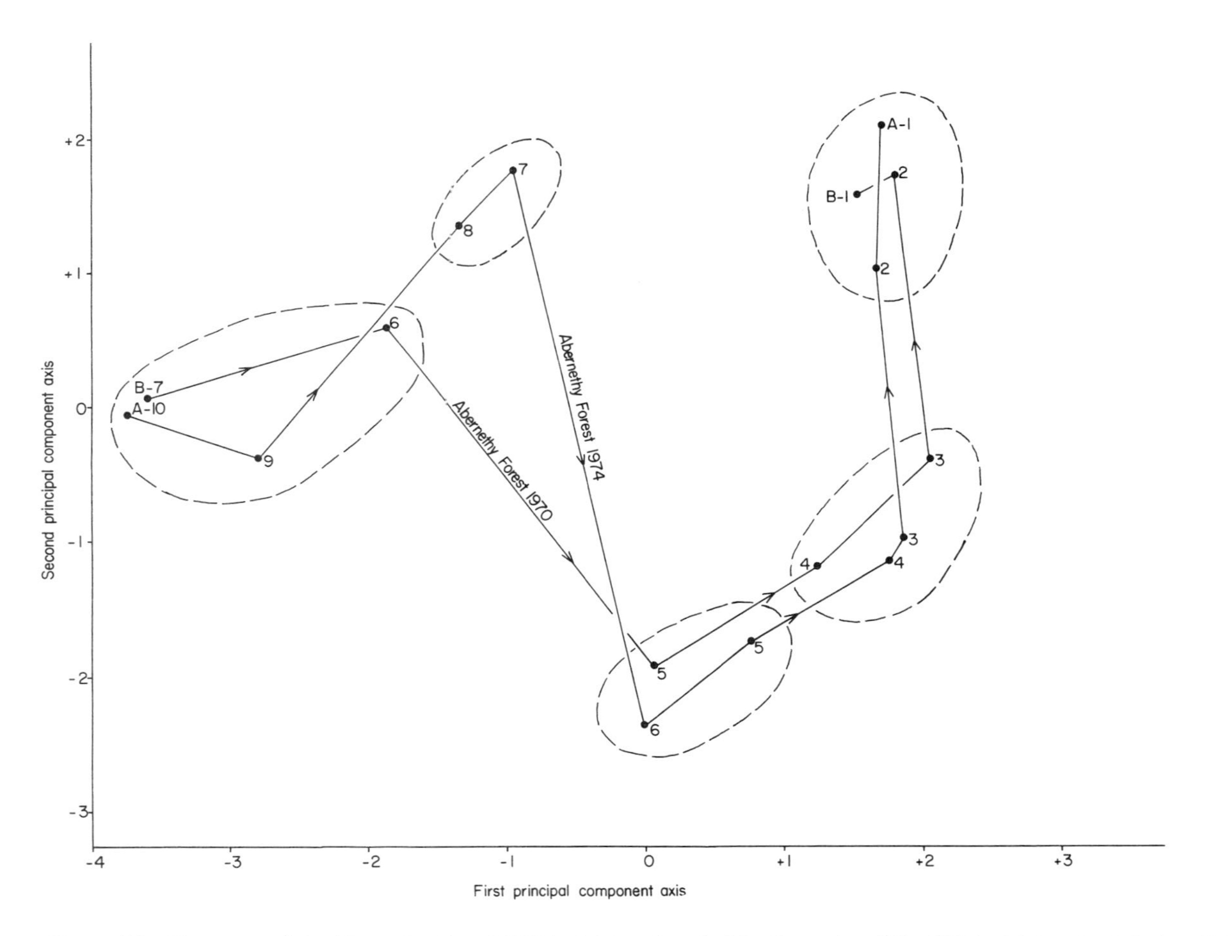

Figure 4.1 The positions of the 10 samples of the Abernethy Forest 1974 test data set, and of the 7 samples of the 1970 test data set, plotted on their first and second principal component axes. Information on partitioning the data according to the minimum sum-of-squares criterion is also shown. The samples from the two sequences are joined up in stratigraphical order.

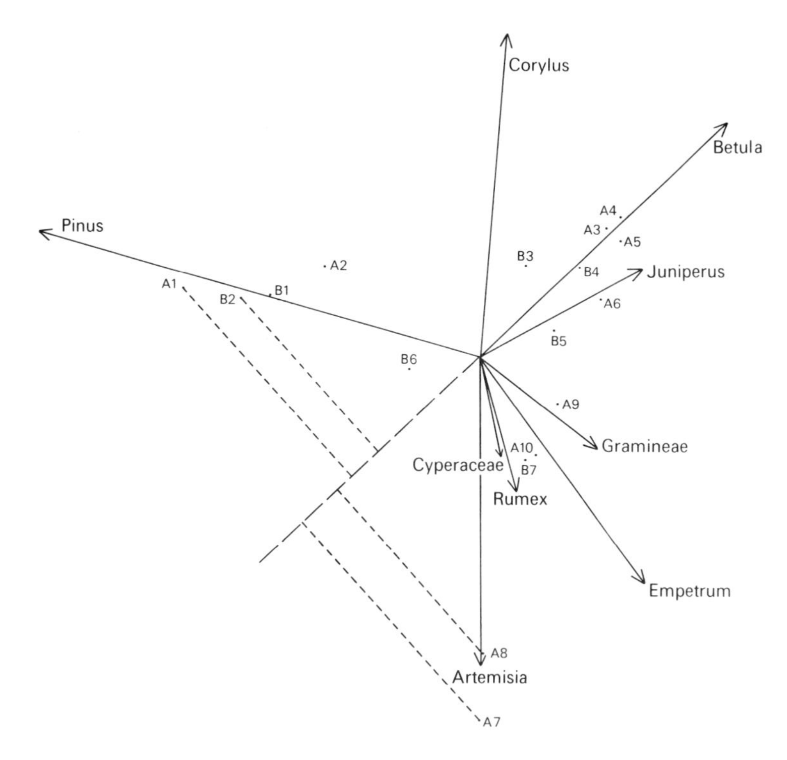

Figure 4.2 A biplot of the 17 samples from the two Abernethy Forest test data sets. The vectors for the three most variable taxa, *Pinus, Betula,* and *Artemisia,* have each been scaled down to one-third of their lengths. Also shown are the projections of several sample points on to the *Betula* vector.

As an illustration, Figure 4.2 shows a biplot of the set of 17 spectra from the two Abernethy Forest test data sets. An outline of the manner in which properties 1–3 above are realised is given below; mathematical details are presented by Gabriel (1971) and Gordon (1981, Chapter 5). The approximations mentioned below are due to the fact that this is a two-dimensional summary of nine-dimensional data; the results are exact in the higher dimensional space, but because Figure 4.2 captures 76.4% of the variability, the main features of the data should be accurately portrayed.

Property 1 The points representing the 17 spectra are such that the distance between any pair of points is an approximation to the Mahalanobis distance between the corresponding pair of spectra. (Mahalanobis distance is a generalisation of ordinary distance which takes into account the correlations between the taxa; see Subsection 5.3.3). It can be seen that the relative positions of the points are similar to those given in Figure 4.1. In other words, the precise standardisation which is carried out on the data has not obscured the resemblances between the spectra.

Property 2 Each taxon is represented by a vector. The amount of variability between samples in each taxon is indicated by the length of the corresponding vector, this length being an approximation of the standard deviation of the pollen values for the taxon in question. For convenience of representation, the lengths of the three longest vectors—corresponding to the most variable taxa, *Pinus, Betula,* and *Artemisia*—have each been scaled down by a factor of 3.

In addition, the correlation coefficient between two taxa is given by the cosine of the angle between the corresponding pair of vectors. Thus, taxa whose behaviour is similar are represented by vectors pointing in similar directions. For example, grasses, sedges, and other herbs are seen to vary in a manner which is similar to one another but largely complementary to the behaviour of *Pinus*.

Property 3 The interaction between spectrum and taxon is obtained from the perpendicular projection of each point on to each vector (extending the vector forwards and backwards in the same direction in space if necessary). This is illustrated for some of the points on the *Betula* vector in Figure 4.2. A spectrum which contains relatively large amounts of pollen from a specified taxon will be represented by a point located in the region of the space defined by large positive values of the vector representing that taxon. For example, the projections of the points onto the *Betula* vector are ranked in the following order: A4, A5, A3, B4, A6, B3, B5, A9, A10, B7, A2, B6, B1, B2, A1, A8, A7. It can be seen from Tables 1.3 and 1.4 that this closely approximates the ordering of the *Betula* proportions in the 17 spectra. Differences are due to the fact that a nine-dimensional data set has been approximated in two dimensions; in the higher-dimensional space, not only the ordering but also the proportions themselves would be recovered exactly.

As an example of the use which can be made of biplots, we can note from Figure 4.2 that, for example, spectra A1, A2, B1, and B2 have relatively high contributions of *Pinus* pollen, and that spectra A7 and A8 have relatively high contributions of *Artemisia* pollen. It would be unwise to attempt too detailed an interpretation of a biplot, particularly if only a small proportion of the variability has been captured in the first two dimensions, but it can present a useful summary of the main features of a data set.

Further discussion of the results of analysing the test data sets by geometrical methods is deferred until after a description of clustering procedures.

4.4.2 Partitioning the Data Set

An alternative classificatory approach is to seek to partition the data set into, say, g groups so as to optimise some mathematical criterion. A commonly used one is the minimum sum-of-squares criterion, that is, we wish to divide the data into g groups (where g is assumed to be known) so as to minimise the total within-group sum-of-squared distances about the centroid. With p_{ik} and q_{ik} defined as in Section 3.4, we seek that division of the data which minimises

$$\sum_{i} \sum_{k=1}^{9} (p_{ik} - q_{ik})^2,$$

where i is summed over the 17 spectra.

The situation differs from that described in Section 3.4 in that there are now no stratigraphical constraints on the set of allowable partitions. This greatly increases the number of possible partitions (see Fortier and Solomon, 1966), and will ensure that an examination of all of them to identify the optimal one would take too long to be computationally feasible for all but the smallest data sets. Various approximating algorithms have been proposed, and these are critically reviewed by Gordon and Henderson (1977). The most commonly used of these is an agglomerative algorithm, which proceeds as follows. We start with all the spectra in separate, single-member groups; at each stage we amalgamate those two groups which lead to the minimum increase possible at that stage in the total within-group sum-of-squares, and continue the process until all the objects are in a single group. This algorithm is referred to as Ward's method (after Ward, 1963), as optimal agglomeration by Orloci (1967a), or as minimum variance cluster analysis by Pritchard and Anderson (1971); an efficient version of the algorithm is described by Wishart (1969).

The results of using this agglomerative algorithm can be summarised by a dendrogram, such as the one shown in Figure 4.3, where the abscissa gives a measure of dissimilarity; the earlier an amalgamation occurs, the more similar the relevant spectra are in overall pollen composition. This procedure avoids difficulties associated with specifying a value for g, the number of groups which are of interest, by obtaining a partition into g groups for all values of g (a partition can be obtained by sectioning the dendrogram at some value of the abscissa, e.g., at the position indicated by the dashed line). However, two points should be noted. Firstly, the partition of the data into g groups produced by this algorithm is only an approximation to the optimal (minimum sum-of-squares) partition into g groups. Secondly, this algorithm imposes a hierarchy on the set of partitions (e.g., the partition into three groups must be nested within the partition into two groups, and this structure may be unwarranted for some data sets (see L. Fisher and Van Ness, 1971; Van Ness, 1973).

In the discussion given in Section 2.5, it was stressed that imposing a hierarchy on data may give misleading results in some cases. Nevertheless, a dendrogram can often give a useful and informative summary at several different levels. Figure 4.3 shows the sum-of-squares dendrogram for the two test data sets considered together, with no stratigraphical constraints imposed on the classification. Sectioning this dendrogram at the place indicated by the dashed line yields five groups, and the extent of these groups is also shown by the dashed lines in Figure 4.1. An investigation of the data by several other of the sum-of-squares approxi-

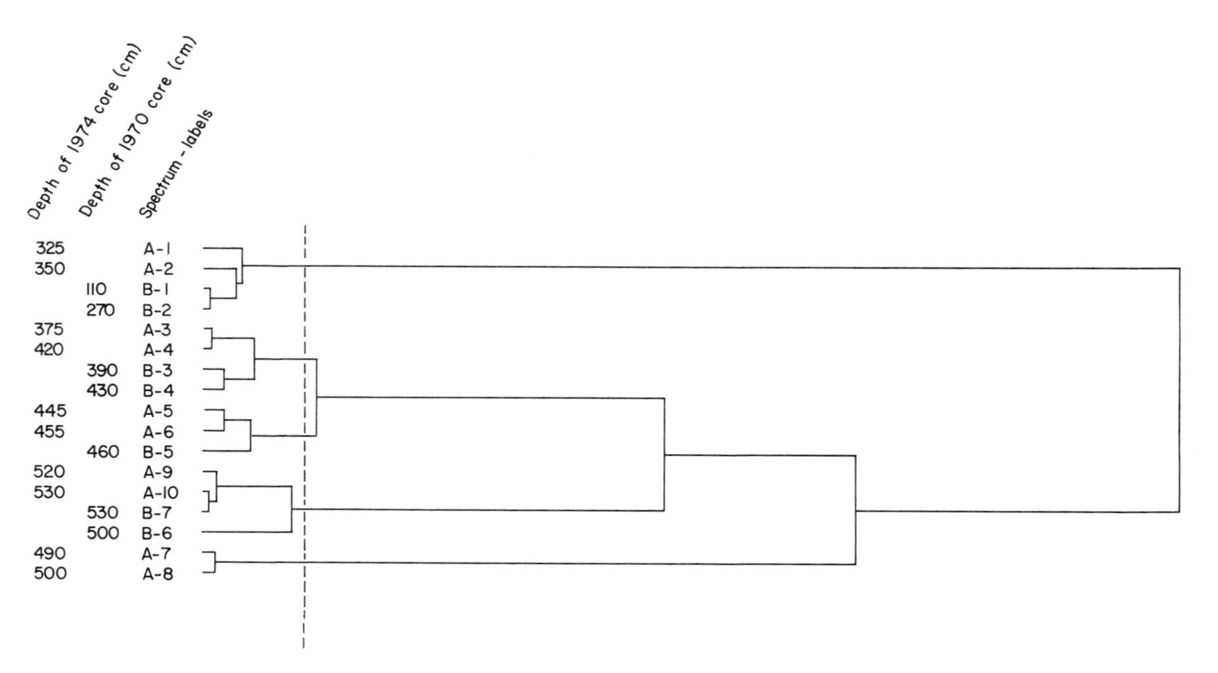

Figure 4.3 Sum-of-squares dendrogram for the Abernethy Forest 1974 and 1970 test data sets. For spectrum labels, A signifies the 1974 core and B the 1970 core.

mating algorithms described in Gordon and Henderson (1977) suggests that the overall optimal sum-of-squares partition into five groups is as shown in Figures 4.1 and 4.3.

An examination of Figures 4.1 and 4.2 lends weight to the suggestion that it would not be a gross distortion of the data to divide them into the five groups shown. The results are also in broad agreement with those described in the previous section. For example, the 'zone' (A3, A4) was seen to be more homogeneous than the 'zone' (B3, B4), but when these zones were compared an *H*-match resulted, and this is reflected in the proximities of these points in the geometrical representations.

The data sets used to illustrate the use of geometrical representation and partitioning techniques in comparing pollen diagrams contain a very small number of samples compared with 'real' pollen analytical data sets. However, if we were to regard the test data sets as complete data from two sequences or sites, the results obtained would suggest that five groups of spectra can be distinguished. The composition of these is as follows: (A1, A2, B1, B2), (A3, A4, B3, B4), (A5, A6, B5), (A9, A10, B6, B7), (A7, A8). As four of these five groups contain pollen spectra from both cores, regional pollen zones could be delimited on the basis of these four groups. The remaining group contains spectra from the Abernethy Forest 1974 core only, and might be of only local occurrence. Clearly, data from other sequences and sites are required to establish whether spectra such as A7 and A8 are unique to the Abernethy Forest 1974 core.

4.5 Comparison of Sequences by Slotting

4.5.1 The Statistic

In contrast to the previous two sections, the method described below makes direct use of the ordering of the spectra within each core. We can regard the two test data sets as two sequences of objects (spectra), namely,

S_1: A1, A2, A3, A4, A5, A6, A7, A8, A9, A10;
S_2: B1, B2, B3, B4, B5, B6, B7.

We can slot these two sequences into a single joint sequence which preserves the ordering within each of the two original sequences, for example,

$S_1 \cup S_2$: B1, A1, B2, A2, B3, A3, B4, A4, A5, B5,
A6, A7, A8, B6, A9, B7, A10.

There are, however, almost 20,000 different ways in which the sequences S_1 and S_2 can be slotted together. We wish to ensure that similar objects from the two sequences appear close together in the joint sequence, subject to the ordering condition described above. This objective is achieved by defining a statistic to measure the 'discordance' of a slotting, and obtaining that slotting which minimises the measure of discordance.

The measure of the discordance between sequences to be described below assumes that the relevant differences between each pair of objects in the two sequences can be summarised by a measure of pairwise dissimilarity d. A matrix of dissimilarities between the spectra in the two test data sets has already been evaluated in Section 2.6 (the values contained within the central box in Table 2.2), and will be used to illustrate the method. In addition, 'dummy' objects A0 and A11 (in general, A$(M + 1)$ if there are M objects in S_1) are introduced respectively before A1 and after A10 by identifying A0 with A1, and A11 with A10. By convention, A0 and A11 (A$(M + 1)$) will always bracket the objects of S_2. Dummy objects B0 and B8 (in general, B$(N + 1)$) are introduced in a similar manner for S_2. The introduction of the dummy objects is an expedient for enclosing the sequences of objects. Dissimilarities relating to the dummy objects are given in the first and last row and column of Table 2.2.

An object from one sequence may be said to fit satisfactorily between a pair of objects in the other sequence if it is sufficiently similar to both of them. Accordingly, the measure of discordance $\sigma(S_1, S_2)$ is defined as the sum (over all objects in each sequence) of the dissimilarities between an object and the two objects in the other sequence which bracket it. This concept is illustrated in Figure 4.4, which describes the slotting $S_1 \cup S_2$ defined above, but in which each object is linked to the pair of objects in the other sequence which brackets it. In this figure, B-A-B links are shown by unbroken lines, and A-B-A links are shown by broken lines. For example, A5 lies between B4 and B5, and is linked to both of them by unbroken lines in Figure 4.4; these links represent a contribution to the measure of discordance of $d(\text{A5}, \text{B4}) + d(\text{A5}, \text{B5})$.

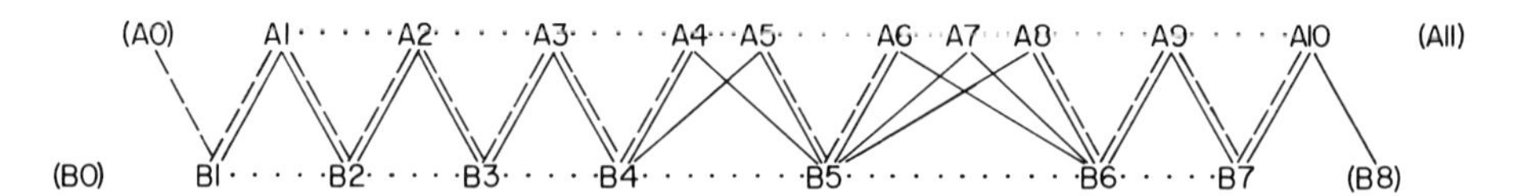

Figure 4.4 Slotting of the sequences S_1 (A1, A2, ..., A10) and S_2 (B1, B2, ..., B7), illustrating the contributions to the measure of discordance $\sigma(S_1, S_2)$ and the 'length' of the sequences, $\mu(S_1, S_2)$.

In some situations, we wish to compare a given sequence with each of a set of sequences, and it is desirable to standardise $\sigma(S_1, S_2)$ in some way. We can define the 'length' of the sequences, $\mu(S_1, S_2)$, by

$$\mu(S_1, S_2) \equiv \sum_{i=1}^{M-1} d(\text{A}(i), \text{A}(i + 1)) + \sum_{j=1}^{N-1} d(\text{B}(j), \text{B}(j + 1)),$$

that is, by the sum of all the dissimilarities between the pairs of objects shown linked together by dotted lines in Figure 4.4. Then, a standardised measure of the discordance is

$$\psi(S_1, S_2) \equiv \frac{\sigma(S_1, S_2) - \mu(S_1, S_2)}{\mu(S_1, S_2)}.$$

Other measures of discordance are described by Gordon (1973b) and Delcoigne and Hansen (1975).

It might be useful to enter one caveat about the slotting method at this stage: the method will always produce a slotting of the sequences, no matter how slight the resemblance between them. The value taken by $\psi(S_1, S_2)$ gives an indication of the overall similarity of S_1 and S_2, lower values of ψ suggesting a closer resemblance. Further comments on the assessment of results, and discussion of other properties of the method are postponed until Sections 4.6 and 4.8.

4.5.2 A Dynamic Programming Algorithm

There is a large number of different ways in which two sequences, even of moderate length, can be slotted together. Fortunately, we do not have to examine all possible slottings in order to identify the one with minimum discordance; Delcoigne and Hansen (1975) noted that one can build the solution up recursively using the optimality principle of dynamic programming (Bellman and Dreyfus, 1962, p. 15). Let $F(i, j)$ denote the minimum contribution which it is possible to make to the discordance when the part-sequence (A0, A1, ..., A(i)) and (B0, B1, ..., B(j)) are slotted together. For example, it can be shown that the dissimilarities contributing to $F(2, 2)$ are shown by the links in Figure 4.5. Thus, the B-A-B contributions (unbroken lines) total

$$\begin{aligned} &d(\mathrm{A1}, \mathrm{B1}) + d(\mathrm{A1}, \mathrm{B2}) + d(\mathrm{A2}, \mathrm{B2}) + d(\mathrm{A2}, \mathrm{B3}) \\ &\quad = 0.454 + 0.375 + 0.478 + 0.963 \\ &\quad = 2.270 \end{aligned}$$

(values are taken from Table 2.2). Similarly, the A-B-A contributions (broken lines) total $d(\mathrm{A0}, \mathrm{B1}) + d(\mathrm{A1}, \mathrm{B1}) + d(\mathrm{A1}, \mathrm{B2}) + d(\mathrm{A2}, \mathrm{B2}) = 1.761$. Since the slotting shown in Figure 4.5 is the optimal slotting of the part sequences (A1, A2) and (B1, B2), it follows that

$$F(2, 2) = 2.270 + 1.761 = 4.031.$$

The optimal slotting of the whole sequences is built up recursively for successively larger values of i and j until $F(M, N)$ is reached. For example, suppose we wished to evaluate $F(2, 3)$. At this stage in the recursion, we would already know that $F(2, 2) = 4.031$, and that $F(1, 3) = 5.049$. The last object in the joint slotting of (A1, A2) and (B1, B2, B3) (see Figure 4.6) must be either (a) A2, or (b) B3. In case (a), the contribution to the discordance will be

$$F(1, 3) + d(\mathrm{A2}, \mathrm{B3}) + d(\mathrm{A2}, \mathrm{B4}),$$

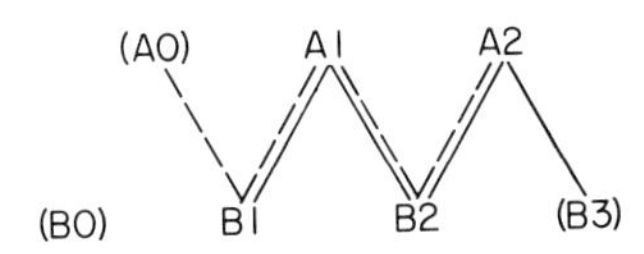

Figure 4.5 Contributions to the measure of discordance given by the optimal slotting of the part-sequences (A1, A2) and (B1, B2).

Figure 4.6 Construction of the slotting of the part-sequences (A1, A2) and (B1, B2, B3) when the last object in the joint slotting is (a) A2 and (b) B3.

the first term because we wish to incorporate the best slotting possible of (A1) and (B1, B2, B3), and the last two terms to give the extra contribution by adding A2 last, as shown by the links in Figure 4.6a.

By a similar argument, the contribution to the discordance in case (b) will be

$$\begin{aligned} &F(2, 2) + d(\text{A2, B3}) + d(\text{A3, B3}) \\ &= 4.031 + 0.963 + 0.505 \\ &= 5.499. \end{aligned}$$

This value is less than the contribution made in case (a) (5.049 + 0.963 + 1.124 = 7.136), and so $F(2, 3) = 5.499$, and the optimal slotting of the part-sequences (A1, A2) and (B1, B2, B3) is that implied by Figures 4.5 and 4.6b.

A geographical analogy might be useful in helping to illustrate the manner in which the optimal splotting of the two test data sets can be effected. Figure 4.7 gives a map of the mythical dictatorship of Yaweno. A strictly organised people, the Yawenonians have 88 towns, named (i, j) (for $i = 0, 1, \ldots, 10; j = 0, 1, \ldots, 7$) and situated on a regular grid as shown in Figure 4.7. There are tollroads connecting some of the towns, but traffic may only move along a road in the direction indicated by the arrow, that is, only in a southward or eastward direction. The charge for using each tollroad is given by the amount printed above the road in Figure 4.7. Visitors to Yaweno must enter the country at town (0, 0) and leave from the rather overcrowded town (10, 7). Which route through Yaweno incurs the minimum cost in toll charges?

This problem can be solved recursively by finding the cheapest route from (0, 0) to each town in turn, in the order (1, 0), (2, 0), (3, 0), ..., (10, 0), (0, 1), ..., (10, 1), (0, 2), ..., (10, 2), ..., (0, 7), ..., (10, 7). The total cost of reaching town (i, j) ($F(i, j)$, say) is shown in the box under the 'name' of the town; for example, the total cost of reaching town (2, 3) is $F(2, 3) = 5.499$. This can be verified by noting that only two roads enter town (2, 3), from the north and the west, and using the second of these would ensure a smaller total cost (4.031 + 1.468 = 5.499, compared with 5.049 + 2.087 = 7.136). This is exactly analogous to the slotting problem, with the charge for using a road being given by the sum of a pair of dissimilarities (as illustrated in Figure 4.6) corresponding to the increased contribution to the discordance yielded by adding another object to the joint slotting.

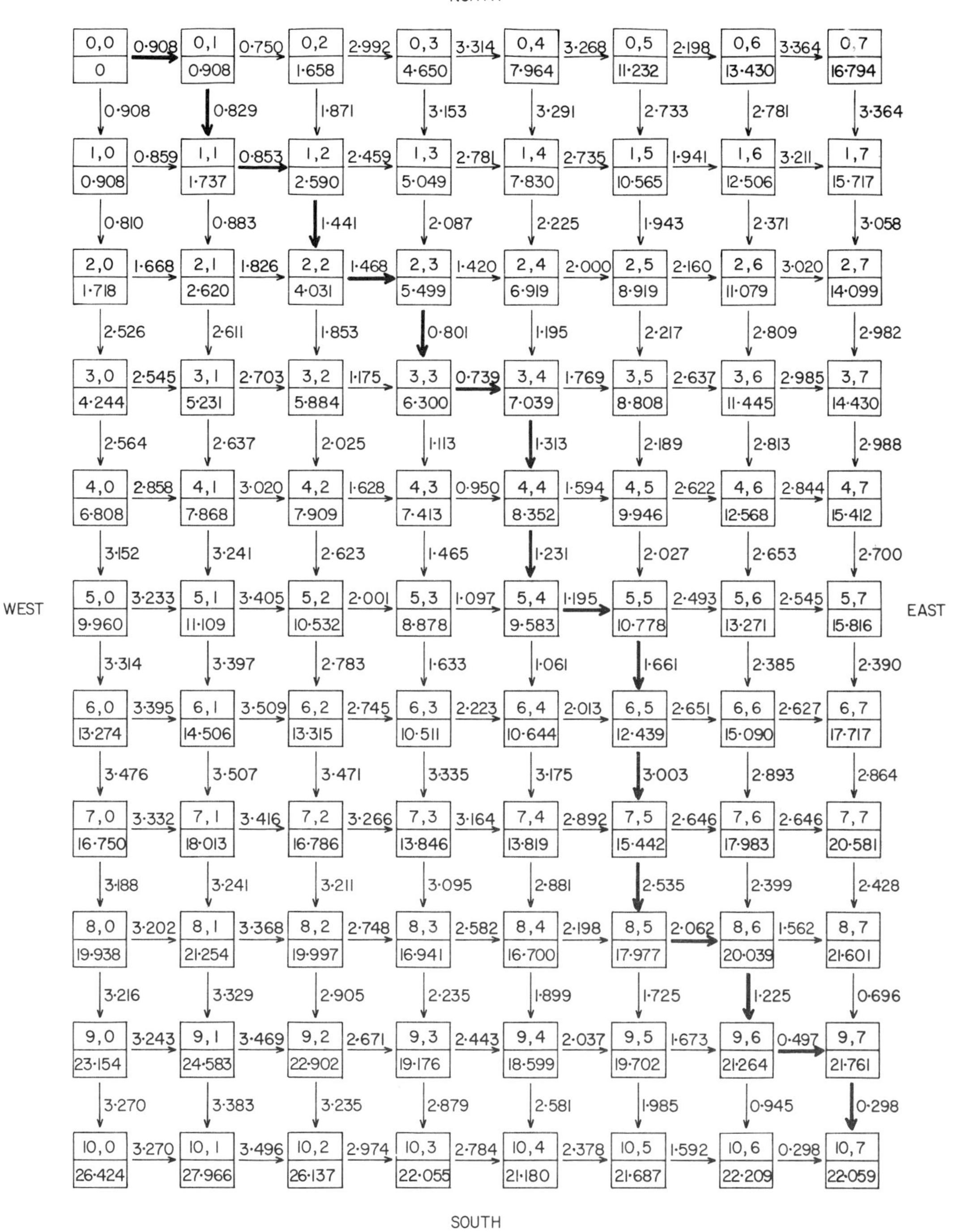

Figure 4.7 A map of the mythical dictatorship of Yaweno, showing the optimal route through the country from (0, 0) to (10, 7). See text for further explanation.

Once $F(10, 7)$ has been evaluated, the optimal route to town (10, 7) from town (0, 0) can be obtained by tracing back along one of the pairs of roads entering a town, choosing the road which led to the minimum total cost of reaching the town. This route has been emphasised by heavier arrows in Figure 4.7; the corresponding optimal slotting is, in fact, the one shown in Figure 4.4.

The correspondence can be stated formally as follows. We start from (0, 0) (no objects yet slotted). Using a southbound road is equivalent to bringing an object from S_1 into the joint slotting, whereas using an eastbound road is equivalent to bringing in an object from S_2. We build up the slotting in the order specified by the optimal route. A listing of a FORTRAN IV program SLOTSEQ implementing this algorithm is given by Gordon (1980c).

It has been seen, therefore, that when the two test data sets are optimally slotted together in the manner shown in Figure 4.4,

$$\sigma(S_1, S_2) \equiv F(10, 7) = 22.059.$$

By summing the diagonal entries in Table 2.1, the 'length' of S_1 is seen to be $0.533 + 1.138 + 0.193 + \cdots + 1.258 + 0.375 = 6.219$. Similarly, the 'length' of S_2 can be shown to be 4.106, hence

$$\mu(S_1, S_2) = 6.219 + 4.106 = 10.325,$$

and the standardised measure of discordance, to four decimal places,

$$\psi(S_1, S_2) = \frac{22.059 - 10.325}{10.325} = 1.1365.$$

4.5.3 Discussion

It is appropriate at this point to make some comments about the slotting method. Firstly, the optimal slotting need not be uniquely defined; there can be more than one way of slotting the sequences together which lead to the minimum value of the discordance. (Alternatively, there can be more than one equally expensive route between the towns (0, 0) and (M, N).) If such a multiple solution exists, which is not the case for these data, it is indicated by the program SLOTSEQ (Gordon, 1980c).

Secondly, in order to trace back the route taken, SLOTSEQ requires to hold the cost (F) and dissimilarity (d) matrices in store. This will limit the size of problem which this program can analyse without recourse to external store, but most computers should be able to compare two sequences with more than 100 objects in each sequence. An alternative algorithm is described by Gordon and Reyment (1979); this can handle much larger sequences, but requires increased effort on the part of the user.

Thirdly, there is no mathematical barrier to slotting together more than two sequences simultaneously (Gordon, 1973b; Gordon and Birks, 1974), but computer storage requirements again limit the size of more general slotting problems which can be handled in this manner. An alternative approach would be to regard one of the sequences under study as the 'master' sequence, and to slot each of the other sequences against it in turn, possibly incorporating them into a new extended master sequence at each stage. With such a strategy, however, the precise form of the results is likely to be influenced by the choice of master sequence and the order in which the other sequences are merged into it.

Finally, we can impose constraints on the set of allowable slottings (alternatively, the set of allowable routes for travelling between towns (0, 0) and (M, N)). This option is likely to be of less interest in Quaternary palynological studies where one is most commonly concerned with the overall palynological similarities between the two sequences, and where one does not wish to invoke *at this stage* any assumptions based on sediment lithology or assumed time equivalence. In other fields of study, however, the possibility of imposing such constraints could be a useful additional strategy when comparing sequences. For example, when comparing physical logs from boreholes, it is relevant to know that 'objects' from different sequences occurred above or below a persistent marker bed, such as a coal bed, tephra layer, or bone bed. A description of three different types of constraint which can be imposed is given by Gordon and Reyment (1979) and Reyment (1980a).

Christopher (1978) has developed a graphical method for comparing two stratigraphical sequences on the basis of similarities in pollen and spore assemblages in connection with his work on Upper Cretaceous stratigraphy of South Carolina. His method, called *similarity coefficient matrix contouring,* involves contouring a matrix of similarities between all pairs of samples in the two sections to be compared. The samples are retained in their stratigraphical order within the two sections. In common with other contouring methods (see Subsection 5.3.2), the results are affected by the amount of 'smoothing' carried out. The resulting contour map is assessed by eye; areas which are similar in the two sections will appear as 'hills', and areas where there are rapid changes in pollen and spore

Table 4.3
Optimal Slotting of the Abernethy Forest Test Data Sets[a]

Abernethy Forest 1974 Core		Abernethy Forest 1970 Core	
Depth (cm)	Spectrum	Spectrum	Depth (cm)
		B1	110
325	A1		
		B2	270
350	A2		
		B3	390
375	A3		
		B4	430
420, 445	A4, A5		
		B5	460
455–500	A6–A8		
		B6	500
520	A9		
		B7	530
530	A10		

[a] $\psi = 1.137$.

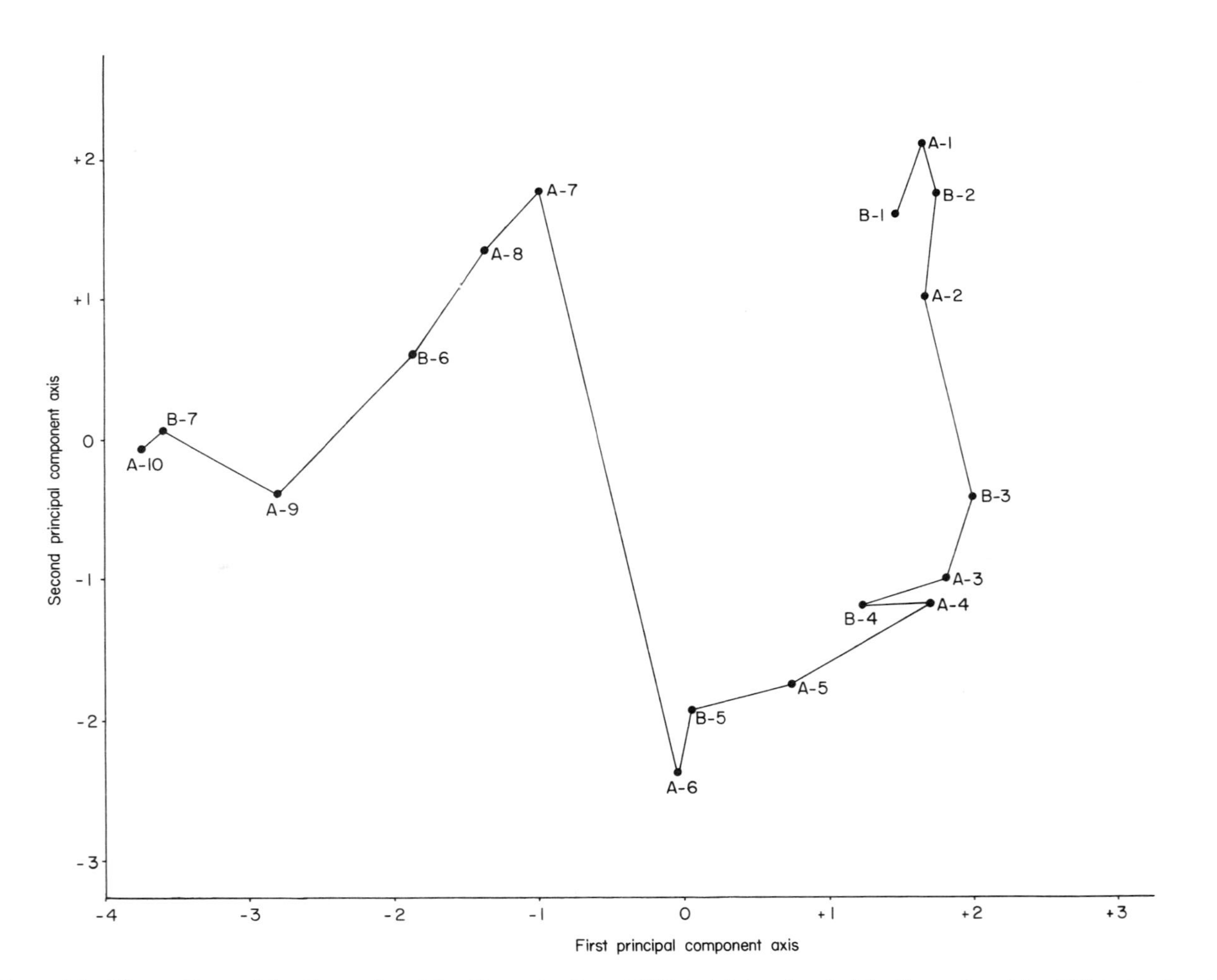

Figure 4.8 The positions of the 10 samples of the Abernethy Forest 1974 test data set and of the 7 samples of the 1970 test data set plotted on their first two principal component axes. The samples are joined up in the order suggested by the sequence-slotting results.

composition will be depicted by regions of closely spaced contours, indicating steep gradients. The results obtained by Christopher (1978) proved useful in highlighting groups of samples of similar composition in the two sequences.

Results of slotting the two Abernethy Forest test data sets are shown in Table 4.3 and Figure 4.8. In Figure 4.8, the 17 spectra are displayed on the first two principal component axes of Figure 4.1, but the spectra are linked together in the order suggested by the joint slotting (Table 4.3) rather than in stratigraphical order of the two sequences separately. The slotting results generally confirm the conclusions of the earlier techniques for comparing the pollen sequences, in that the slotting method has positioned together spectra of similar composition from the two different sequences. The order of the spectra suggested by the joint slotting and shown in Figure 4.8 could, if desired, now be used to carry out a constrained classification study of all 17 spectra, using the methods described in Chapter 3. Such methods would partition the combined sequence of spectra into a series of zones. If such a zone contained a selection of spectra from each of the two original sequences, it could be regarded as a *regional* pollen zone, based on the information from both sequences, rather than solely a local zone derived from a single sequence.

It will prove useful in the analysis of the complete Abernethy Forest data sets to be described in the next section if the general shape of the 'slotting' and 'time' curves in Figures 4.1 and 4.8 are borne in mind.

4.6 Numerical Comparisons of Abernethy Forest

All the numerical comparison methods described in Sections 4.3–4.5 have been applied to compare the two complete data sets from the Abernethy Forest 1970 and 1974 cores. The results of these comparisons are described and discussed below.

4.6.1 Zone-by-Zone Comparisons

The zones for comparison of the two sequences were selected on the basis of consistencies between the results of the various numerical zonation methods presented in Section 3.7 and on the basis of the original zonations for each sequence delimited by H. H. Birks (1970) and H. H. Birks and Mathewes (1978). (See Figures 3.8 and 3.9 for the pollen diagrams and zonation analyses of the complete data sets.) Local pollen zones consistently recognised in the 1974 sequence (Figure 3.8) contain levels 1–15 (zone AFP-6), 16–26 (zone AFP-5), 35–41, 42–45 (zone AFP-2), and 46–49 (zone AFP-1). Several different zonations can be suggested for the part of the sequence containing levels 27–41 (cf. Figure 3.8). For the zone-by-zone comparisons, the original zonation of zones AFP-4 (levels 27–33) and AFP-3 (levels 34–41) were used, along with the following numerically defined local zones: levels 16–32 (optimal sum-of-squares partition zone v and SPLITLSQ), 27–32 (SPLITINF ≡ zone AFP-4 lacking level 33), 16–29 and 30–34 (BARRIER), 33–34 (optimal partition zone iv), and 33–41 (SPLITINF).

As discussed in Section 3.7 the results of the numerical zonations of the Abernethy Forest 1970 sequence (Figure 3.9) are consistent between themselves, but the numerical results correspond only in part with the original zonation of H. H. Birks (1970). Several zones were thus used for comparisons with the 1974 sequence. The zones selected contained the following levels: 1–3 (zone AF-5), 4–23 (zone AF-4), 24–29 (zone AF-3), 30–33 (zone AF-2), 34–41 (zone AF-1), 1–17 (BARRIER), 1–18 (SPLITLSQ, and optimal partition zone v), 18–23 (BARRIER), 19–23 (SPLITINF, SPLITLSQ, and optimal partition zone iv), 24–30 (optimal partition zone iii), 25–30 (BARRIER, SPLITINF), 31–34 (BARRIER, SPLITINF, and optimal partition zone ii), and 35–41 (BARRIER, optimal partition zone i, SPLITINF, and SPLITLSQ).

The information radius H of each zone in each sequence is given in Table 4.4 along with a summary of the H-matches (see Section 4.3) between zones in each sequence. In addition 'near-matches' defined by

$$H(1, 2) \leq 1.1(H_1 + H_2)$$

are also summarised in Table 4.4.

The effects of excluding levels that are transitional in their pollen composition are clearly shown by the information radii for the zones. For example, removing level 33 from local zone AFP-4 to form a zone containing levels 27–32 reduces the information radius from 0.166 to 0.072. Removing level 33 from the zone containing levels 33–41 reduces the information radius from 0.336 to 0.159. If level 34 is also removed to form a zone containing levels 35–41, the information radius is reduced further to 0.063. It is clear that high within-zone heterogeneity, as measured by H, can often result from one or two transitional samples.

The results of the zone-by-zone comparisons generally follow the major pollen stratigraphical features of the two sequences. The local zone containing spectra dominated by *Pinus* pollen in the 1974 core (levels 1–15, zone AFP-6) matches, on the H-match criterion, the zones dominated by *Pinus* pollen in the 1970 core (1–3, 1–17, 1–18, 4–23, 18–23). The local zones characterised by high *Betula* and *Corylus/Myrica* pollen percentages and low *Pinus* pollen frequencies in the 1974 sequence (16–26, 16–29, 16–32) are equated, by the H-match criterion, with zones AF-3 and AF-2 in the 1970 sequence as well as with zones containing levels 24–30, 25–30, and 31–34. Inspection of the original pollen diagrams (Figures 3.8 and 3.9) shows that in the 1970 sequence the phase of high *Juniperus* pollen (zone AF-2) also contains 20% or more of *Corylus/Myrica* pollen, suggesting that zone AFP-5 in the 1974 sequence is similar not only to zone AF-3 but also to zone AF-2 in the 1970 sequence.

Table 4.4

Summary of Zone-by-Zone Comparisons between the Abernethy Forest 1970 Core and the Abernethy Forest 1974 Core[a]

	1970 Core													
	AF-5			AF-4			AF-3			AF-2		AF-1		
1974 Core	1–3	1–17	1–18	4–23	18–23	19–23	24–29	24–30	25–30	30–33	31–34	34–41	35–41	H
AFP-6 1–15	**	**	**	**	**									0.081
AFP-5 16–26							**	**	**	*	**			0.064
16–29							**	**	**	**	**			0.126
16–32							**	**	**	**	**	*		0.217
AFP-4 27–33										**	**			0.166
27–32										*	**			0.072
30–34		*	**	**						**	**	**	**	0.454
33–34		**	**	**						**	**	**	**	0.367
33–41												**	**	0.336
AFP-3 34–41													*	0.159
35–41														0.063
AFP-2 42–45												**	**	0.119
AFP-1 46–49												*	*	0.123
H	0.103	0.145	0.142	0.184	0.108	0.088	0.118	0.113	0.068	0.121	0.157	0.298	0.239	

[a] H-matches are shown by **, near-matches are shown by * (see text for definition of matches).

Local pollen zone AFP-4 and levels 27–32 in the 1974 sequence, with their high *Betula* and *Juniperus* pollen percentages, match zone AF-2 and levels 31–34 in the 1970 sequence on the *H*-match criterion (see Table 4.4). Some local zones in the 1974 sequence, such as those containing levels 30–34, 33–34, and 33–41, have high within-zone information radii (0.454, 0.367, 0.336, respectively), and hence high within-zone heterogeneity. These zones are characterised by high but varying percentages of *Juniperus*, Gramineae, *Artemisia*, *Betula*, and *Empetrum* pollen. They are, rather surprisingly, matched, on the *H*-match criterion, with spectra dominated by *Pinus* pollen but with occasional high frequencies of Gramineae, *Betula,* and *Empetrum* pollen. These results indicate the importance of using zones with low within-zone heterogeneity as the basic units of comparison, even if samples at or near the zone boundaries have to be excluded (see Gordon and Birks, 1974).

Using the *H*-match criterion, there is no zone in the 1970 sequence that matches the very distinctive *Artemisia*-dominated local pollen zone AFP-3 in the 1974 sequence. When the criterion is relaxed to the 'near-match' criterion, zone AFP-3 is matched with the basal seven spectra of the 1970 sequence. When the uppermost spectrum in local zone AFP-3 is removed, thereby reducing the within-zone information radius from 0.159 to 0.063, the resulting zone (levels 35–41) has no match at all in the 1970 sequence. Local zone AFP-2, with its high *Betula*, *Empetrum*, and Gramineae pollen values matches zone AF-1 and levels 35–41 in the 1970 sequence, whereas the basal zone in the 1974 sequence (zone AFP-1) only has 'near-matches' with zones AF-1 and levels 35–41 in the 1970 sequence.

In summary, the zone-by-zone comparisons indicate that zones AFP-1 and AFP-3 in the 1974 sequence cannot be convincingly matched with any zones in the 1970 sequence (cf. H. H. Birks and Mathewes, 1978). Zone AFP-2 is, however, similar to zone AF-1. Zone AFP-4 is most similar to zone AF-2, zone AFP-5 matches zones AF-3 and AF-2, and zone AFP-6 closely matches zones AF-4 and AF-5 and the composite zone of zones AF-4 and AF-5. This general correspondence between the two sequences can provide a basis for delimiting a single set of pollen assemblage zones for the two sequences.

4.6.2 Classification Study

The combined 1974 and 1970 Abernethy Forest data sets were analysed by both geometrical and clustering procedures. The ordination results described below are obtained from principal components analyses of the correlation matrices of the nine pollen taxa included in Figures 3.8 and 3.9. A biplot analysis suggested similar conclusions, and these results are not presented here.

It is convenient to begin by analysing the Abernethy Forest 1974 data on their own. The results of the principal components analysis of these data are given in Table 4.5, and the positions of the 49 samples on the first two principal component axes are shown in Figure 4.9. This geometrical representation in two dimensions

Table 4.5
Results of Principal Components Analysis of the Complete Abernethy Forest 1974 Data: Component Loadings for the First Four Principal Components

	Component			
	1	2	3	4
Betula	0.360	−0.495	0.180	−0.025
Pinus	0.203	0.561	−0.354	0.351
Corylus/Myrica	0.366	−0.145	0.555	0.131
Juniperus	−0.004	−0.477	−0.542	−0.186
Empetrum	−0.239	−0.278	−0.294	0.154
Gramineae	−0.442	−0.234	0.055	0.199
Cyperaceae	−0.476	−0.022	0.177	0.091
Artemisia	−0.215	0.241	0.143	−0.818
Rumex acetosa-type	−0.412	0.032	0.315	0.289
Eigenvalue	3.513	1.811	1.191	1.036
Percentage of total variance	39.03	20.12	13.22	11.52
Cumulative percentage of total variance	39.03	59.15	72.37	83.89

accounts for just over 59% of the original variability. The definitions of these principal components in terms of the original variables (Table 4.5) are similar to those presented in Section 4.4 for the analysis of the two test data sets (Table 4.2). Small differences between the two analyses are that *Empetrum* exerts rather less influence and *Betula* rather more influence on the first component of the complete 1974 data set than in the analysis of the two test data sets and that *Empetrum* exerts rather more influence on the second component of the 1974 data set than it does on the second component of the two test data sets.

The 49 samples plotted in Figure 4.9 appear to fall into five or six distinct groups. If the minimum sum-of-squares criterion is adopted, the optimal partition into six groups is given by the groups labelled I, IIa, IIb, III, IV, and V on Figure 4.9. The optimal partition into five groups is the same except that groups IIa and IIb are united. These two partitions are obtained by the agglomerative algorithm and by several other approximating algorithms, strongly suggesting that these groups correspond to the overall optimal partitions according to the sum-of-squares criterion. The 10 spectra which were selected to form the test data set from the 1974 core (A1–A10) are circled and labelled on Figure 4.9. It is interesting to note that when the samples are joined up in stratigraphical order, as in Figure 4.9, the partitions respect the stratigraphical ordering of the spectra, with the exception of levels 33 and 34. The composition of the groups of spectra delimited by the partitions in Figure 4.9 is summarised and compared with the site

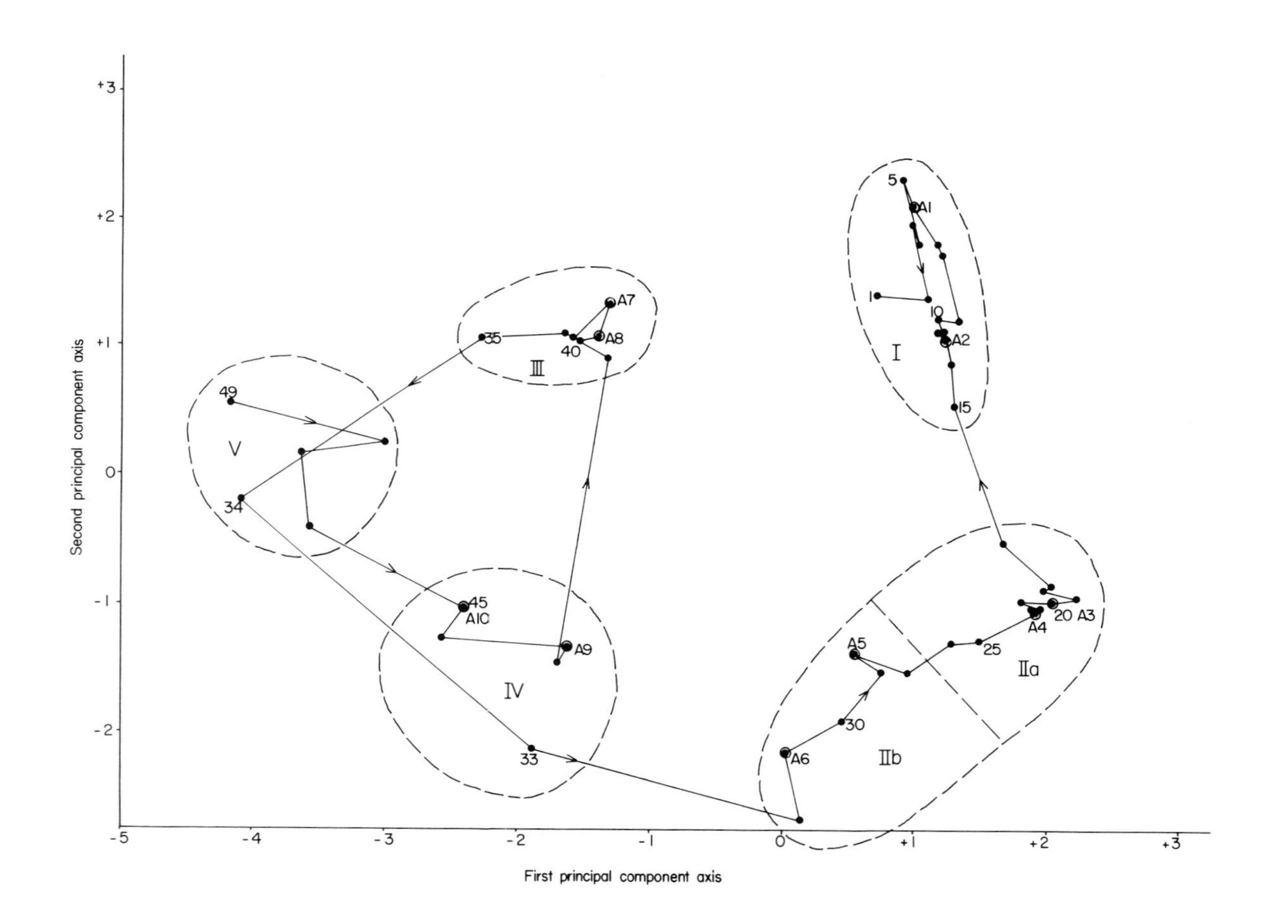
Second principal component axis
First principal component axis
+3
+2
+1
0
-1
-2
-5
-4
-3
-2
-1
0
+1
+2
+3
I
IIa
IIb
III
IV
V
5
A1
1
10
A2
15
20 A3
A4
25
A5
30
A6
33
34
35
40
A7
A8
A9
45
A10
49

Table 4.6

Comparison of the Sum-of-Squares Partition of the Abernethy Forest 1974 Data Set into Six Groups with the Local Pollen Zones

Group	Constituent spectra		Local zones
	Numbers	Depth (cm)	
I	1–15	300–362	AFP-6
IIa	16–26	365–430	AFP-5
IIb	27–32	435–460	AFP-4 minus spectrum 33
III	35–41	475–510	AFP-3 minus spectrum 34
IV	33, 42–45	465, 515–530	AFP-2 plus spectrum 33
V	34, 46–49	470, 535–550	AFP-1 plus spectrum 34

zones AFP-1–AFP-6 in Table 4.6; it can be seen that there is a satisfactorily close correspondence.

In palynological terms the partition into groups I, IIa, and IIb and groups III, IV, and V on the first principal component emphasises the distinction between spectra dominated by *Pinus*, *Betula*, and/or *Corylus/Myrica* pollen (spectra with positive scores on the first component; groups I, IIa, and IIb in Figure 4.9) and spectra dominated by herb pollen such as Gramineae, *Artemisia*, and *Rumex acetosa*-type (spectra with negative scores on the first component; groups III–V in Figure 4.9). The second principal component has high negative contributions from *Betula*, *Juniperus*, and *Empetrum* and high positive contributions from *Pinus* and *Artemisia*. This component is thus contrasting pollen spectra dominated by *Betula* (zone AFP-5), *Betula* and *Juniperus* (zone AFP-4), and *Betula* and *Empetrum* pollen (zone AFP-2) (groups IIa, IIb, and IV in Figure 4.9, all with negative scores on the second component) with spectra dominated by either *Pinus* pollen (zone AFP-6 and group I) or *Artemisia* pollen (zone AFP-3 and group III), all with positive scores on the second principal component. Group V, containing spectra of zone AFP-1 and spectrum 34, has high Gramineae and *Rumex acetosa*-type pollen and is clearly separated on the first principal component but has scores near 0 on the second component.

The interpretation of the results of a principal components analysis of the two full data sets is more complicated. The 49 samples of the 1974 core and the 41 samples of the 1970 core are plotted on the first two principal components of the

Figure 4.9 The positions of the 49 samples of the Abernethy Forest 1974 data set plotted on their first and second principal component axes. The samples are joined up in stratigraphical order. The samples that comprise the 1974 test data set are circled and labelled. The partitions based on the minimum sum-of-squares criterion into five (I–V) and six (I, IIa, IIb, III, IV, V) groups are also shown.

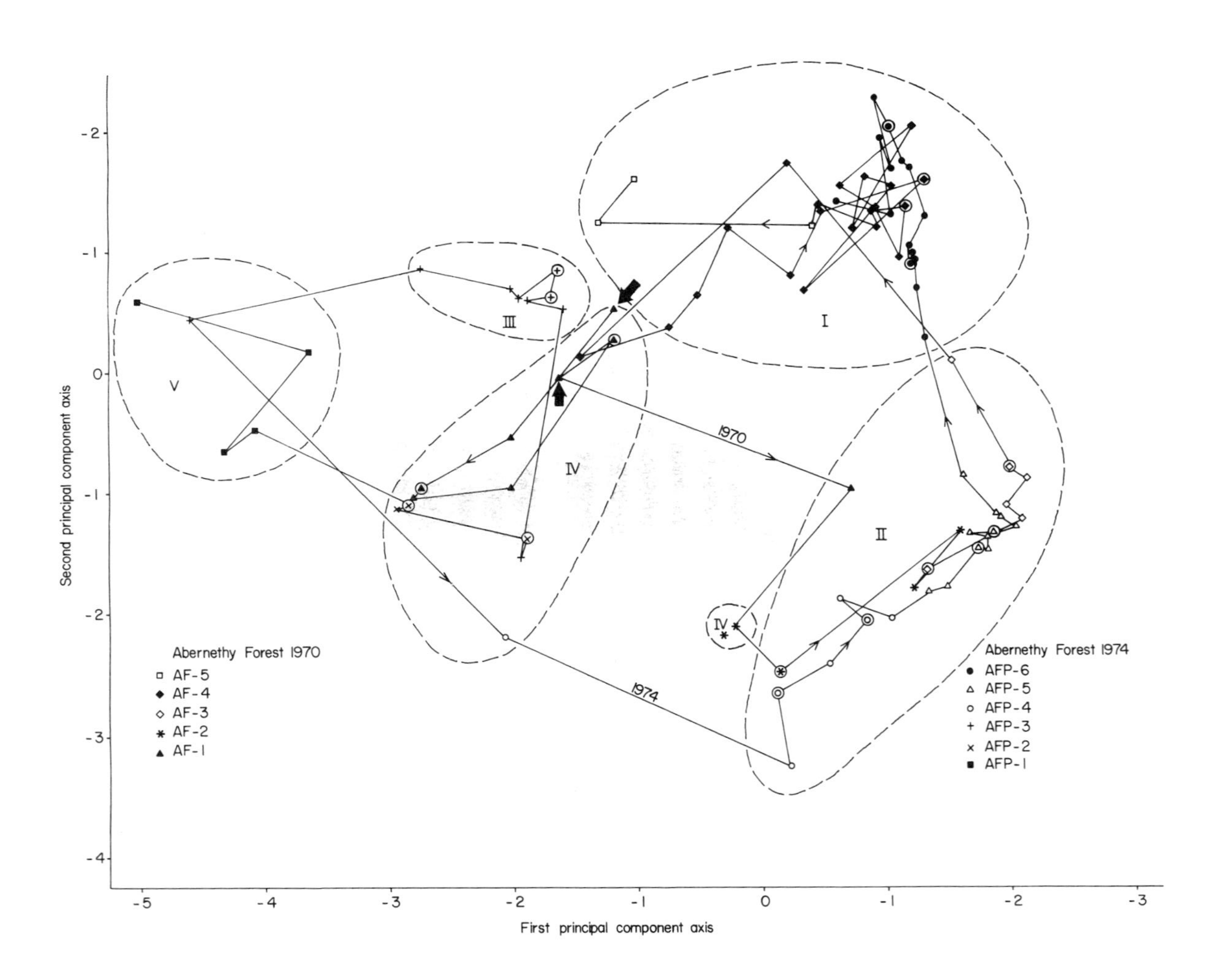
Abernethy Forest 1970
AF-5
AF-4
AF-3
AF-2
AF-1
Abernethy Forest 1974
AFP-6
AFP-5
AFP-4
AFP-3
AFP-2
AFP-1
I
II
III
IV
V
1970
1974
First principal component axis
Second principal component axis

Table 4.7

Results of Principal Components Analysis of the Abernethy Forest 1974 and the Abernethy Forest 1970 Data: Component Loadings for the First Five Principal Components

	Component				
	1	2	3	4	5
Betula	0.319	−0.520	−0.109	0.069	−0.039
Pinus	0.189	0.614	0.385	−0.068	−0.092
Corylus/Myrica	0.356	−0.137	−0.276	0.555	0.079
Juniperus	0.004	−0.477	0.307	−0.493	−0.406
Empetrum	−0.282	−0.239	0.371	−0.008	0.813
Gramineae	−0.438	−0.180	0.146	0.204	−0.201
Cyperaceae	−0.471	0.078	−0.037	0.191	−0.185
Artemisia	−0.225	0.100	−0.705	−0.506	0.181
Rumex acetosa-type	−0.437	−0.033	−0.117	0.324	−0.225
Eigenvalue	3.126	1.909	1.068	1.003	0.735
Percentage of total variance	34.73	21.21	11.87	11.15	8.17
Cumulative percentage of total variance	34.73	55.94	67.81	78.96	87.13

combined data in Figure 4.10. Samples in the two sequences are joined up in stratigraphical order. This representation accounts for just under 56% of the variability. Spectra that comprise the two test data sets used earlier in this chapter and also in Chapter 3 are circled, to allow comparison with Figures 4.1, 4.2, and 4.8. The definitions of the first five principal components in terms of the original pollen types are given in Table 4.7.

The first two components of this analysis (see Table 4.7) are very similar to those defining the first two axes of Figure 4.9 (see Table 4.5). The first component in both analyses is defined by high positive loadings for *Betula*, *Pinus*, and *Corylus/Myrica* and by high negative loadings for Gramineae, *Empetrum*, Cyperaceae, *Artemisia*, and *Rumex acetosa*-type. The second component in both analyses has high negative loadings for *Betula*, *Corylus/Myrica*, *Juniperus*, *Empetrum*, and

Figure 4.10 The positions of the 49 samples of the Abernethy Forest 1974 data set and the 41 samples of the Abernethy Forest 1970 data set plotted on the first and second principal component axes derived from a combined analysis. The samples are joined up in stratigraphical order. The samples that comprise the two test data sets are circled. The local pollen zones for each sequence are shown, as is the optimal sum-of-squares partition of the data into five groups found by the approximating algorithms described in Gordon and Henderson (1977).

Gramineae and high positive loadings for *Pinus* and, to a lesser extent, *Artemisia*. The ecological interpretation of these two components of the combined data is thus identical to that presented above for the first two components of the 1974 data set.

Principal components of pollen stratigraphical data can be viewed as 'composite curves' (see H. J. B. Birks, 1974; H. J. B. Birks and Berglund, 1979), and can be of considerable use in portraying the major stratigraphical patterns in the data set and in indicating gradients within the data (see Adam, 1974). It is, however, relevant to make two caveats. Firstly, this assumes implicitly that trends in the data are *linear* trends, and that it is relevant to interpret axes or directions in the plot. There are examples of plots which show clear *non-linear* trends in the configuration of points (e.g., H. J. B. Birks and Berglund, 1979), these trends being not readily interpretable in terms of linear directions in the space (I. C. Prentice, 1980).

Secondly, if two successive principal components account for the same amount of variability, there is an indeterminacy in their direction, as any pair of orthogonal axes in the same plane would serve equally well (Morrison, 1976, pp. 277, 287). Because of multinomial sampling errors associated with pollen counts, one should therefore be cautious about basing too many conclusions on components accounting for about the same amount of variability. In the principal components analysis of the combined full data sets (see Table 4.7) the third component is dominated by *Artemisia* and accounts for 11.87% of the total variability. The fourth component accounts for 11.15% of the variability but it contrasts *Juniperus* and *Artemisia* with *Corylus/Myrica*. In terms of the amount of variability explained, the third and fourth components would appear to have changed place between Tables 4.5 and 4.7, but because of the similar amount of variability explained by these two components, it would be safer simply to state that the components together effectively distinguish between spectra of zone AFP-4 and spectra 31–33 of zone AF-2 on the one hand and zone AFP-5 (less spectra 25 and 26) and zone AF-3 plus spectra 30 and 34 of the 1970 core on the other. In palynological terms this is primarily the distinction between the rapid phase of *Juniperus* pollen abundance at the beginning of the post-glacial and the succeeding phase of *Corylus/Myrica* pollen abundance. Both phases have high *Betula* pollen percentages but differ in the relative importance of *Juniperus* and *Corylus/Myrica* pollen. The very distinctive *Artemisia* pollen-dominated zone in the 1974 core (zone AFP-3 less spectrum 34) is also strongly separated from the rest of the data, reinforcing some differences on the first two component axes. The fifth component, accounting for 8.17% of the total variability, is largely dominated by *Empetrum* (see Table 4.7). This component separates out the spectra containing high *Empetrum* pollen percentages (mainly zones AFP-2 and AF-1) and highlights very effectively the unusually high value of *Empetrum* pollen in spectrum 15 of the 1970 core (see Figure 3.9).

The first five components account for over 87% of the total variability, and there are diminishing returns from investigating higher components. It is relevant in this context to stress the distinction between statistical significance and biological significance (Blackith and Reyment, 1971, p. 91): the last few principal axes may often make a contribution which is statistically significant, but which does not add much to the ecological interpretation of the data. A useful rule-of-thumb when using principal components analysis of a correlation matrix between variables is to consider only those components which have eigenvalues of 1.0 or more (see Jeffers, 1967; Isebrands and Crow, 1975). In this case only the first four principal components, accounting for nearly 79%, would be considered.

It can be seen that the plot of the combined 1970 and 1974 data sets on the first two principal component axes (Figure 4.10) does not portray the joint data in distinct well-separated groups to the same extent that the 1974 data on their own were displayed in Figure 4.9. An informal assessment of Figure 4.10 (which, after all, only accounts for just over half the observed variability in the data given in Table 4.7) could be in danger of biassing the assessment to accord with one's own preconceived views about how the data are interrelated. Partitioning the same data by the sum-of-squares criterion can help to guard against the incursion of potentially erroneous preconceptions. Table 4.8 summarises the minimum sum-of-squares partition into five groups (see also Figure 4.10). This is the optimal partition found by the approximating algorithms described in Gordon and Henderson (1977). The agglomerative algorithm produced a partition with a higher sum-of-squares than the partition presented in Table 4.8, but the compositions of the two partitions were very similar, differing only in the location of the following spectra from the 1970 core: 20 (340 cm) and 21 (350 cm) (both moved from group I to group IV in the agglomerative results), and 33 (470 cm) (starred in Figure 4.10, moved from group IV to group II). It is, of course, possible that none of these partitions represents the overall minimum sum-of-squares partition. Partitioning the combined data into more than five groups divides the large cluster II, but not in a manner analogous to the IIa/IIb division illustrated in Figure 4.9; the division of cluster II is more by the year of the data sets. This division appears to reflect the rather different proportions of *Betula* and *Corylus/Myrica* pollen in the two data sets.

Assimilation of the principal components results shown in Figure 4.10 may be helped by remembering the configurations and the form of the 'time curve' in the two earlier principal components plots (Figures 4.1 and 4.9). Interpretation is also helped if instead of considering all the samples shown in Figure 4.10, the mean coordinates or component scores of the samples within each of the local pollen zones in each sequence on the first two principal components are calculated. These means and the ranges of the scores of the samples in each zone are plotted in Figure 4.11, and the zones of each sequence are joined up in stratigraphical order (see Ritchie, 1977; Ritchie and Yarranton, 1978a, 1978b; H. J. B. Birks and Berglund, 1979, for similar plots). If the ranges of component scores of spectra comprising local pollen zones in different sequences overlap markedly, it suggests

Table 4.8

Optimal Partition into Five Groups of the Combined Abernethy Forest 1974 and 1970 Data Found by the Approximating Sum-of-Squares Algorithms[a]

Group	Constituent spectra {depth in cm. (number)}		Local pollen zones	
	1974	1970	1974	1970
I	300–362 (1–15)	0–350, 370 (1–21, 23)	AFP-6	AF-5, AF-4 (−360 cm)
II	365–460 (16–32)	380–460, 480 (24–32, 34)	AFP-5, AFP-4 (−465 cm)	AF-3, AF-2 (−470 cm, +480 cm)
III	475–510 (35–41)	—	AFP-3 (−470 cm)	—
IV	465, 515–530 (33, 42–45)	360, 470, 490–550 (22, 33, 35–41)	AFP-2 (+465 cm)	AF-1 (−480 cm, +360 cm, +470 cm)
V	470, 535–550 (34, 46–49)	—	AFP-1 (+470 cm)	—

[a] Sum-of-squares algorithms described in Gordon and Henderson (1977).

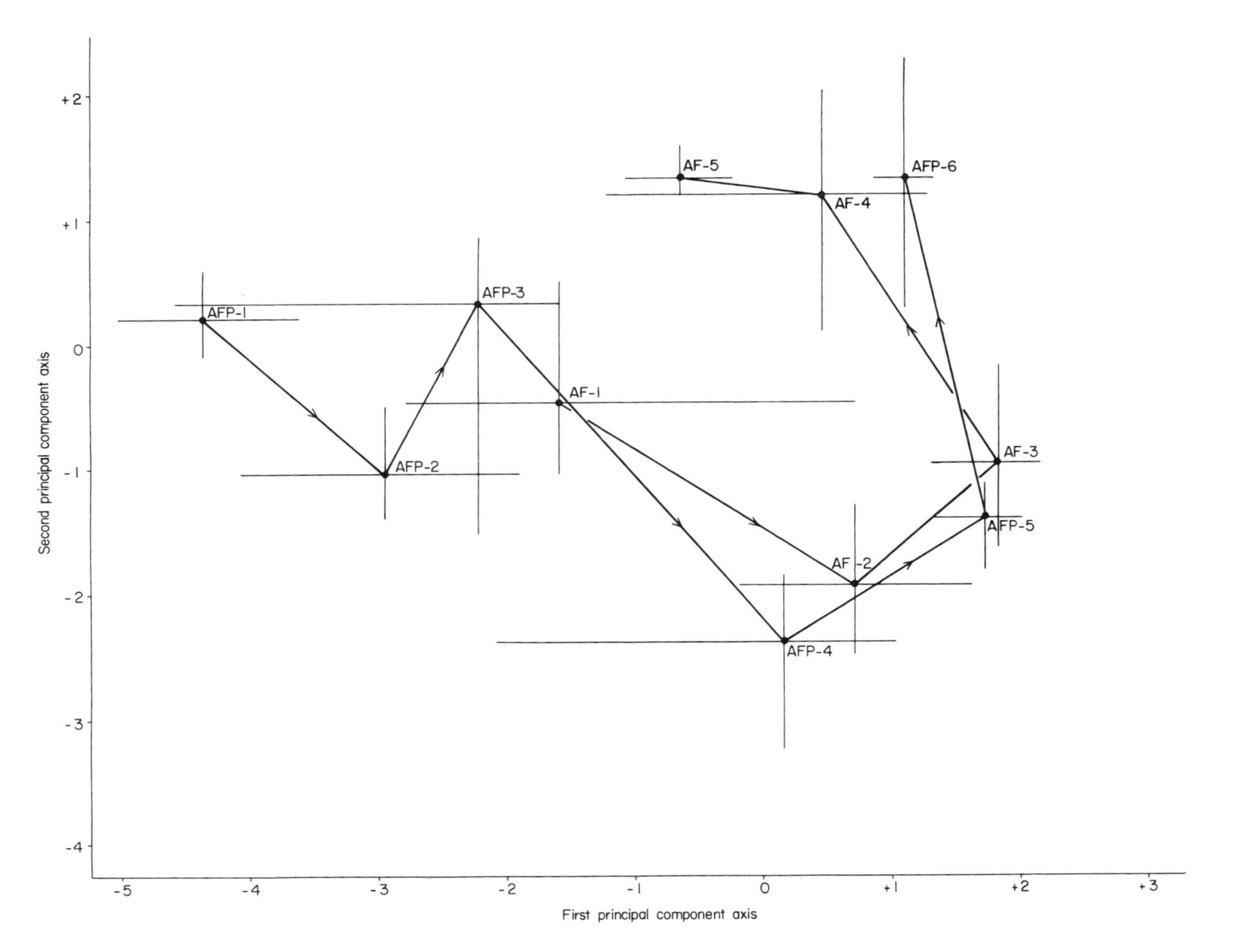

Figure 4.11 Plot of the mean principal component scores for each local pollen zone and their ranges on the first and second principal component axes for the Abernethy Forest 1974 data (zones AFP-1 to AFP-6) and the Abernethy Forest 1970 data (zones AF-1 to AF-5). See text for further explanation.

that the zones are of similar pollen composition and hence may belong to the same regional pollen assemblage zone. Examination of the two principal components plots of the combined data set (Figures 4.10 and 4.11) suggests a broad similarity between zones AFP-6 and AF-4; between zones AFP-5 and AF-3; between zones AFP-4 and AF-2; and between zones AFP-2 and AF-1. These conclusions agree fairly well with the results from partitioning the combined data (Table 4.8 and Figure 4.10), suggesting that four regional pollen assemblage zones can be delimited on the basis of the two Abernethy Forest cores. The principal components results suggest that zones AFP-1 and AFP-3 of the 1974 core do not have correlative zones in the 1970 core. There is, however, very considerable range in the component scores of the spectra within zone AFP-3, especially spectra 42 (515 cm) and 34 (470 cm) (see Figure 4.10). These two spectra cause some overlap between spectra from zone AFP-3 and zone AF-1. The large effect of such transitional samples indicates the advisability of considering whether such spectra should be included in the definition of local pollen zones, especially when summary diagrams like Figure 4.11 are being constructed.

Besides emphasising the points of similarity in the pollen composition of the two sequences, the principal components results, particularly Figure 4.10, and the partitioning results highlight several interesting differences between the two cores. The three spectra (0, 30, 50 cm) that comprise the uppermost zone (AF-5) in the 1970 sequence do not correspond very closely with any of the other spectra in group I, mainly because of the high Gramineae and Cyperaceae pollen values in zone AF-5. Spectra 35 and 41 (490 and 550 cm, respectively) in the 1970 core (marked with small arrows in Figure 4.10) appear more similar to some of the spectra of group III (zone AFP-3) than they are with other spectra of zone AF-1 in group IV, reflecting the high but isolated *Artemisia* pollen percentages in spectra 35 and 41.

In general, the results of these two classificatory approaches, one using partitionings of the combined data sets and the other using a low-dimensional geometrical representation of the data, are consistent. This consistency gives one greater confidence in the numerical results as a whole than is possible when only one numerical technique is used.

4.6.3 Sequence Slotting

Table 4.9 displays the result obtained when the 1970 and 1974 Abernethy Forest sequences were slotted together so as to minimise the statistic ψ defined in Section 4.5. A discussion of some properties of the slotting method is deferred until the final section, but it is relevant to draw attention at this point to the phenomenon of 'blocking'. This refers to several consecutive 'objects' in a sequence occurring next to one another in the joint slotting without the interposition of any objects from the other sequence, for example, objects 1–13 from the 1970 sequence and objects 1–8 of the 1974 sequence in Table 4.9. Such blocking often indicates that there is an inexact correspondence of pollen composition as one moves down the two sequences; even if there were a broad comparability between

Table 4.9
Optimal Slotting of the Complete 1974 and 1970 Abernethy Forest Data[a]

Abernethy Forest 1974			Abernethy Forest 1970		
Local zones	Depth (cm)	Spectra	Spectra	Depth (cm)	Local zones
			1–3	0–50	AF-5
	300–335	1–8	4–13	60–210	
	340–357	9–13	14–16	230–270	AF-4
AFP-6			17–23	290–370	
	360	14	24	380	
	362	15			
	365	16	25–27	390–410	
	367–372	17–19	28	420	AF-3
AFP-5	375–420	20–24	29	430	
	425, 430	25, 26			
	435–445	27–29	30, 31	440–450	
	450, 455	30, 31	32	460	AF-2
AFP-4	460	32	33, 34	470, 480	
	465	33			
AFP-3	470, 475	34, 35	35	490	
	480–510	36–41			
	515	42	36	500	
AFP-2	520, 525	43, 44	37, 38	510, 520	AF-1
	530	45	39	530	
AFP-1	535–545	46–48	40, 41	540, 550	
	550	49			

[a] $\psi = 2.232$.

the respective zones in each sequence, the object-to-object variation down the sequences could differ.

The results are broadly in agreement with the two earlier studies. One would be hesitant to make too direct a comparison at the start of the two sequences, but thereafter the correspondence is fairly close: zone AFP-5 with zone AF-3; zone AFP-4 with zone AF-2; and most of the members of zone AF-1 with zone AFP-2. From our experience of comparing sequences, a value of 2.232 for ψ indicates a reasonable correspondence between the two sequences. These results suggest that no satisfactory matches can be proposed for zone AFP-3 or zone AF-5.

4.6.4 Conclusions

The various numerical approaches to comparing the two Abernethy Forest data sets provide broadly consistent results. The results suggest that zone AFP-6 of the 1974 core can be matched with zone AF-4 of the 1970 core. Both zones are characterised by very high percentages of *Pinus* pollen, and can be correlated with the *Pinus* regional pollen assemblage zone of H. H. Birks (1970). Zone AFP-5 is fairly consistently matched with zone AF-3. Although both zones have high percentages of *Betula* and *Corylus/Myrica* pollen, the relative proportions of these types differ in the two sequences. The reasons for this difference are not known, but possible explanations include differential sedimentation of *Corylus/Myrica* or *Betula* pollen in different parts of the Abernethy Forest channel, or different identification criteria for these triporate pollen types by the pollen analysts concerned. The former explanation is the more likely; and, interestingly, Pennington (1964, 1970) has noted differences in the proportions of *Corylus/Myrica* pollen in different parts of the same lake basin at sites in the English Lake District. The mechanism for such differential pollen sedimentation is not known (cf. M. B. Davis, 1968, 1973; M. B. Davis and Brubaker, 1973; Ammann-Moser, 1975; Ammann, 1979), but its operation may be important at several sites (see Sønstegaard and Mangerud, 1977). Despite these differences in the pollen proportions, zones AFP-5 and AF-3 appear to belong to the same regional pollen assemblage zone, the *Betula–Corylus/Myrica* zone of H. H. Birks (1970).

Zones AFP-4 and AF-2 are consistently matched together. Both are characterised by high *Betula* and *Juniperus communis* pollen frequencies, and they belong to the *Betula–Juniperus* regional zone of H. H. Birks (1970).

Convincing matches between the basal three zones of the 1974 core and zones of the 1970 core are virtually absent except for a tendency for zone AFP-2 to be matched with some spectra of zone AF-1. No matches can be suggested for the very distinctive zones AFP-3 and AFP-1. H. H. Birks (1970) commented, in her description of the broad Gramineae–*Rumex–Artemisia* regional pollen zone with which zone AF-1 was equated, that when further studies of late-glacial deposits were made in the area, it might be necessary to subdivide the zone. The more detailed work by H. H. Birks and Mathewes (1978) confirmed that prediction. A three-fold subdivision of the late-glacial pollen stratigraphy of this area of Scotland can now be consistently recognised, with a lowermost *Rumex acetosa*-type–Gramineae zone, an *Empetrum–Betula* zone, and an uppermost *Artemisia* zone. As discussed by H. H. Birks and Mathewes (1978), these zones can be widely distinguished in late-glacial pollen sequences in the Eastern Highlands of Scotland.

No convincing match can be found in the 1974 core for zone AF-5 in 1970 core. This is not surprising as zone AF-5 is only found in the uppermost 50 cm of the 1970 core, whereas the pollen analyses of the 1974 core only cover the depth interval 300–550 cm.

4.7 Other Examples of Numerical Comparisons

4.7.1 Stratigraphical Studies

Gordon and Birks (1974), H. J. B. Birks and Berglund (1979), and H. J. B. Birks (1981a) present and discuss several examples of the application of the numerical approaches considered above for comparing a range of pollen stratigraphical sequences as a tool in delimiting regional pollen assemblage zones. Gordon and Birks (1974) compared the late-Wisconsin pollen sequence at Wolf Creek, central Minnesota (H. J. B. Birks, 1976b), with the pollen stratigraphy at Horseshoe Lake, central Minnesota (Cushing, 1967b), the type locality and section for the late-Wisconsin regional pollen assemblage zones for central Minnesota. Such a comparison was particularly important, as the Wolf Creek data were collected subsequent to the establishment of the regional pollen zones by Cushing (1967b). It was clearly valuable to be able to compare the Wolf Creek sequence, and any other new sequences, with the existing regional pollen stratigraphy, especially at the type section at Horseshoe Lake. The results that Gordon and Birks (1974) obtained for zone-by-zone comparisons, for a combined ordination of the two sequences, and for sequence-slotting were very consistent. The results indicated that the three local pollen zones at Wolf Creek could be matched with the regional pollen assemblage zones at Horseshoe Lake. The sequence-slotting results are shown in Figure 4.12. These results differ slightly from those presented in Figure 9 of Gordon and Birks (1974), which were obtained using the approximating algorithm described in Gordon (1973b). The results given in Figure 4.12 show the optimal slotting, obtained by the dynamic programming algorithm of Delcoigne and Hansen (1975), described in Section 4.5. The radiocarbon dates obtained for the zone boundaries at the two sites are also shown. It should be emphasised that each of these dates has a laboratory counting error of between ±160 and ±400 radiocarbon years. In addition, the dates are liable to further, at present largely unknown and intractable, errors resulting from the radiocarbon dating of different limnic sediments. Despite these limitations, there is a satisfactory correspondence between the slotting based on similarities in pollen composition and the radiocarbon dates for the zone boundaries. There is, however, quite extensive 'blocking' in the joint slotting. For example, spectra 12–26 from the Wolf Creek core appear in a single block of spectra without the interposition of any spectra from Horseshoe Lake. This is explicable by the fact that these Wolf Creek spectra are all very similar in pollen composition. The significance of such blocking in sequence-slotting is discussed further in the final section.

H. J. B. Birks (1981a) compared three late-Wisconsin pollen sequences from northeast Minnesota by means of a combined principal components analysis of the three sequences (as in Section 4.4). The results obtained provided a useful insight into the similarities among the pollen spectra and showed how the local pollen zones at Kylen Lake, a site studied in the area between 1970 and 1977, could be compared with the regional pollen zones previously delimited and defined for northeast Minnesota by Cushing (1967b).

Gordon and Birks (1974) also compared four Holocene (post-glacial) pollen sequences studied by McAndrews (1966, 1967) along a west–east transect near Itasca in northwest Minnesota. This transect crossed the major vegetational formations of the area (prairie, oak savanna, deciduous forest, and mixed coniferous–deciduous forest). The numerical approaches applied to these data included zone-by-zone comparisons after each site had been subjected to numerical zona-

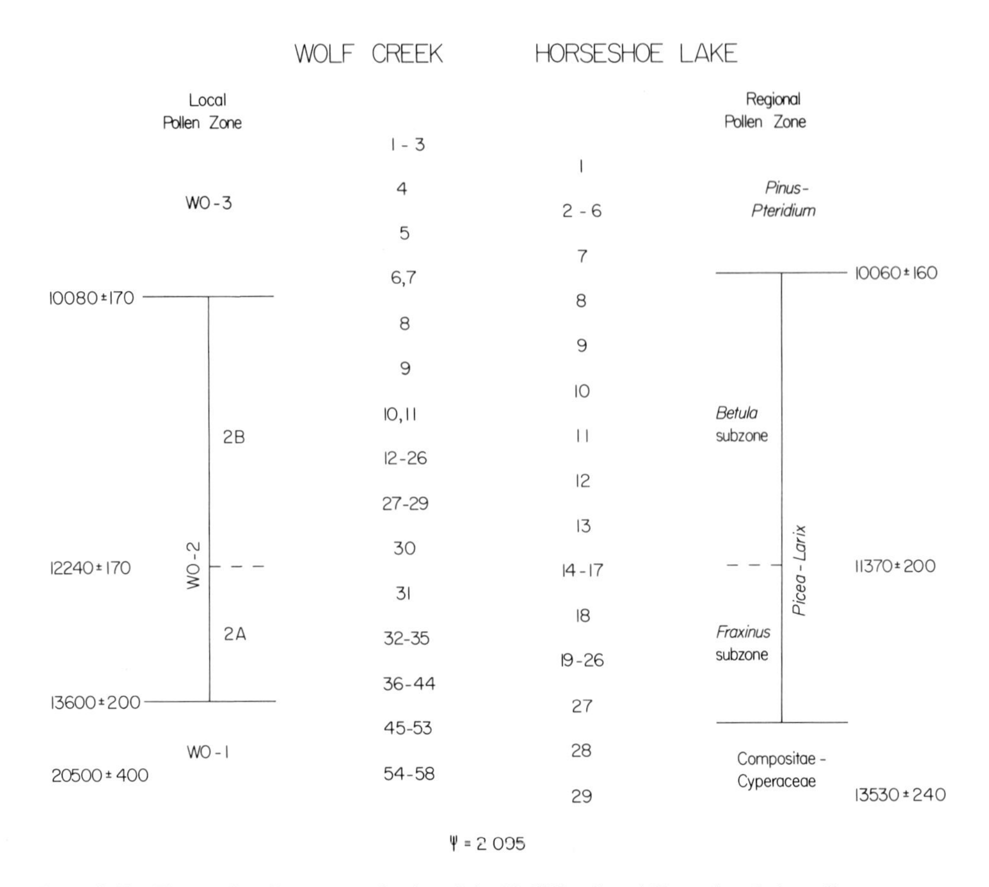

Figure 4.12 The results of sequence-slotting of the Wolf Creek and Horseshoe Lake pollen sequences ($\psi = 2.095$). Radiocarbon dates for the pollen zone boundaries are also given, expressed as radiocarbon years before present (B.P.).

tion methods similar to those described in Chapter 3, and sequence-slotting. The numerical results obtained indicate that some of the pollen assemblages occurred at all the sites along the transect, implying formerly widespread vegetation types, whereas some of the pollen assemblages could only be matched at adjacent sites along the transect. In general, the results confirmed McAndrews' conclusions, but the numerical results differed in some minor details. The results not only highlighted the similarities between the sequences, but also emphasised important differences between sites. These differences appear to reflect major differences in the pollen stratigraphy and hence the vegetational history of the sites along the Itasca transect, and emphasise the complex vegetational development of this ecotonal area.

In their re-appraisal of the Holocene pollen stratigraphy of southern Sweden using numerical methods, H. J. B. Birks and Berglund (1979) compared two long sequences from Blekinge in southeast Sweden to derive regional pollen assemblage zones for this phytogeographically distinct area of Sweden. They also compared one of the Blekinge sequences with a long sequence from a different phytogeographical area (Scania) in southern Sweden. Initially each sequence has divided into a series of local pollen zones on the basis of four different methods of numerical zonation. The two Blekinge sequences were compared directly by sequence-slotting and by a combined principal components analysis of the sequences (as in Section 4.4). The results obtained were consistent, and indicated strong similarities in pollen composition between the two sequences. On the basis of these results, Birks and Berglund delimited a series of regional pollen assemblage zones for Blekinge. When pollen sequences from Scania and Blekinge were compared numerically, the combined principal components analysis showed very clearly that there were virtually no similarities in pollen composition between the sequences. This conclusion was confirmed by the high value of ψ (7.42) obtained when the two sequences were slotted together. These results indicate that, contrary to previous practice in southern Sweden, two separate series of regional pollen assemblage zones are required for Blekinge and Scania, if the criterion for zonation is to be overall similarity in pollen composition. The numerical comparisons emphasised the strong phytogeographical and ecological distinctions between the two areas that exist today and that must have existed in the past.

Other recent numerical comparisons of late- and post-glacial pollen sequences have usually involved combined principal components analysis of two or more sequences; see, for example, Björck (1979, 1981; several sequences from southern Sweden), Björck and Persson (1981; two sequences from northeast Greenland), Lamb (1982; seven sequences from Labrador), and Wiltshire and Moore (1983; two sequences from central Wales). Sequence-slotting has been used by Ritchie (1982) to compare sequences from the northern Yukon, Northwest Territories, and Alaska.

There has, to date, been only one application of numerical techniques to the critical problem of comparing pollen sequences from previous interglacial deposits. This is rather surprising as pollen stratigraphy is the major means of comparing and correlating interglacial sequences (see, e.g., West, 1977), as no absolute chronology is available for terrestrial interglacial deposits. The current range of radiocarbon dating is about 70,000 years (Muller, 1977; Grootes, 1978; Maugh, 1978a, 1978b), whereas the last interglacial is generally regarded as ending at about 115,000 years ago (Woillard and Mook, 1982).

H. J. B. Birks and Peglar (1979) used principal components analysis to compare pollen profiles from two interglacial peat beds in the Shetlands. The rationale of their approach was that a geometrical disposition of samples from different profiles, as suggested by principal components analysis, would allow the detection of groups of samples of similar pollen composition that may originate from the same interglacial stages, as well as the recognition of groups of samples of dissimilar pollen composition that may originate from different interglacial stages. The positions of the individual samples from the Fugla Ness interglacial peat bed

(H. J. B. Birks and Ransom, 1969) and from the Sel Ayre peat section (H. J. B. Birks and Peglar, 1979) on the first and second principal component axes are shown in Figure 4.13. The first component effectively distinguishes between pollen spectra from the two sections, whereas the second component highlights changes in pollen composition within each sequence. These results suggest that the sequences differ in their pollen composition, and thus that they may derive from different interglacial stages. H. J. B. Birks and Peglar (1979) present botanical reasons for correlating the Sel Ayre profile with the Ipswichian (last) interglacial and Fugla Ness with the Hoxnian interglacial. A principal components analysis of the two Shetland interglacial sequences and of a long Flandrian (post-glacial) sequence from Shetland (Johansen, 1975) was also made by H. J. B. Birks and Peglar (1979). The results show that the three sequences are distinct in their pollen composition, with no overlap between the sequences when the individual samples are plotted on the first and second principal component axes. There is clearly considerable potential for applying this type of approach to interglacial pollen stratigraphical and correlation problems elsewhere in the world.

4.7.2 Comparisons for Non-Stratigraphical Purposes

The major aim of all the numerical comparisons of pollen sequences discussed so far has been entirely stratigraphical, namely, the detection and delimitation of regional pollen assemblage zones or the comparison and biostratigraphical correlation of interglacial sequences. Numerical methods for comparing pollen sequences can also aid the pollen analyst in summarising the major patterns of stratigraphical change within and between pollen sequences from specific geographical or ecological areas. Such a summary can contribute greatly to our understanding of vegetational dynamics and patterns of vegetational changes over long time-periods.

Ritchie and Yarranton (1978a, 1978b) (see also Ritchie, 1977) have used principal components analysis as a means of comparing several pollen sequences in Canada. After a pilot study using four sites (Ritchie and Yarranton, 1978a), they (1978b) analysed pollen stratigraphical data from 14 sites situated today within the boreal forest, aspen parkland, or prairie–grassland of central Canada. Each sequence was zoned independently into local pollen zones using the sequential correlation method of Yarranton and Ritchie (1972) (see Section 3.2). The sequences were then compared by means of a principal components analysis of the mean pollen composition of each of the 65 local zones at the 14 sites.

Plots of the component scores for the zones on the second and third principal component axes at four representative sites are shown in Figure 4.14. The zones at each site are joined up into stratigraphical order. In addition, Ritchie and Yarranton positioned modern surface pollen spectra from 136 lakes in central Canada on these principal component axes (see Section 6.4), as a means of comparing modern and fossil pollen spectra. Three of the sites shown in Figure 4.14,

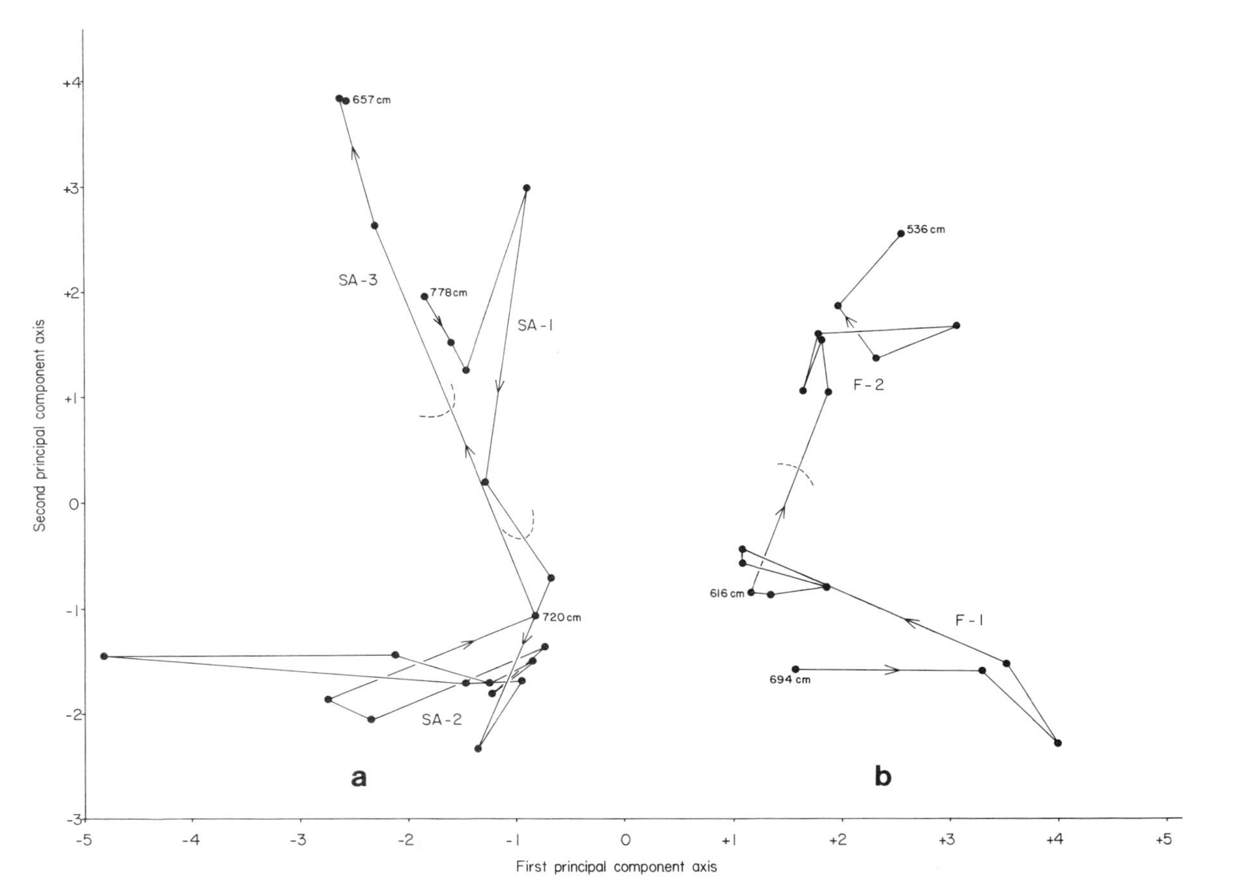

Figure 4.13 The positions of the pollen samples from the (a) Sel Ayre and (b) Fugla Ness interglacial sequences on their first and second principal component axes. The samples are joined up in stratigraphical order from bottom to top of each sequence; local pollen zones for each sequence are also shown (After H. J. B. Birks and Peglar, 1979.)

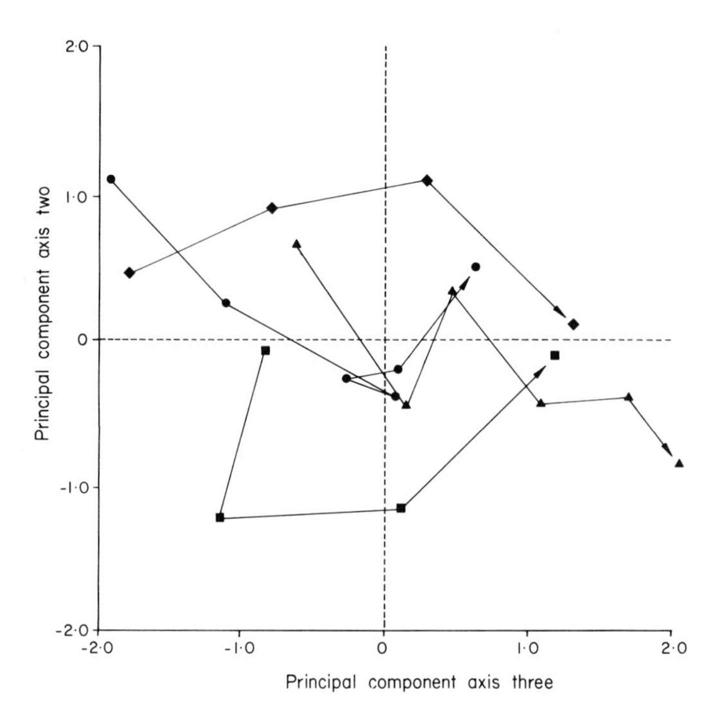

Figure 4.14 The position of the pollen zones, as defined by the mean pollen composition of each, for four sites in central Canada—Flin Flon (◆), Lofty Lake (●), B Lake (■), and Glenboro (▲)—on the second and third principal component axes. The zones are joined up in stratigraphical order from bottom to top of each sequence. Flin Flon, Lofty Lake, and B Lake are situated today in the boreal forest, whereas Glenboro is in the aspen parkland (After Ritchie and Yarranton, 1978b.)

Flin Flon, Lofty Lake, and B Lake, are situated today within the boreal forest. The principal components results show very clearly that the pollen stratigraphy and hence the vegetational history of these sites have been very different in the past (see also Ritchie and Yarranton, 1978a). Flin Flon appears to have been surrounded by closed forest throughout the post-glacial period, whereas the forest history of the B Lake area has been interrupted in mid-post-glacial times by herb–grassland development. The pollen record at Lofty Lake occupies an intermediate position between Flin Flon and B Lake, with some increased representation of grassland taxa in the mid-post-glacial, but not as marked an increase as at B Lake.

Such comparisons of pollen sequences provide a useful summary of the major patterns of variations within a large body of pollen stratigraphical data. As Ritchie and Yarranton emphasise, the results indicate that present-day vegetation types have developed to their present structure and composition through markedly different vegetational histories, a conclusion that is not predictable from present-day ecological observations. This general approach has considerable potential in the comparison and synthesis of pollen stratigraphical sequences over large geographical areas.

Tallis and Johnson (1980) adopted a broadly similar approach for comparing pollen sequences from peat deposits that had accumulated within landslip hollows in the southern Pennine hills of northern England. Twelve pollen profiles were initially zoned into 27 local pollen zones using Yarranton and Ritchie's (1972) sequential correlation method. The similarities between the mean pollen composition of these zones were then displayed by means of principal coordinates analysis (see Subsection 5.3.4) using Gower's (1971) similarity coefficient. Four groups of zones from the basal peats were recognised from the scaling results. The zones had characteristic pollen compositions which could be correlated with radiocarbon-dated pollen profiles from elsewhere in the southern Pennines. Tallis and Johnson were thus able to suggest possible dates for the inception of the landslips.

Turner and Hodgson (1979) compared relative pollen spectra from 41 sites in the northern Pennines by calculating mean pollen proportions for a specific time interval in the early post-glacial at each site. The means for the seven major taxa at the 41 sites were then analysed by principal components analysis of the correlation matrix of the means. The results showed considerable differences between sites that were related, in part, to altitude, soil type, and geology. This fine-scale local variation in pollen assemblages was also detected by a sum-of-squares partitioning of the same data. Turner and Hodgson (1983) similarly compared mid-post-glacial pollen assemblages from 38 sites in the same area using principal components analysis. They demonstrated that, as in the early post-glacial, the major patterns of variation in pollen assemblages were related to altitude, geology, soils, and climate.

The methods of analysis used by Ritchie and Yarranton (1978a, 1978b), Tallis and Johnson (1980), and Turner and Hodgson (1979, 1983) have certain disadvantages. The use of local zone averages raises the same problems as zone-by-zone comparisons using information radii (Section 4.3), namely, that zones must be specified at the outset by the investigator and that an incorrect choice of zones can result in misleading results. Clearly, this latter problem can be largely overcome by the careful and critical use of several numerical zonation techniques on each sequence and the selection of consistently delimited zones as the zones for subsequent analysis.

A further limitation of these studies is that analysis of the *mean* pollen composition of the zones fails to take into account the inherent within-zone variability in pollen composition. In some situations the within-zone variation may be as great as the variation between zones at different sites (cf. Turner and Hodgson, 1983). A relevant method for analysing such data is canonical variates analysis (Blackith and Reyment, 1971), which will be illustrated in Chapters 5 and 6. This method displays geometrically the variation between predetermined groups of objects (in this case, each group consists of fossil pollen spectra from a single local pollen zone) in a way which discriminates as effectively as possible between the different groups. As far as we are aware, it has only been used once for comparing fossil pollen sequences (Turner and Hodgson, 1979). It has been used more frequently, however, to compare groups of modern surface pollen samples (see Subsection 5.3.3) and to compare groups of modern and fossil pollen spectra (see Subsection 6.4.3). It has also been employed to compare, for example, the morphology of particular fossils within specific stratigraphical zones of the Cretaceous and Tertiary (Malmgren, 1974; Reyment, 1966, 1980a; Reyment, Hayami, and Carbonnel, 1977).

4.7.3 Difference Diagrams

In some pollen analytical studies, particularly those concerned with fine-scale vegetational differentiation in relation to specific environmental factors such as soil, the major aim of comparing pollen sequences is not primarily to detect regional similarities, but to highlight local differences between profiles. Various methods for detecting such differences are considered below.

Andersen (1978a) compared two post-glacial sequences from Denmark, one from a large lake where the pollen stratigraphy is assumed to reflect primarily regional vegetational history, the other from a very small hollow where the pollen record reflects local vegetational history. After correlating the two sequences with reference to major regional pollen stratigraphical changes, Andersen compared the two profiles by plotting together the percentage curves for the major taxa at the two sites on a common age rather than depth scale. By superimposing one pollen profile on another in this way, Andersen was able to display differences in the pollen composition at the two sites, which he interpreted as reflecting differences in local and regional vegetation.

Jacobson (1975, 1979) developed so-called difference diagrams for comparing pollen sequences from nearby sites situated on contrasting soil types in central Minnesota. The sites were initially correlated on the basis of radiocarbon dates and regional pollen stratigraphical changes. Groups of 2 to 4 stratigraphically adjacent samples were then amalgamated and their pollen compositions averaged. Difference diagrams between pairs of sites were then constructed, using both pollen percentage data and pollen-accumulation rates. If y_{1ik} and y_{2ik} denote the pollen-accumulation rates of taxon k at the sampling position corresponding to time i at sites 1 and 2, respectively, the difference in accumulation rates is

$$\Delta y_{ik} \equiv y_{1ik} - y_{2ik}.$$

For comparing pollen proportions p_{1ik}, p_{2ik} at time i at the two sites, Jacobson defined the difference in proportions to be

$$\Delta p_{ik} \equiv \log(p_{1ik}/p_{2ik}).$$

Both Δy_{ik} and Δp_{ik} were plotted stratigraphically (varying i) for each taxon k.

Some pollen types showed little difference in their percentages or accumulation rates at the two sites, suggesting that these taxa were either both growing locally at the two sites or were part of the regional vegetation. Other pollen types, such as *Pinus strobus*, *Pinus banksiana/Pinus resinosa*, and *Quercus* had higher percentages and/or accumulation rates at one of the two sites, suggesting local occurrence on a particular soil type.

Bradshaw (1978, 1981b) constructed similar stratigraphical difference diagrams for pollen percentage data from two sites on contrasting soils in East Anglia. The measure of difference he used was

$$\Delta p'_{ik} \equiv p_{1ik} - p_{2ik}.$$

By means of difference diagrams, he was able to demonstrate marked vegetational differences related to different soil types throughout much of the post-glacial. Heide (1981) used a similar measure of difference for contrasting two sites in Wisconsin.

The approach pioneered by Jacobson (1975, 1979) represents a useful and important means of highlighting and displaying differences between pollen sequences (see Jacobson and Bradshaw, 1981). As described above, the approach requires a reliable and independent time control, to ensure that similar periods of time are being compared. There is the further point that the observed data will be subject to statistical counting errors, as discussed in Chapter 2. In that one is examining the *difference* between pairs of readings, each of which will be subject to counting errors, the observed difference may possess a relatively large standard error; a useful precaution would be to plot not only the observed difference, but also its associated standard error, as is done for conventional pollen diagrams by, for example, Pennington and Bonny (1970), Maher (1972a, 1972b, 1977), Sercelj and Adam (1975), and Adam (1975). As further pollen analytical studies are made of local vegetational differentiation and of the contrast between local and regional vegetational history, new methods for comparing sequences and for quantifying the differences between pollen diagrams may be required (see I. C. Prentice, 1982a). Clearly this is an area for future research and development.

4.8 Properties of the Numerical Methods of Comparison

The general advantages and disadvantages of numerical comparisons can be summarised under the same kinds of heading as those given in Section 3.8 and will not be repeated here. It seems appropriate, however, to comment separately on the strengths and weaknesses of each of the three basic methods described in Sections 4.3, 4.4, and 4.5, and illustrated in Sections 4.6 and 4.7.

4.8.1 Zone-by-Zone Comparisons

The zone-by-zone comparison method requires that the zones to be investigated must be specified at the beginning of the analysis. An injudicious choice of local zones can give rise to misleading conclusions. On the other hand, this biostratigraphical unit is generally the temporal scale at which the palaeoecologist operates in regional studies. In such studies, one is usually more interested in environmental conditions prevailing over a reasonably long period of time; a correspondence between two spectra from different cores generally gives much less information. A further point is that amalgamation of the original spectra allows one to carry out broader-scale studies.

One drawback is that the ordering of the spectra within the zone is not used; this can give relevant information on vegetational changes during the period of time when the pollen was being deposited. For example, the fact that the proportion of one pollen type steadily increases within a zone and oscillates in another zone could be of great interpretative importance. However, our experience is that pollen analysts generally make little use of such information of within-zone variation in the *early* stages of data handling, although more use is made of it when detailed interpretations particularly of population change, are attempted later; virtually all published descriptions and definitions of pollen zones (e.g., Cushing, 1967b; H. J. B. Birks, 1973b) reduce the observed data to a set of means and ranges for each principal taxon.

4.8.2 Classification Methods

The classification methods do not require zones to be specified but allow them to emerge during the analysis. Unconstrained classifications (in contrast to those described in Chapter 3) can assist in the identification of similar spectra from different parts of the same core, or from different cores. Spectra which differ markedly from their neighbours are highlighted, and can be excised from potential local zones. The results of the classification studies enable groups of similar spectra to be identified, and their similarities with one another noted. Again, one is not using at this stage the information available from the ordering of the spectra. In addition, the number of spectra which can be analysed will be smaller than by the zone-by-zone comparison method. However, classification is probably the most widely used of the three methods described in this chapter (see also Chapters 5 and 6).

Geometrical representation procedures or ordination techniques within the general field of classification are now fairly widely used in Quaternary pollen analysis for detecting and summarising major trends in pollen stratigraphical data from one sequence (e.g., Adam, 1970, 1974; H. J. B. Birks, 1974; Pennington and Sackin, 1975; H. J. B. Birks and Berglund, 1979) or more than one sequence (e.g., Gordon and Birks, 1974; Ritchie and Yarranton, 1978a, 1978b; H. J. B. Birks and Berglund, 1979; H. J. B. Birks and Peglar, 1979; Björck, 1979, 1981; H. J. B. Birks, 1981a; Lamb, 1982). These scaling methods have also been used in the analysis of surface-pollen data, and examples of such studies, involving the methods of principal components analysis, principal coordinates analysis, non-linear mapping, and non-metric multidimensional scaling, will be presented in Section 5.3. Comparative comments on different ordination methods are postponed to that section.

4.8.3 Sequence Slotting

In contrast to the other approaches, the sequence-slotting method makes use of the ordering of the spectra within the cores; like the classification approach, it does not require an initial choice of zones. It can identify groups of spectra from different cores that are similar in pollen composition, making use of the entire data set at the same time. The method, however, is only feasible for studying a few sequences at a time, and (unlike the other two methods) cannot be used to compare modern pollen spectra with fossil spectra.

The value of ψ gives an indication of how similar the two pollen sequences are, smaller values indicating a closer fit. Gordon (1982a) describes how one could test the null hypothesis that the second sequence was obtained by re-sampling at various positions down the first stratigraphical column. This is, however, too restrictive a hypothesis for stratigraphical pollen data, and a less formal approach is adopted here; larger values of ψ are regarded with an increasing degree of suspicion. From our experience with the analysis of various stratigraphical pollen data sets, a value for ψ of 2.232 (which is obtained for the slotting displayed in Table 4.9) may tentatively be described as indicating a very reasonable fit.

Our experience has suggested that the slotting method is fairly sensitive to different developments in the separate sequences, and can be influenced to some extent by sampling fluctuations if the counts in each spectrum are small. The presence of 'blocking', described in Sections 4.6 and 4.7, is often a sign that there are differences between the sequences. A complicating factor is that blocking sometimes occurs when there is very little difference between several consecutive objects within a core (see Gordon and Birks, 1974, p. 238), though this generally results in smaller values of ψ than occur when large differences are present.

The method, then, does appear to be most robust when comparing sequences which are *fairly* similar, but not too similar, and this desideratum may limit its usefulness in pollen analytical studies. However, there are many palaeoecological situations in which it is relevant to compare sequences in this manner, and in which the method may have a contribution to make; an application to comparing physical logging data is described by Gordon and Reyment (1979). The method has also been used by Reyment (1978a, 1978b, 1980a) to slot together Cretaceous borehole sequences from Morocco and Nigeria; these sequences were defined either on the basis of morphological characteristics of ostracod carapaces or foraminifer tests, or on the ostracod composition of the sequences. The method has also been used to compare magnetic susceptibility curves from cores of lake sediment, ammonite suture lines (Reyment, personal communication), frequencies of foraminifers and mean diameters of their proloculi in bore holes (Reyment, 1982), and surface pollen data collected along transects.

It is possible that the slotting method would be more robust if slotting were carried out to optimise a statistic other than ψ; investigation of this has not yet borne fruit.

4.8.4 Conclusions

Lest these comments be interpreted as over-critical, we emphasise that we believe that the three approaches can all be of assistance in the comparison of pollen data, in the detection and delimitation of regional pollen zones, and (the first two methods) in the comparison of fossil and modern pollen spectra. In certain situations, one method might be more appropriate than the others, but when all three methods are applied to the same data sets, they can complement one another, and give the investigator more confidence that the results suggested are not solely an artifact of the method employed.

The methodology presented in Chapter 3 and in this chapter is directly applicable to the International Geological Correlation Programme's project 158B on the palaeohydrology of the Northern Hemisphere described by Berglund and Digerfeldt (1976) and Berglund (1979a; 1983). Numerical methods allow the independent zonation of sequences from specific reference areas, the comparison of sequences from different reference areas within the same type region, and the comparison of sequences between one type region and another (I. C. Prentice, 1982b). Comparison and correlation of sequences between regions and the construction of correlation schemes linked to a radiocarbon chronology can provide a rigorous basis for subsequent, more detailed palaeoecological studies. In addition, the construction of difference diagrams for carefully selected pairs of sites in contrasting ecological settings provides a powerful means of reconstructing regional and local vegetational history, and of detecting vegetational differentiation in both time and space.

CHAPTER 5

The Analysis of Modern Pollen Data

5.1 Introduction

Pollen analysis of sediments of late-Quaternary age provides the basic data for the reconstruction of the past floras, plant populations, and vegetational communities of an area surrounding the site of study. The interpretation of such data in terms of past floras, populations, and communities has long been recognised as a difficult and complex problem (see von Post, 1918, 1967; Godwin, 1934; Erdtman, 1943; Fagerlind, 1952; M. B. Davis, 1963; Janssen, 1970; D. Walker, 1972). It requires not only a thorough knowledge of the present-day ecology of the taxa involved (Faegri, 1966; Janssen, 1970, 1981b), but also information on the relationships between modern, pollen spectra and the composition of the vegetation from which the spectra are derived.

A quantitative relationship is assumed to exist between the number of pollen grains of a taxon deposited in the sediment at a site and the number of individuals of that taxon in the vegetation surrounding the site (see Section 1.1 and von Post, 1918, 1967; Fagerlind, 1952; Cushing, 1963; M. B. Davis, 1963; Leopold, 1964; Mosimann and Greenstreet, 1971; D. G. Green, 1983). Such a relationship is undoubtedly very complex, being influenced by a large number of interacting factors. Among these are the genetic, physiological, climatic, and ecological factors controlling the flowering and pollen production of the individual plant; the abundance of the taxon within the vegetation; the structure of the community in which the taxon occurs; the mode of pollen dispersal; the meteorological factors influencing pollen transportation; the physical, chemical, and biological conditions affecting pollen sedimentation and preservation at the site of deposition; the size and morphometry of the basin of deposition; and local site factors such as the presence or absence of fringing vegetation. H. J. B. Birks and H. H. Birks (1980) discuss what is known about the processes of pollen production, dispersal, sedimentation, and preservation in relation to Quaternary pollen analysis. The interactions of these factors are so complex (see Tauber, 1965, 1967, 1977; Cushing, 1967a; Havinga, 1967; M. B. Davis, 1968, 1973; R. B. Davis *et al.*, 1969; M. B. Davis *et al.*, 1971; Berglund, 1973; M. B. Davis and Brubaker, 1973; Peck, 1973; Andersen, 1974, 1980b; Bonny, 1976, 1978, 1980; Jacobson and Bradshaw, 1981) and, as yet, so poorly understood that no satisfactory quantitative *explanatory* models of modern pollen production, dispersal, sedimentation, and preservation, and hence of modern pollen representation have yet been devised (cf. Kabailiene, 1969; Tsukada, 1981, 1982c) that can assist in the reconstruction of past floras, populations, and/or communities from fossil pollen spectra (A. M. Solomon and Harrington, 1979).

In view of these complexities, most Quaternary pollen analysts have studied the final product of these processes only, namely, *modern pollen spectra* preserved in recent sediments such as moss polsters, surficial lake-muds, and surface soils, in relation to the surrounding flora and vegetation (Wright, 1967) without considering in detail the various intermediate processes that have formed and influenced the composition of the spectra. Studies of modern pollen preserved in these 'surface samples' provide, in many ways, the simplest and certainly the soundest and most repeatable approach to the reconstruction of past plant populations and communities as well as palaeoenvironments (D. Walker, 1972). Surface-sample studies are an essential stage in the interpretation of fossil pollen assemblages. Although the desirability of studying surface samples was emphasised by von Post (1918, 1967), Godwin (1934), and Erdtman (1943), it has only been comparatively recently that the full potential of surface samples has been explored. Wright (1967), M. B. Davis (1969b), Livingstone (1969), T. Webb (1971), D. Walker (1972), Richard (1976), A. M. Solomon and Harrington (1979), H. J. B. Birks and H. H. Birks (1980), and I. C. Prentice (1982a) review current approaches.

Modern pollen data, like fossil stratigraphical data, are complex, quantitative, and multivariate. Unlike stratigraphical data, modern data do not have any stratigraphical context. However, they usually have some spatial context, either in terms of geographical location, ecological or vegetational setting, or position along vegetational or altitudinal transects. In order to derive empirical relationships between modern pollen and contemporary vegetation, surface-sample data must be compared with the vegetation from which the modern pollen spectra originate. Such relationships are also complex, and numerical methods for establishing them have been increasingly used during the 1970s. This chapter discusses the various numerical approaches currently available for the analysis of modern pollen data and for modelling and establishing contemporary pollen–vegetation relationships relevant to the interpretation of fossil pollen stratigraphical data.

5.2 Numerical Approaches to the Analysis of Modern Pollen Data

There are two main approaches to the use of modern pollen spectra in the interpretation of Quaternary fossil pollen assemblages: the comparative approach and the representation-factor approach (see Livingstone, 1969; M. B. Davis, 1969b; D. Walker, 1972; H. J. B. Birks, 1973b; I. C. Prentice, 1982a). In the so-called comparative approach (H. J. B. Birks and H. H. Birks, 1980) or analogue method (I. C. Prentice, 1982a), the aim is to establish similarities between modern and fossil spectra, so that the latter can be interpreted in terms of modern vegetational analogues. Initially, modern vegetation types are characterised by their contemporary pollen spectra (e.g., Wright *et al.,* 1967; Lichti-Federovich and Ritchie, 1968; Mott, 1969; R. B. Davis and Webb, 1975; T. Webb and McAndrews, 1976; I. C. Prentice, 1978). Numerical methods are particularly useful for detecting, summarising, and displaying major patterns of variation within modern pollen data (see Figure 5.1). Numerical methods can also be used to compare contemporary pollen data and the vegetation types from which the surface samples were collected, and thus to characterise modern vegetation types in terms of contemporary pollen assemblages. Various numerical methods for analysing modern pollen and vegetational data are discussed in Section 5.3. If modern pollen–vegetation relationships or 'corresponding patterns' (*sensu* T. Webb, 1974a) (see Figure 5.1) are established, comparisons on the basis of overall similarity can then be made between fossil pollen spectra and contemporary pollen assemblages from known vegetation types (e.g., P. S. Martin and Gray, 1962; Leopold, 1964; McAndrews, 1966, 1967; M. B. Davis, 1967a, 1969b; Wright *et al.,* 1967). If similarities exist between modern and fossil assemblages, it is likely that the spectra were produced by broadly similar vegetation; a modern vegetational analogue can thus be proposed for the past vegetation. Such comparisons provide a factual and repeatable basis for vegetation reconstruction. Quantitative methods for comparing modern and fossil spectra are discussed in Section 6.4.

In the second major approach to the analysis of surface-pollen data, the aim is to derive modern quantitative pollen-representation factors, so-called correction factors, that can be used to estimate the abundance of individual taxa in the past (see Figure 1.1). Such factors are based on mathematical models of contemporary pollen–vegetation relationships, whose parameters are estimated solely from the available surface-pollen data and modern vegetation data (see Figure 5.1; Andersen, 1970; Parsons and Prentice, 1981; T. Webb, Howe, Bradshaw, and Heide, 1981; I. C. Prentice, 1982a). Representation factors attempt to quantify the relationship between pollen values (expressed either as proportions, pollen concentrations, or accumulation rates of different taxa) in modern spectra and the quantitative abundance of the same taxa in the surrounding vegetation. The abundance and population size of the different taxa in the past can then be estimated directly by assuming that the contemporary representation factors for each taxon are

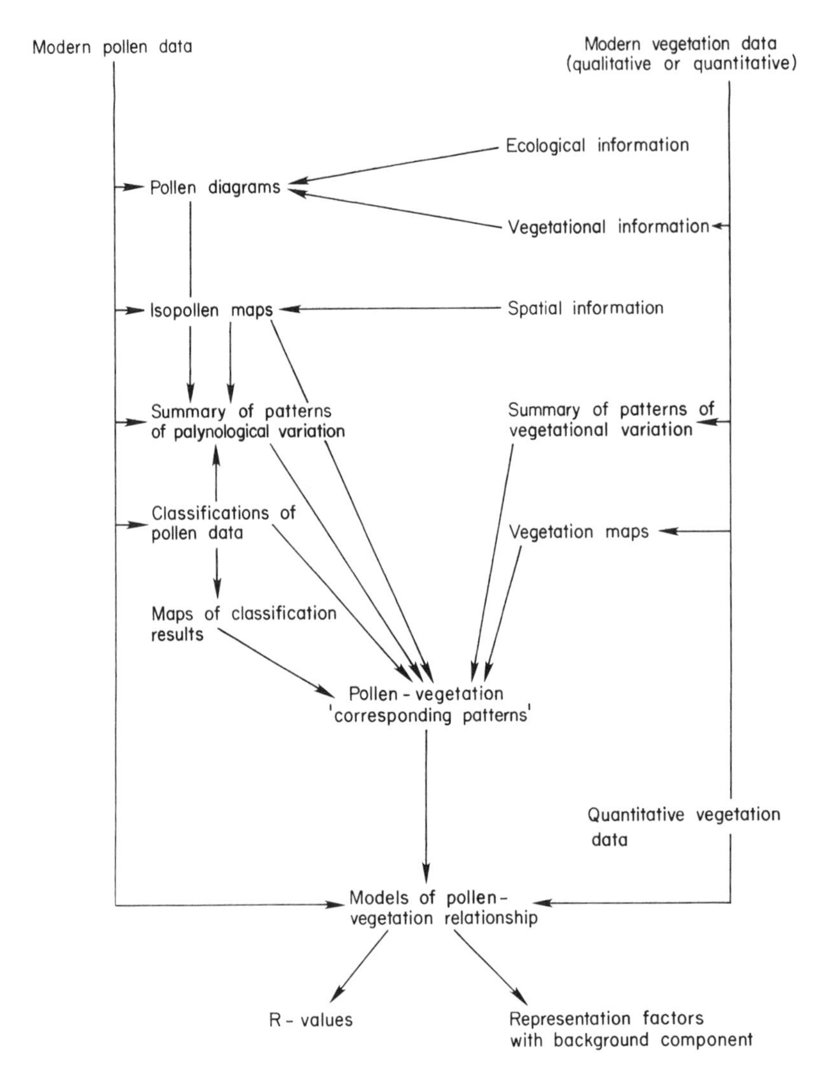

Figure 5.1 Stages in the analysis of modern pollen data for the derivation of contemporary pollen–vegetation 'corresponding patterns' and modelling of modern quantitative pollen–vegetation relationships.

invariant in time and space, and are thus applicable to fossil pollen spectra preserved at different sites. By combining these individual estimates of past populations and by assuming that the fossil pollen spectra are derived from a specified and constant source area, it is possible to reconstruct the past vegetation in quantitative terms (see Livingstone and Estes, 1967; Livingstone, 1968; Donner, 1972; Andersen, 1973, 1975, 1978a, 1980c; Bradshaw, 1981b; Heide, 1981). Numerical methods for estimating modern pollen-representation factors are discussed in Section 5.4, and the application of modern representation factors to the reconstruction of past populations and communities is considered in Section 6.3.

In this and the next chapter, a modern pollen spectrum is assumed to be the proportions or percentages of pollen types within the pollen sum which have been estimated by identifying and counting every grain in a pollen count of a surface sample (see Section 1.1 and Figure 1.1). In some instances, pollen concentrations or accumulation rates have been estimated for surface-mud samples (e.g., M. B. Davis *et al.*, 1973; R. B. Davis and Webb, 1975; I. C. Prentice, 1978; Brush and De Fries, 1981; Elliott-Fisk, Andrews, Short, and Mode, 1982; Lamb, 1982, 1984). Such estimates are simpler to analyse mathematically than relative percentage data because of the absence of constraints on the amounts of different taxa (Mosimann and Greenstreet, 1971). However, absolute pollen data from within and between lakes frequently show great variability (e.g., M. B. Davis *et al.*, 1973; Pennington, 1973, 1979; M. B. Davis and Ford, 1982) as a result of complex and poorly understood processes of pollen recruitment and sedimentation in lakes. Such data can only usefully be compared at a coarser level than percentage data (T. Webb, Laseski, and Bernabo, 1978); numerical analysis of modern absolute data is thus often not appropriate (cf. Lamb, 1982, 1984).

Simplifying assumptions are required in both the comparative approach and the representation-factor approach. The most important and most critical assumption concerns the definition of the pollen-source area of the vegetation around the surface-sample sites. It is essential to know what area of vegetation should be compared with a modern pollen spectrum in either approach. For this reason, the broad spatial relationships between modern pollen and contemporary vegetation must be established (see Figure 5.1) before pollen–vegetation relationships can be modelled, pollen-representation factors estimated, and quantitative population and vegetational reconstructions derived.

The strengths and weaknesses of the comparative and the representation-factor approaches in reconstructing past vegetation are discussed thoroughly by Wright (1967), M. B. Davis (1969b), Livingstone (1969), D. Walker (1972), A. M. Solomon and Harrington (1979), H. J. B. Birks and H. H. Birks (1980), and I. C. Prentice (1982a). It is not strictly correct to describe the comparative approach as a qualitative approach and the representation-factor approach as a quantitative approach (cf. Parsons and Prentice, 1981), because the comparison of modern and fossil spectra, and hence the reconstruction of past vegetation, can also be done statistically (see Section 6.4 and I. C. Prentice, 1982a) even though the past vegetation is not reconstructed in quantitative terms.

The success which has been achieved in the reconstructions and the sophisticated interpretations of vegetational history through the use of surface-sample data and pollen-representation factors (see H. J. B. Birks and H. H. Birks, 1980, for examples) clearly demonstrates the importance of studying modern pollen assemblages in relation to contemporary vegetation. Studies of modern pollen–vegetation patterns and relationships provide a factual basis for the reconstruction of past vegetation through the extrapolation of modern analogues backwards in time, namely, methodological uniformitarianism (see Section 1.1). These studies can also provide important insights into the accuracy with which pollen data reflect vegetation patterns, and into the geographical and vegetational scales at which pollen data most closely correspond with and therefore characterise vegetation. Modern pollen–vegetation studies can provide important checks on the sensitivity of pollen analysis as a tool for characterising vegetation, and can show the potential sensitivity and resolution of pollen analysis when applied to specific palaeoecological problems. Studies of modern pollen spectra are thus an important means of testing some of the basic assumptions of the pollen analytical method (T. Webb, 1974a; T. Webb, Laseski, and Bernabo, 1978).

The various stages in the analysis of modern pollen data and the derivation of 'corresponding patterns' between pollen and vegetation and quantitative models of pollen–vegetation relationships are summarised in Figure 5.1.

5.3 Presentation and Comparison of Modern Pollen Spectra from Different Vegetation Types

There are several ways of presenting modern pollen data. Firstly, if surface pollen spectra can be ordered according to their position along transects or classified into vegetation types or broad geographical regions, the modern pollen data can be displayed as a pollen diagram (see Figure 5.1) in which the ordering or grouping of the surface samples replaces the stratigraphical positioning of samples in a conventional pollen diagram. Secondly, modern pollen values of each taxon can be mapped and isofrequency contours drawn for individual pollen types, so-called isopollen maps (see Figure 5.1). These two approaches are described in Subsections 5.3.1 and 5.3.2.

A third approach uses multivariate statistical methods to analyse and summarise modern pollen data and to aid comparison between surface-pollen spectra and contemporary vegetation. This broad approach is summarised under three headings, in Subsections 5.3.3, 5.3.4, and 5.3.5. The first of the multivariate statistical methods is canonical variates analysis, which utilises information about the vegetation types from which the surface samples were collected in order to provide *a priori* groupings of pollen samples. The rationale is to compare and discriminate between these groups of modern spectra solely on the basis of their pollen composition. If the between-group variation in pollen composition is greater than the within-group variation, we can conclude that the groups of modern pollen spectra—and hence the vegetation types from which the samples were collected—differ in their overall pollen composition. This method is discussed in Subsection 5.3.3.

The second of the multivariate statistical methods is numerical classification, which seeks to detect prominent patterns or structure within the pollen data solely on the basis of the pollen composition of the modern samples, without any reference to the vegetation. When a grouping of pollen samples has been established, the vegetation data can be examined to see how closely the structure within the pollen data reflects the vegetation patterns. This allows us to investigate the rôle which vegetation plays in influencing the structure of the pollen data, without the *a priori* assumption of a direct rôle, which is made in canonical variates analysis. Numerical classification of modern pollen data is reviewed in Subsection 5.3.4.

Finally, various ways of measuring the degree of resemblance or 'corresponding patterns' between modern pollen and contemporary vegetation are discussed in Subsection 5.3.5. A range of methods is available, depending on the amount of detail in which the vegetational data have been described.

5.3.1 Display of Modern Pollen Data

Modern pollen data can most conveniently be represented in a pollen diagram (of the resolved bar-histogram type; see Section 1.3) when there are some independent criteria for ordering or grouping the individual samples. Suitable criteria can include position along a transect (e.g. Janssen, 1966, 1973, 1981a; Caseldine and Gordon, 1978; Caseldine, 1981), or *a priori* groupings of samples into vegetation types (e.g., Bent and Wright, 1963; H. J. B. Birks, 1973a, 1973b, 1977b, 1980; Flenley, 1973; O'Sullivan, 1973; Ritchie, 1974; A. M. Davis, 1980), vegetation–landform units (e.g., McAndrews, 1966; Lichti-Federovich and Ritchie, 1968; Janssen, 1981a), or broad geographical or ecological regions (e.g., M. B. Davis, 1967a; Wright *et al.*, 1967; Mott, 1969; I. C. Prentice, 1978) from which the surface samples were collected (see Figure 5.1).

If there is a large number of samples from each vegetation type or region, the mean values of the individual pollen types can be calculated and plotted. Figure 5.2 shows the mean percentages of all non-aquatic taxa with values of 2.5% or more of total pollen in one or more samples from surficial sediments of 131 lakes situated within 11 distinctive vegetation–landform units in the western interior of Canada. The vegetation–landform units range from northern low-arctic, dwarf-shrub tundra; through the boreal coniferous forest of central Manitoba; to the deciduous forests, grasslands, and aspen parklands of southern Manitoba and Saskatchewan (see Figure 5.3). These data, collected by Lichti-Federovich and Ritchie (1968), are one of the most comprehensive sets of surface-pollen counts that have ever been collected and analysed in a consistent and detailed way. They are used in this chapter and in Chapter 6 to illustrate the use of various numerical methods in the analysis of surface-sample data (see also H. J. B. Birks, Webb, and Berti, 1975). Details of the original data, site locations, and vegetation–landform units from which the samples were collected are given by Lichti-Federovich and Ritchie (1968).

Figure 5.2 shows a continuum in the mean pollen percentages from the tundra to the grassland samples. There are high mean values of *Betula* and Cyperaceae pollen in the tundra samples; high *Alnus* and *Picea* pollen percentages in the

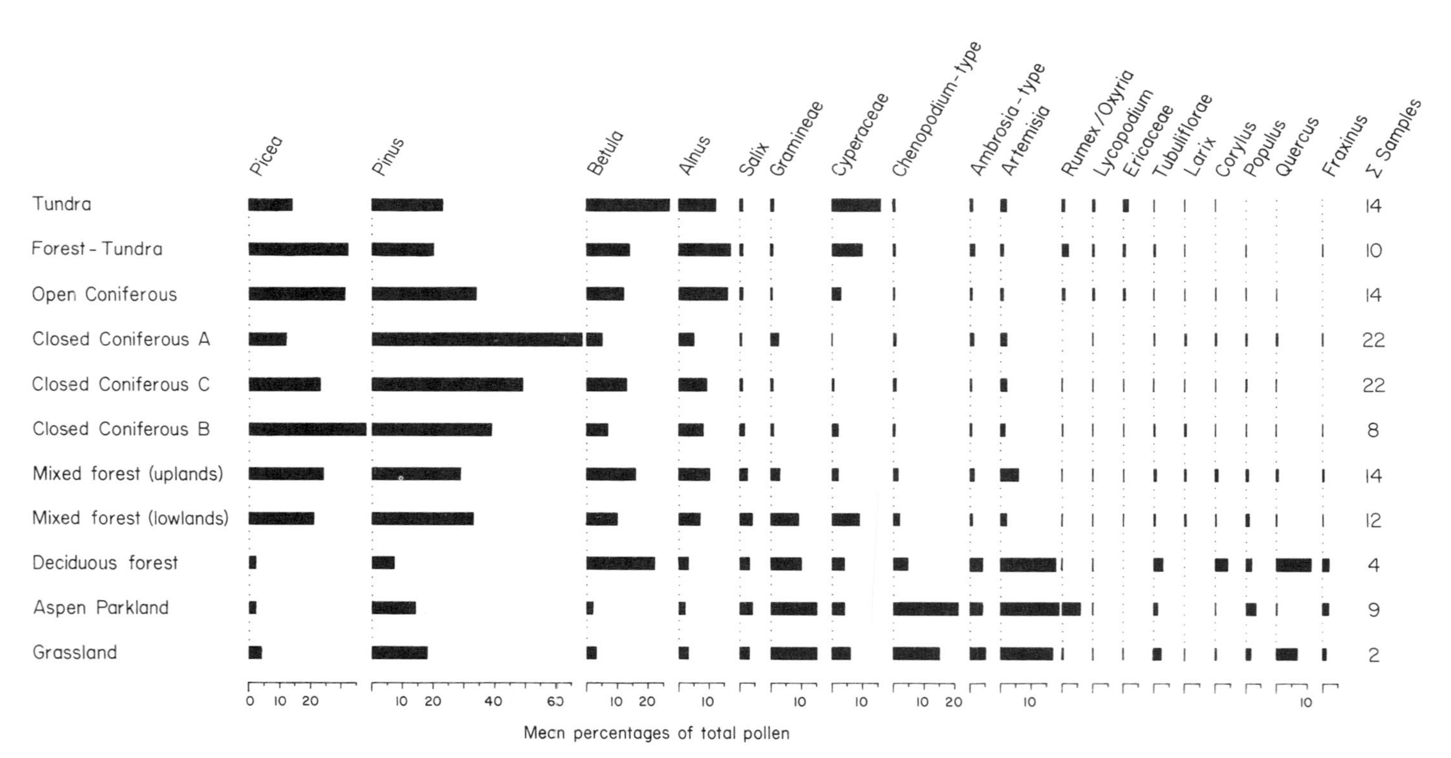

Figure 5.2 Mean pollen spectra for each of the 11 vegetation–landform units in the western interior of Canada shown on Figure 5.3. (Data are from Lichti-Federovich and Ritchie, 1968.)

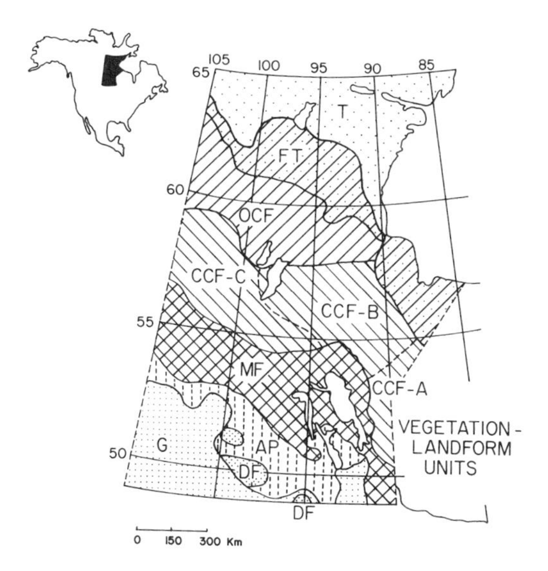

Figure 5.3 Location of vegetation–landform units in the western interior of Canada from which Lichti-Federovich and Ritchie (1968) collected surface-sample data. The units are tundra (T); forest–tundra (FT); open coniferous forest (OCF); closed coniferous forest (CCF), subregions A, B, and C; mixed deciduous–coniferous forest (MF); aspen parkland (AP); deciduous forest (DF); and grassland (G). The MF unit is subdivided into upland and lowland regions by Lichti-Federovich and Ritchie (1968). (After H. J. B. Birks, Webb, and Berti, 1975.)

forest–tundra samples; high mean values of *Pinus* pollen in samples from the closed coniferous forest units; and high values of *Chenopodium*-type, *Artemisia,* and Gramineae pollen in the southern deciduous forest, aspen parkland, and grassland samples. This summary suggests that differences exist among the mean spectra from some of the vegetation–landform units. It fails, however, to display the variation in pollen composition within the units or to illustrate whether the differences in pollen composition between units are abrupt or gradual. Some information about the extent of variation within groups can be presented by plotting not only the mean percentages but also the maximum and minimum percentages for each taxon in the vegetational units (Caseldine and Gordon, 1978). However, as each taxon is displayed separately, this can still mask the nature of the differences in pollen composition between vegetational units.

5.3.2 Mapping of Modern Pollen Data

A second approach to displaying and summarising modern pollen data is to map the pollen values for each taxon (see Figure 5.1). The raw data will be of the form $\{(x_i, y_i, p_i), i = 1, \ldots, n\}$, where p_i denotes the proportion of the pollen sum at the ith site which belongs to the taxon of interest, and the ith site has geographical coordinates x_i(longitude) and y_i(latitude); thus, p_i is viewed as the height of the function 'proportion of pollen' at the point (x_i, y_i).

Mapping procedures summarise the main spatial patterns and variations in the set of geographically located modern pollen values $(p_1, p_2, \ldots, p_n)$ to enable them to be compared more readily with contemporary vegetation patterns. Several methods exist for mapping quantitative data (see Krige, 1966; Harbaugh and Merriam, 1968; J. C. Davis, 1973; Unwin and Hepple, 1974; J. C. Davis and McCullagh, 1975; Webster, 1977; Sibson, 1981), and the results can be presented in several different formats. The p values can be approximated by a smooth surface, which can be presented as a three-dimensional diagram, drawn in perspective. Alternatively, the height above the (x, y) plane can be categorised into a small number of classes, and the codes for these classes plotted on the (x, y) plane; the codes for each class can be represented by a different colour, or by a different symbol on a line-printer. Once this categorisation has been done, 'isopollen' contours between different height classes can be drawn, although many methods of analysis obtain contours directly without obtaining, in an intermediate step, a smooth surface approximating the p values.

Szafer (1935) was the first palynologist to construct pollen contour or 'isopollen' maps as a means of illustrating the distribution patterns of selected pollen types at certain time intervals during the post-glacial (see Section 1.3). More recently, T. Webb (1973, 1974a, 1974b, 1974c), H. J. B. Birks, Webb, and Berti (1975), R. B. Davis and Webb (1975), T. Webb and McAndrews (1976), Bernabo and Webb (1977), I. C. Prentice (1978, 1983a), T. Webb, Yeracaris, and Richard (1978), T. Webb *et al.* (1983), Lamb (1982, 1984), Elliott-Fisk *et al.* (1982), P. A. Delcourt, H. R. Delcourt, and Davidson (1983), and Huntley and Birks (1983) have constructed isopollen maps of modern pollen data. Two major techniques have been used for the mapping of modern pollen data: interpolation procedures (e.g., T. Webb, 1974a, 1974b, 1974c; R. B. Davis and Webb, 1975) and trend-surface analysis (e.g., T. Webb and McAndrews, 1976).

For many purposes it is important to display and summarise the modern data in as much detail as the data permit, that is, with the minimum amount of spatial smoothing. Contour maps of pollen values based on simple interpolation between neighbouring sample points can provide useful summaries of the data. For small data sets (less than about 75 samples), contouring can often be done readily by visual interpolation (e.g., I. C. Prentice, 1978). For larger data sets, the assistance of computer programs or packages is invaluable. Such programs usually estimate the value of p at each point on a regular finely spaced grid in the (x, y) plane, using for the estimation a weighted combination of the values at nearby sample points. The number of sample points used in each estimation and the distance between them (which influences the amount of smoothing and level of detail to be presented in the map) can often be specified by the investigator (e.g., Dudnik, 1971); alternatively, an initial assessment of the data can indicate the relevant weights to be used in the estimation (Delfiner and Delhomme, 1975). The estimated pollen values at the grid points can then be categorised and printed, and contours drawn, in the manner described above.

Figure 5.4 shows isopollen maps for six pollen types in the Lichti-Federovich and Ritchie (1968) data set (from H. J. B. Birks, Webb, and Berti, 1975), plotted on a base map of the major vegetation–landform units of Figure 5.3. These maps show peak values of Cyperaceae and *Betula* pollen in the tundra. There are isolated high values (>5%) of Cyperaceae pollen farther south. These presumably reflect local growth of sedges around some of the lakes from which surface samples were collected. *Picea* pollen dominates with values of 30 to 40% over a broad zone (*ca*. 500 km) south of the tundra. *Pinus* pollen is predominant, with values of 30 to 60% over a 500-km region to the south of the zone of high *Picea* pollen values. To the south and west of the areas of high *Pinus* and *Picea* pollen values, there is a sharp gradient in conifer pollen with percentages falling from 30 to 15% or less in about 200 km. This sharp fall is matched by a sharp rise in the pollen percentages of herbs such as *Artemisia* and *Chenopodium*-type in the south and west.

Isopollen maps such as these reveal broad correspondences between the distribution of selected pollen types and the distribution of major vegetation types (see also H. J. B. Birks, Webb, and Berti, 1975; R. B. Davis and Webb, 1975; T. Webb and McAndrews, 1976; I. C. Prentice, 1978, 1983a; T. Webb, Yeracaris, and Richard, 1978; T. Webb *et al.*, 1983; Lamb, 1982, 1984). The maps place the palynological patterns of Figure 5.2 in a geographical context. In addition, the maps show the nature of the gradients in pollen percentages among the various vegetation units. Most of these gradients are gradual (e.g., between tundra and forest–tundra, forest–tundra and open coniferous forest), but steep gradients exist in the south and west. Such maps can also indicate where it might be profitable to take further samples, in order to resolve uncertainties about the behaviour of the contour lines, although it may prove difficult to find suitable sites exactly where desired.

McAndrews and Power (1973), McAndrews and Adams (1974), T. Webb and McAndrews (1976), and McAndrews and Manville (1980) have used trend-surface analysis to display patterns within modern pollen data. These authors suggest that local variations in pollen frequencies can obscure the broad-scale, regional patterns. They recommend the use of trend-surface analysis to highlight the regional patterns of interest and to smooth over local anomalies between samples.

Trend-surface analysis is broadly similar to a simple regression analysis, with the difference that two independent variables (x for longitude and y for latitude) are used to yield a regression surface rather than a regression curve. This surface is described by a linear combination of various terms in x and y; most commonly, this is an rth-degree polynomial. Thus, the pollen value p at the point (x, y) is estimated by

$$\hat{p} = a_1x^r + a_2y^r + a_3x^{r-1}y + \cdots + a_{q-2}x + a_{q-1}y + a_q,$$

where $q = \frac{1}{2}(r + 1)(r + 2)$ and $a_1, \ldots, a_q$ are regression coefficients. Gittins (1968), J. C. Davis (1973), Unwin (1975), and Mather (1976) give detailed accounts of the mathematics of trend-surface analysis. For a specified value of r, the aim is to find the coefficients $a_1, \ldots, a_q$, leading to fitted values $\hat{p}_i$ at (x_i, y_i), which ensure that

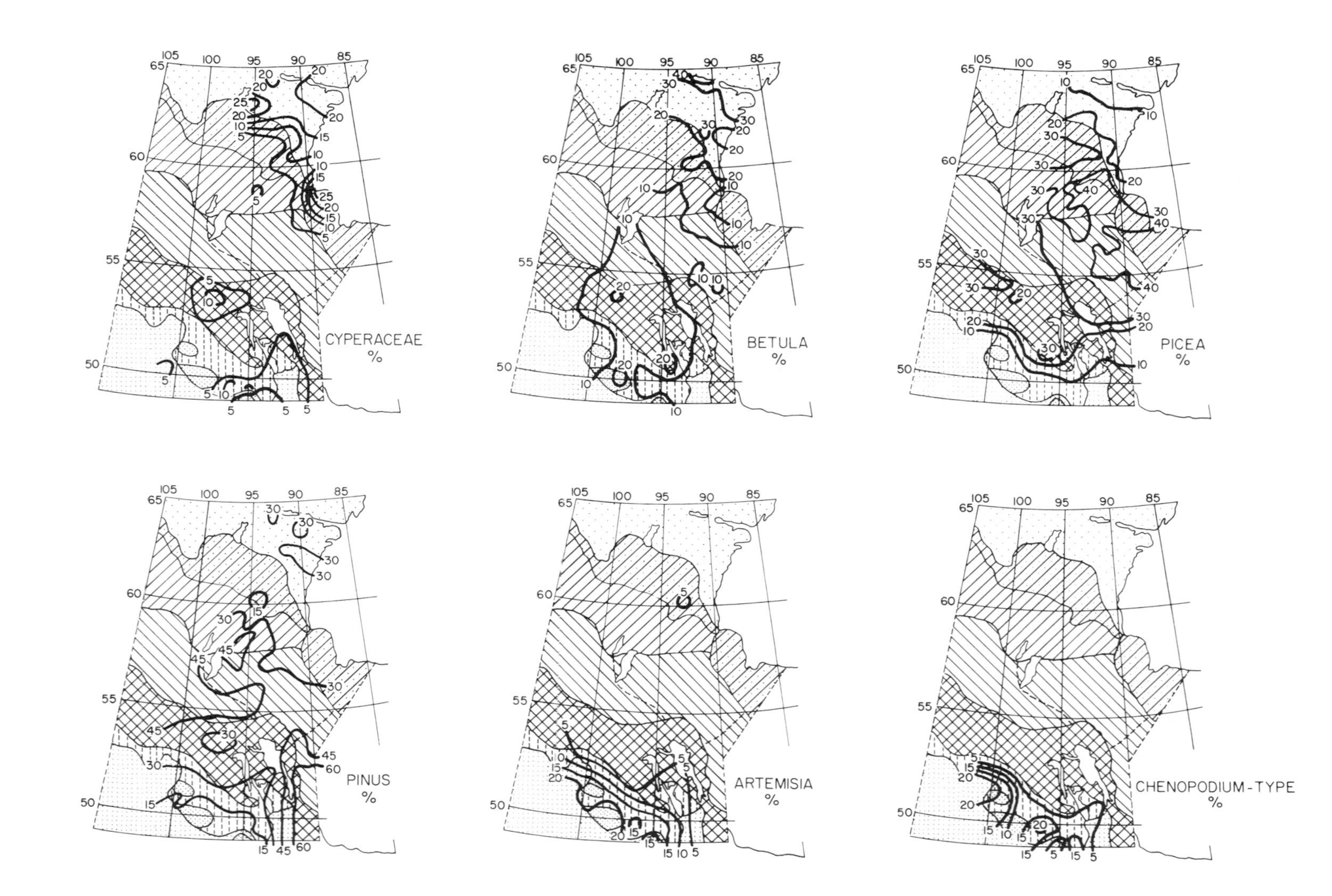
CYPERACEAE
%
BETULA
%
PICEA
%
PINUS
%
ARTEMISIA
%
CHENOPODIUM-TYPE
%

the sum of the squares of the differences $p_i - \hat{p}_i$ is minimised. As the degree r of the polynomial is raised, the surface more closely fits the data and smooths out less and less of the local variation. The choice of value for r is thus somewhat arbitrary: J. C. Davis (1973, p. 335) comments that 'we often must rely on experience and intuition' to decide what is a satisfactory fit in a trend-surface analysis. It can be useful and informative to examine the set of residuals $p_i - \hat{p}_i$ to assess the spatial dependence of any features which have been 'smoothed over' by the trend surface.

Some comparative comments follow on these methods of representing modern pollen data. Interpolated isopollen contour maps show major geographical patterns but incorporate various scales of variation, from local to regional. In some instances, local differences may even obscure regional patterns.

In trend-surface analysis, a single surface is fitted to the entire data set, and the general shape of the surface in one area can be strongly influenced by data from a considerable distance away. Polynomial surfaces of low degree smooth over local differences and can emphasise some regional patterns but at the same time may poorly represent the original data. On the other hand, polynomial surfaces of high degree can display wild fluctuations in regions of the plane where there are few data points (Whitten, 1975), and can prove difficult to assess. For these reasons, we feel that trend-surface analysis is unsuitable for the *primary* mapping and display of modern pollen data. It can, however, be useful for specific analyses of surface-pollen data where broad-scale, continental patterns are the major interest (e.g., McAndrews and Manville, 1980).

Various methods for the contouring of spatial data have been proposed which fit a set of smooth surfaces, each surface being based solely on neighbouring sample points; discussions of such methods are presented by Crain (1970) and Sibson (1981). We are not aware of any application of such methods to the analysis of modern pollen data, but they appear to have considerable potential for displaying patterns of variations at a more local scale.

It is relevant to note that isopollen maps, whether prepared by interpolation or by fitting a surface to the sample points, can only show the separate patterns of individual pollen types. Almost all studies to date have considered the analysis of pollen percentage data (cf McAndrews and Power, 1973; I. C. Prentice, 1978; Elliott-Fisk *et al.*, 1982). The constraint that percentages in a spectrum must sum to 100 means that the separate maps for each of the taxa must be related to one another, and one has to assess the set of maps together. One might prefer to remove this constraint by the use of pollen-accumulation rates, but the great variability inherent in such data at present militates against their analysis by mapping techniques (T. Webb, Laseski, and Bernabo, 1978).

Figure 5.4 Isopollen maps for Cyperaceae, *Betula, Picea, Pinus, Artemisia,* and *Chenopodium*-type pollen. Isopolls are contours of equal pollen percentage based on a sum of the 19 pollen types used in the analysis. The isopolls are plotted on a base map showing the vegetation–landform units of Figure 5.3. (After H. J. B. Birks, Webb, and Berti, 1975.)

On the other hand, there is the point that one wishes to see how modern pollen assemblages as a whole reflect contemporary vegetation patterns ('corresponding patterns' in Figure 5.1). It is thus necessary to determine what structure there is within modern pollen data when all the pollen types are considered in all the samples simultaneously, and then to examine whether any structure within the pollen data correspond to patterns within the modern vegetation. Techniques for such analyses are discussed in the next three subsections.

5.3.3 Canonical Variates Analysis

This method displays geometrically the patterns among pre-defined groups of objects (each object being described by several variables) in such a way as to emphasise the overall differences between groups but to allow for the variability within the groups. It reveals the extent to which a grouping of objects based on some external criterion is reflected by the data.

If there are t variables (in our case, pollen types) and h groups (in our case, groups of modern pollen spectra collected from specific vegetation types) and $t > h$, the original data can be represented in a space of t dimensions with each dimension representing a pollen type. Canonical variates analysis obtains the $h - 1$ axes that most effectively maximise the separation between the h groups relative to the within-group variation. These axes will in general not be orthogonal in the original space, but in the transformed space defined by the canonical variates the axes are orthogonal. They are also ordered in the same sense as in principal components analysis, in that they account for progressively less of the between-to-within group variation in the data. The group means and the individual samples can be plotted in this transformed space. In practice, most of the separation between groups is associated with the first few axes, and it is this reduction from t dimensions to the first few axes that is the power of canonical variates analysis.

Healy (1965), Klovan and Billings (1967), Rempe and Weber (1972), Oxnard (1973), and N. A. Campbell and Atchley (1981) give lucid geometrical explanations of the method, whereas Seal (1964), Morrison (1976), Rao (1970), Gnanadesikan (1977), and Maxwell (1977) discuss the detailed mathematics. Reyment and Ramden (1970) and Blackith and Reyment (1971) present a FORTRAN IV program CANVAR to implement canonical variates analysis (see also Reyment, 1980a, p. 156). In biology, canonical variates analysis has been largely applied to morphometric or taxonomic problems (e.g., Jolicoeur, 1959; Reyment, 1969a), but it has also been used in ecology by, for example, Buzas (1967, 1972), R. H. Green (1971, 1974), and Huntley and Birks (1979). Quaternary palynological applications include Adam (1970), H. J. B. Birks, Webb, and Berti (1975), H. J. B. Birks (1976b, 1977b, 1980), Turner and Hodgson (1979), A. M. Davis (1980), Andrews and Nichols (1981), Heide (1981), and Lamb (1982, 1984). Canonical variates analysis has, however, been more widely used in other palaeoecological studies (e.g., Ashton, Healy, and Lipton, 1957; Reyment and Brännstrom, 1962; Reyment, 1963b, 1966, 1979, 1980a; Hemmings and Rostron, 1972; Oxnard, 1972; Brothwell and Krzanowski, 1974; Hills, Klovan, and Sweet, 1974).

Following Gower (1966a, 1967a, 1970), I. C. Prentice (1980) presents canonical variates analysis as a scaling method in which the implicit dissimilarity measure is the square root of Mahalanobis D^2 between groups of pollen spectra. The formula for D^2 is

$$D^2(i, j) = (\bar{\mathbf{p}}_i - \bar{\mathbf{p}}_j)' W^{-1} (\bar{\mathbf{p}}_i - \bar{\mathbf{p}}_j),$$

where $D^2(i, j)$ is the value of D^2 between groups i and j, $\bar{\mathbf{p}}_i \equiv (p_{i1}, p_{i2}, \ldots, p_{it})'$ is the vector of mean pollen values for each pollen type in group i, and W is the within-group dispersion matrix which is estimated by pooling the dispersion matrices of all the groups. If the (k, l)th element of W^{-1} is w^{kl}, this formula can be written as

$$D^2(i, j) = \sum_{k=1}^{t} \sum_{l=1}^{t} (p_{ik} - p_{jk}) w^{kl} (p_{il} - p_{jl}).$$

It should be noted that other authors replace W in the above definition by S, an estimate of the (assumed common) covariance matrix, which is just a constant multiple of W.

I. C. Prentice (1980) describes how D^2 may be regarded as a measure of between-group 'signal' against a background of within-group 'noise'. It seems that D^2 would have considerable potential as a measure of dissimilarity between pollen spectra. However, it requires the groups of spectra to be specified at the outset, and assumes that the dispersion matrix is the same for each group. This assumption, implicit in canonical variates analysis, may not be realistic for many palynological data, but the results obtained often appear to be fairly robust to small departures from this assumption. Alternatively, one can use methods of analysis which reduce the influence that 'atypical' observations have on the results (N. A. Campbell, 1982).

As is the case with most other multivariate statistical procedures described in this book, the strict significance-testing aspects of canonical variates analysis are not important in palynology; the prime usefulness of the approach is its ability to represent the data in ways which enable their structure to be displayed more clearly.

As an illustration of the use of canonical variates analysis as a means of summarising multivariate surface pollen data, the data from the western interior of Canada were analysed using a much modified version of Reyment and Ramden's (1970) CANVAR program. Plots of the canonical group means for the 11 groups of samples on the first and second canonical variate axes are shown in Figure 5.5, and the positions of the 131 individual samples on these two axes are shown in Figure 5.6. As the first two canonical variate axes account for 72% of the between-to-within group variation, these plots represent an efficient summary of the patterns among groups in two dimensions.

These plots show that, although most of the group means are some distance apart along the first two axes (Figure 5.5), there is considerable scatter of individual samples within each group and some overlap between groups (Figure 5.6). The plots emphasise the relative discontinuity in the gradient in pollen composition between samples from the south and west and the rest of the data. North of the aspen parkland, the second canonical variate axis shows group-to-group continuity on a north-to-south line from the tundra (high positive coordinates) to the closed coniferous forest subzone A and the mixed forest region. There are no obvious discontinuities between these groups.

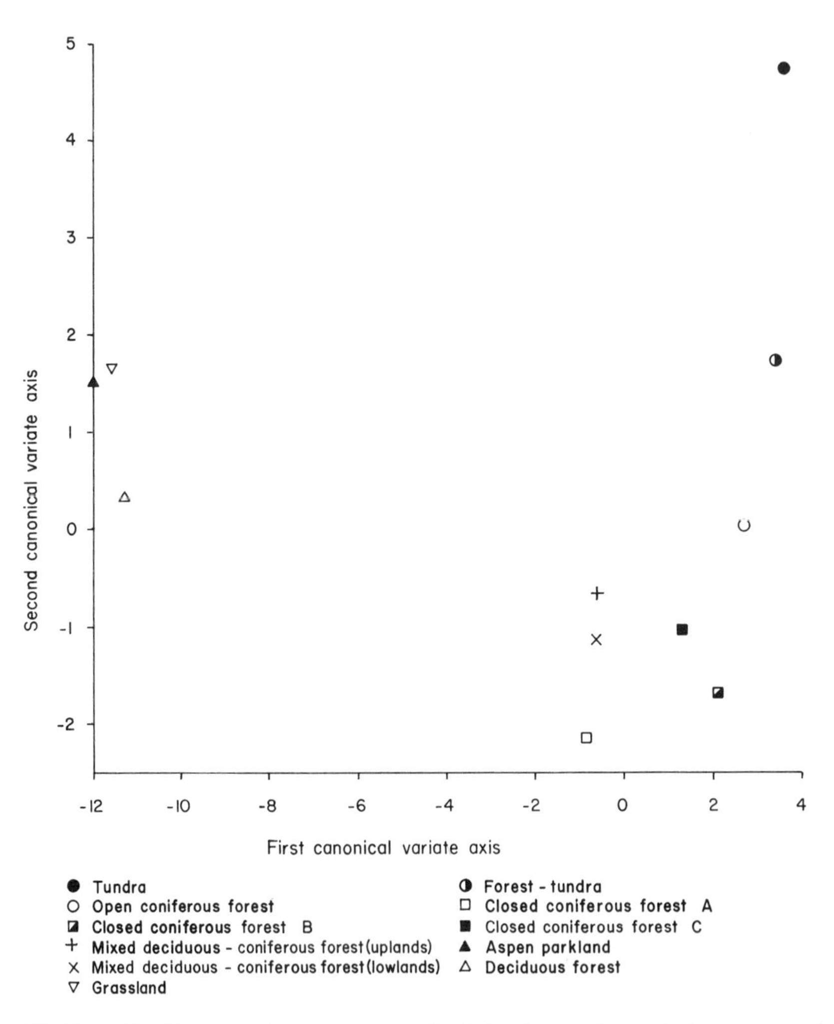

Figure 5.5 Plot of the 11 canonical group means on the first and second canonical variate axes derived from a canonical variates analysis of the surface-sample data from the western interior of Canada.

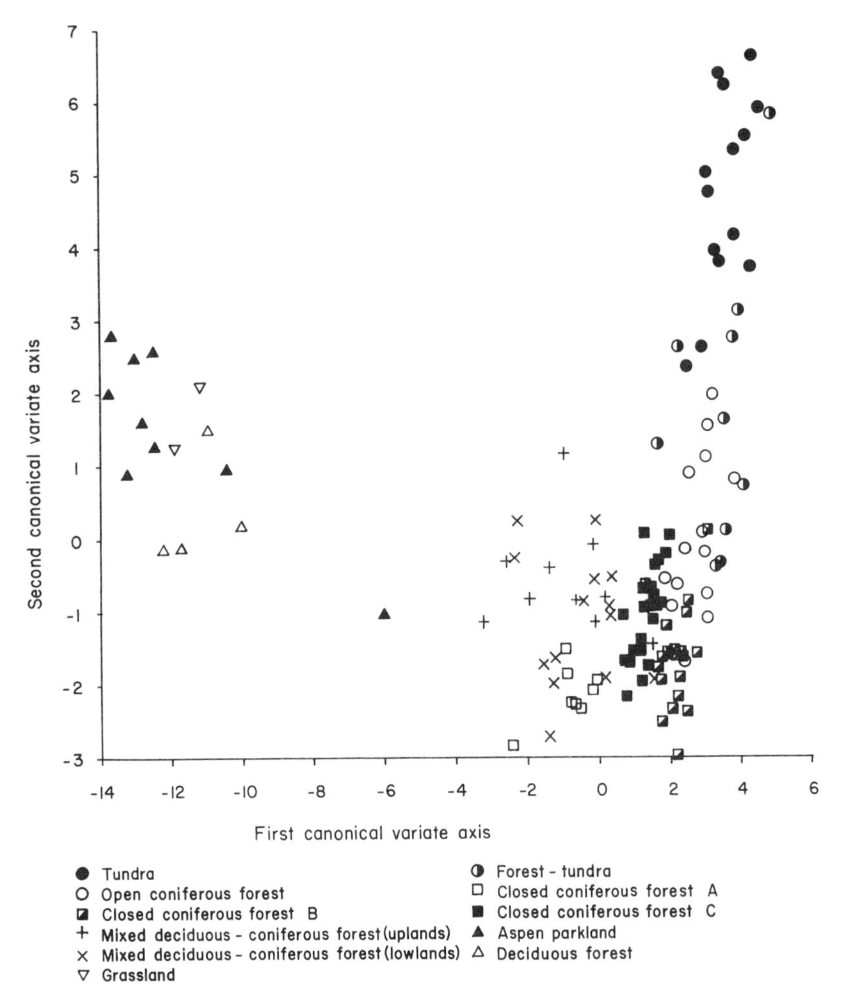

Figure 5.6 Plot of the 131 individual samples on the first and second canonical variate axes of Figure 5.5.

Correlations between the original variables and the canonical variates provide a useful way of interpreting, in palynological terms, the canonical variate axes (see B. K. Williams, 1983). *Picea* and *Alnus* pollen have high positive correlations (>0.5) with axis one, whereas Gramineae, *Chenopodium*-type, *Ambrosia*-type, *Artemisia,* Tubuliflorae, *Populus, Quercus,* and *Fraxinus* pollen all have high negative correlations (<−0.5). These correlations indicate that the first axis discriminates between spectra with high *Picea* and *Alnus* values (i.e., from the tundra, forest–tundra, coniferous forest, and mixed forest units; see Figure 5.2) and spectra with high herb, *Populus, Quercus,* and *Fraxinus* pollen (i.e., southwest samples). *Betula,* Cyperaceae, and Ericaceae pollen and *Lycopodium* spores have high positive correlations with the second canonical variate, in contrast to the high negative correlations of *Picea, Pinus,* and *Larix* pollen. The second axis thus differentiates between those samples with high *Betula,* Cyperaceae, and Ericaceae pollen percentages (i.e., tundra and forest–tundra samples) and samples with either high *Pinus* or high *Picea* pollen values (forest samples).

Further details of canonical variates analysis of these data are given by H. J. B. Birks, Webb, and Berti (1975); other applications of the method to the numerical analyses of surface-sample data include those of Adam (1970), H. J. B. Birks (1977b, 1980), A. M. Davis (1980), Heide (1981), Lamb (1982, 1984), and Elliott-Fisk *et al.* (1982).

Besides being used as a scaling procedure to represent differences between groups of surface-pollen spectra categorised according to some external criterion such as vegetation type, and to determine diagnostic taxa for each group, canonical variates analysis can also be used to address a different type of problem in Quaternary palynology and palaeoecology (I. C. Prentice, 1982b). It can be utilised to distinguish between groups of samples (e.g., surface-pollen spectra) obtained by a classificatory procedure such as numerical partitioning (Subsection 5.3.4) in terms of an additional external set of variables (e.g., environmental variables such as altitude and climatic factors, or site characteristics such as lake size and depth). For example, partitions of modern pollen spectra can be displayed and characterised in terms of site characteristics (cf. H. J. B. Birks, 1977b). We are unaware of any palynological uses of canonical variates analysis for this type of problem, although it has been used to characterise numerically derived clusterings of plant ecological data in terms of environmental variables (e.g., Huntley and Birks, 1979; Gittins, 1979; see also R. H. Green, 1971, 1974).

5.3.4 Classification of Modern Pollen Data

In Section 2.5, we discussed in general terms a group of methods referred to collectively as 'classification methods', in which the aim is to establish whether there is any group structure in the collection of objects under investigation. In this subsection, we describe the use of such methods to analyse modern pollen data solely on the basis of their observed similarities in pollen composition. The numerical classifications obtained can then be compared with the classification based on the vegetation types from which the pollen samples were collected to investigate whether different contemporary vegetation types produce distinctive modern pollen assemblages (see Figure 5.1). If the pollen and vegetation classifications correspond and satisfactory corresponding patterns are found, the modern pollen spectra can be used in the interpretation of fossil pollen assemblages (see Chapter 6) with a greater degree of confidence than would otherwise be possible.

As discussed in Section 2.5, the two main approaches to numerical classification are (1) to obtain a geometrical representation of the objects as points in a low-dimensional space such that similar objects are represented by points which are close together; (2) to partition the objects into a number of disjoint groups, so as to optimise some mathematical criterion. Both approaches will be used in the analysis of the data from the western interior of Canada.

Many different methods have been proposed for relating the dissimilarities between objects to the distances between the corresponding points in a geometrical representation. The biplot, correspondence analysis, and principal components analysis were discussed in Section 4.4. Other geometrical methods, which will be used in this section, are principal coordinates analysis, non-linear mapping, and non-metric multidimensional scaling. As a preliminary to these analyses, some comparative comments follow on ordination procedures.

In Section 4.4, it was convenient to present principal components analysis in terms of a rigid rotation of axes in the original high-dimensional space to new axes, or principal components, so that projection of the points on to the first few principal axes accounted for most of the scatter in the set of points. An alternative characterisation is in terms of the squared interpoint distances. If $d_2(i, j)$, as defined in Section 2.6, represents the squared distance between the ith and jth points in the original space, principal components analysis can be defined by the linear projection of the points from this space into a lower-dimensional space, where they have squared interpoint distances $d_l(i, j)$, which ensures that

$$\sum_i \sum_j (d_2(i, j) - d_l(i, j))$$

is minimised.

In principal coordinates analysis (Torgerson, 1952; Gower, 1966b), this procedure is carried out on a general measure of pairwise dissimilarity, which need not be the squared distance $d_2(i, j)$; I. C. Prentice (1980) discusses the suitability of various measures of dissimilarity for the analysis of pollen percentage data. Several authors (e.g., Kruskal and Carroll, 1969; Sammon, 1969; Kruskal, 1971; A. J. B. Anderson, 1971a, 1971b) have sought configurations of points which optimise other measures of difference between the original pairwise dissimilarities d_{ij} and the interpoint distances in the geometrical representation $\hat{d}_{ij}$, such as

$$\sum_i \sum_j (d_{ij} - \hat{d}_{ij})^2 / d_{ij},$$

the non-linear mapping criterion of Sammon (1969). The method of non-metric multidimensional scaling (Shepard, 1962a, 1962b; Kruskal, 1964a, 1964b) assumes only that there is a monotone relationship between dissimilarities and distances. Although non-metric scaling was used in the first comparison of two pollen sequences by ordination methods (Gordon and Birks, 1974), it does not have any particular advantages in the analysis of the vast majority of pollen stratigraphical data. It does have some important advantages in the analysis of certain ecological (Fasham, 1977; I. C. Prentice, 1977), archaeological (Sibson, 1972a, 1972b), and historical (D. G. Kendall, 1975; Galloway, 1978) data, particularly where principal components analysis or principal coordinates analysis performs poorly. I. C. Prentice (1980) presents a comprehensive review of the use of ordination methods in Quaternary palynology.

Factor analysis *sensu stricto* (Jöreskog *et al.*, 1976) does not appear to have been applied to Quaternary palynological data. Although the aims of factor analytical investigations are sometimes similar to other studies involving ordination techniques, the more formal underlying model of factor analysis, involving specific and common factors, does not appear to be appropriate for palynological data.

It is relevant to note that principal components analysis makes fewer demands on computing resources than the other ordination procedures outlined above. Many of these procedures require the iterative minimisation of a function of nr variables, where r is the number of dimensions in the final configuration and n is the number of samples, and have to carry out this minimisation separately for each value of r. By contrast, principal components analysis and principal coordinates analysis obtain the solution in all numbers of dimensions simultaneously, by the analysis of an $n \times n$ matrix of the pairwise similarities within the set of n objects, or spectra. Further, a principal components analysis can alternatively be implemented via the analysis of a $t \times t$ covariance (or correlation) matrix, where t denotes the number of pollen types represented. This duality between so-called Q-mode ($n \times n$) and R-mode ($t \times t$) methods of conducting a principal components analysis (see Gower, 1966b; Orloci, 1967b, 1973; Jöreskog *et al.*, 1976, Chapter 2; Pielou, 1977, Chapter 21) greatly extends the size of data set which can be analysed in this manner, since the number of pollen types (usually 15–30) is generally considerably less than the number of samples (100–300) to be compared.

Numerous authors have used one or more of these methods in the analysis of surface-sample data. Principal coordinates analysis has been applied, using various dissimilarity measures, to the analysis of modern pollen data by Kershaw (1973), H. J. B. Birks, Webb, and Berti (1975), H. J. B. Birks (1977b), I. C. Prentice (1978, 1982b, 1983a), Lamb (1982, 1984), and Huntley and Birks (1983).

In almost all published applications of principal components analysis to modern pollen data (e.g., T. Webb, 1973, 1974a, 1974b, 1974c; Mack and Bryant, 1974; H. J. B. Birks, Webb, and Berti, 1975; Brubaker, 1975; T. Webb and McAndrews, 1976; Caseldine and Gordon, 1978; T. Webb, Laseski, and Bernabo, 1978; T. Webb, Yeracaris, and Richard, 1978; Bonny, 1978, 1980; Markgraf, D'Antoni, and Ager, 1981; Elliott-Fisk *et al.*, 1982), the pollen percentages were standardised to unit variance for each taxon, a standardisation implicit in the use of a correlation matrix between pollen types in the R-mode implementation of principal components analysis. This standardisation reduces the influence of very variable taxa, and can increase markedly the influence of taxa represented by only a few pollen grains (I. C. Prentice, 1980). In that the proportions in the underlying sediment of such rare taxa will generally be less accurately estimated, this standardisation can have undesirable results, although such effects will be limited by the common practice of not including rare taxa in the numerical analysis (see

Section 2.5). The only studies of modern pollen data of which we are aware, in which untransformed percentage data were subjected to principal components analysis, were conducted by O'Sullivan and Riley (1974) and I. C. Prentice (1982b). The use of other variants of principal components analysis in Quaternary palynology is described by I. C. Prentice (1980).

Correspondence analysis does not appear to have been used to analyse modern pollen data, although it has been used widely in ecology, pre-Quaternary palaeoecology (e.g., Melguen, 1973, 1974; David, Campiglio, and Darling, 1974; David and Dagbert, 1975; Reyment, 1975; Médus and Ipert, 1977; Malmgren, Oviatt, Gerber, and Jeffries, 1978, Cisne and Rabe, 1978; Spicer and Hill, 1979; McNamara and Fordham, 1981), and modern pollen morphology (Hideux and Ferguson, 1976; van der Pluym and Hideux, 1977a, 1977b). It has also been used to analyse modern macrofossil assemblages (H. H. Birks, 1973; Spicer, 1981),

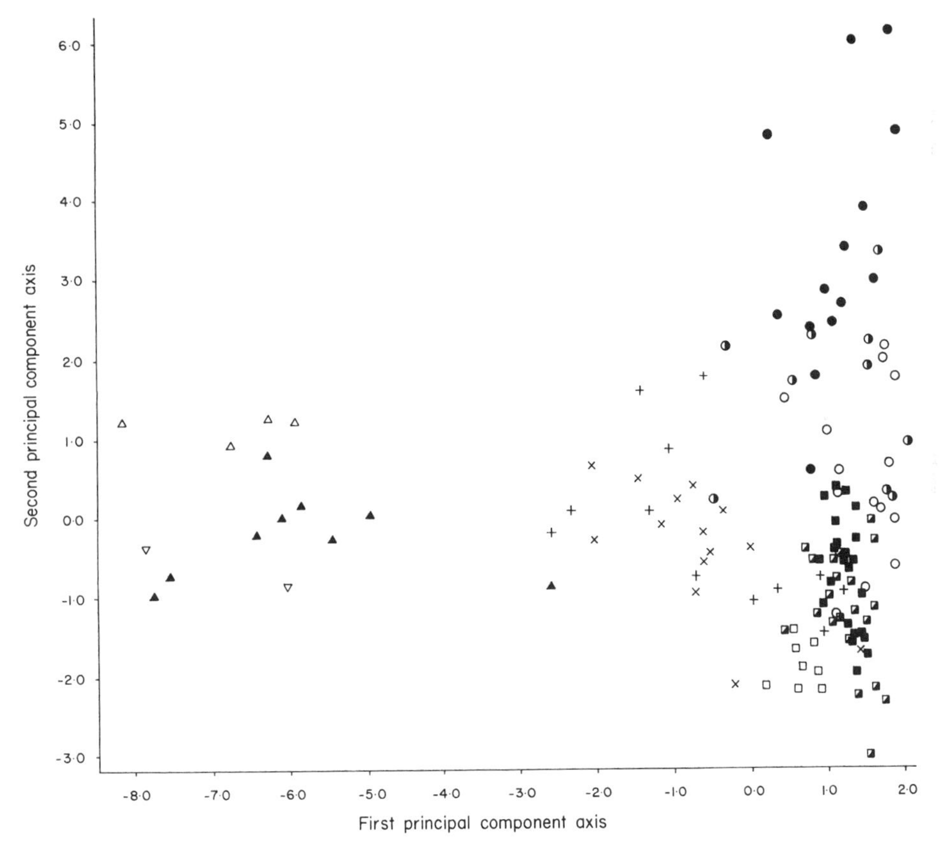

Figure 5.7 Plot of the 131 individual samples on the first and second principal component axes. The coordinates for each sample are the component scores on the first two principal components. The vegetation–landform units from which the samples were collected are also shown. See Figure 5.5 for a key to the symbols used.

Quaternary pollen stratigraphical data (Gordon, 1982c) and post-glacial pollen and diatom stratigraphical data (Elner and Happey-Wood, 1980; Peglar *et al.*, 1984). It warrants use with surface pollen data. Non-metric multidimensional scaling has been applied to modern pollen data by H. J. B. Birks (1973a) and I. C. Prentice (1978). Birks (unpublished data) has used Sammon's (1969) non-linear mapping with limited success in the analysis of several sets of surface-sample data, but we are not aware of any published applications of this method to palynological data.

As an illustration of the use of different geometrical methods to analyse modern pollen data, Lichti-Federovich and Ritchie's (1968) data were analysed by principal components analysis, principal coordinates analysis, non-linear mapping, and non-metric multidimensional scaling. The 19 non-aquatic taxa shown in Figure 5.2 were used. The positions of the 131 surface samples on the first two axes of these four representations are shown in Figures 5.7–5.10. Standard symbols are used to show from which of the 11 vegetation–landform units the samples were collected. The principal components analysis (Figure 5.7) was based on standardised percentage data; the principal coordinates analysis (Figure 5.8) was based on the city-block metric d_1 (see Section 2.6); the non-linear mapping (Figure 5.9) was obtained using the algorithm of Sammon (1969), with simple Euclidean

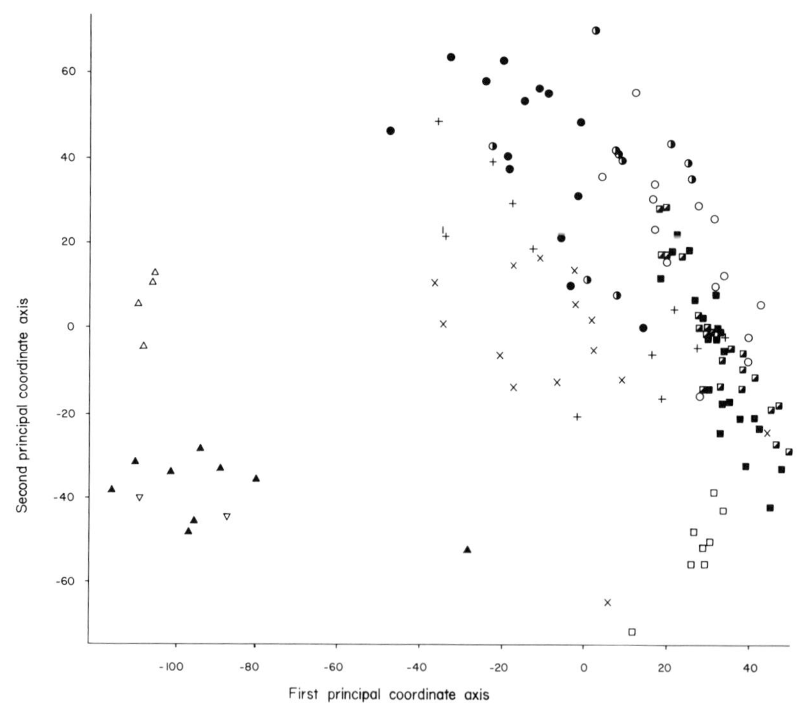

Figure 5.8 Plot of the 131 individual samples on the first and second principal coordinate axes. The vegetation–landform units from which the samples were collected are also shown. See Figure 5.5 for a key to the symbols used.

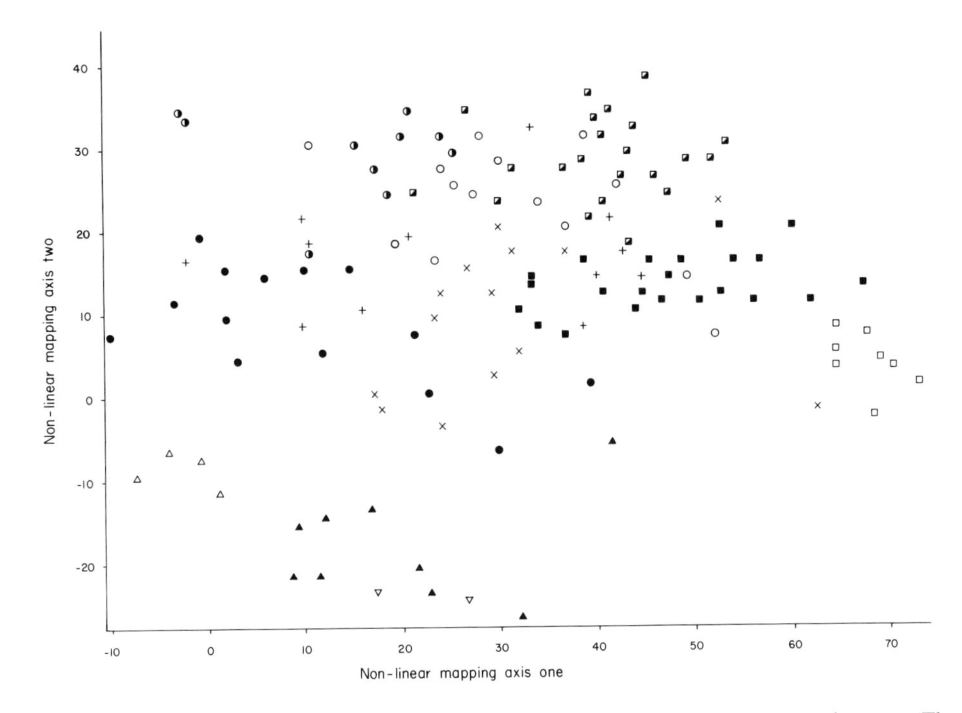

Figure 5.9 Plot of the 131 individual samples on the first and second non-linear mapping axes. The vegetation–landform units from which the samples were collected are also shown. See Figure 5.5 for a key to the symbols used.

distance d_2 (see Section 2.6) as the dissimilarity between samples; and the non-metric multidimensional scaling (Figure 5.10) was based on the city-block metric, and implemented using global scaling and primary treatment of ties (see Sibson, 1972b; I. C. Prentice, 1977, 1980; Gordon, 1981, Chapter 5 for details of these options).

The principal components analysis results (Figure 5.7) suggest a major separation on the first axis (32.6% of the total variation), with samples from the grassland, aspen parkland, and deciduous forest units having high negative scores (−5.0 to −8.0) in contrast to all the other samples with scores between −3.0 and +2.0. Pollen types that have high negative correlations with the first component include Gramineae, *Chenopodium*-type, *Ambrosia*-type, *Artemisia,* Tubuliflorae, *Fraxinus,* and *Populus,* whereas *Picea, Pinus,* and *Alnus* all have high positive correlations.

The sample scores on the first two principal component axes have been mapped in the same manner as separate pollen types in Subsection 5.3.2. A map of the sample scores on component one (Figure 5.11a) shows a sharp gradient in pollen composition between samples in the south and west with high pollen values of prairie herbs (e.g., Gramineae, *Chenopodium*-type, *Artemisia, Ambrosia*-type) and/or deciduous-forest trees (*Quercus, Populus, Fraxinus*) and samples in the

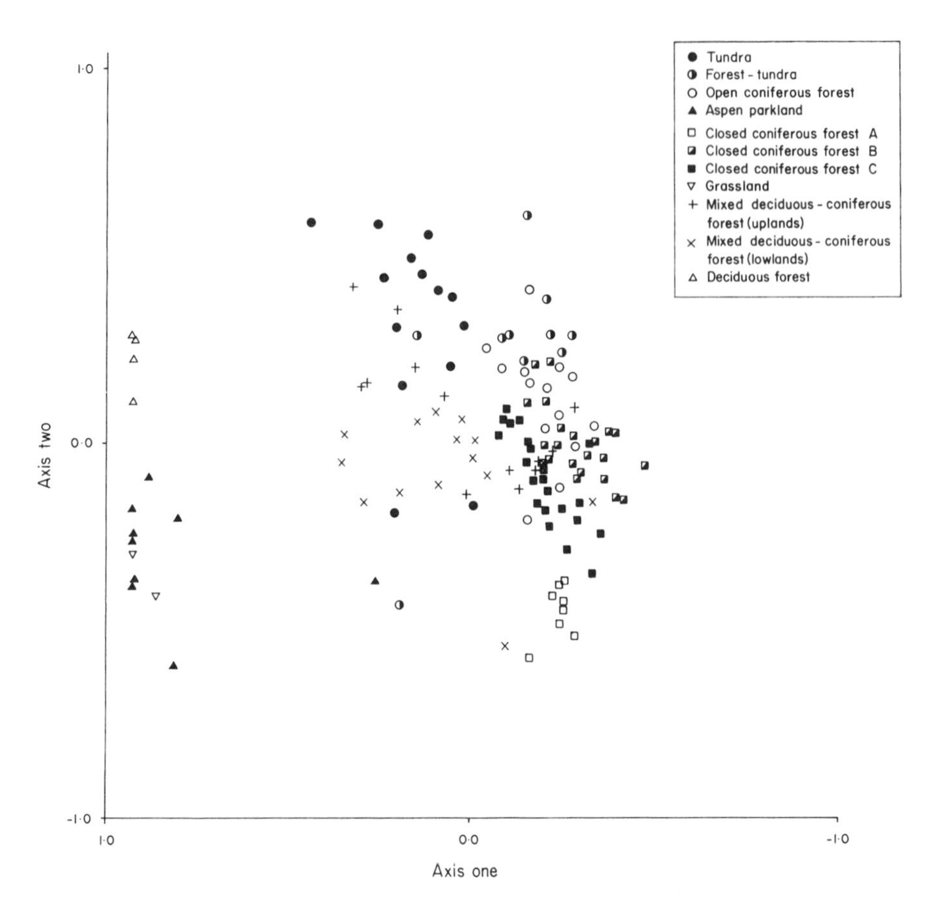

Figure 5.10 Plot of the 131 individual samples on the first and second axes of a two-dimensional non-metric multidimensional scaling. The vegetation–landform units from which the samples were collected are also shown.

north with high values of *Picea, Pinus,* and *Alnus* pollen. The second component (Figure 5.7; 15.2% of the total variability) depicts a gradient between tundra and forest–tundra samples (high positive scores) and closed coniferous forest samples (high negative scores). Samples from the mixed deciduous–coniferous forests occupy intermediate positions on both the first and second axes. A map of the scores for component two (Figure 5.11b) reveals a gradient from the north to the central and southeastern areas. High positive scores are centred on the tundra and forest–tundra regions, scores of 0 to +1 characterise samples from the open coniferous forest unit in the north and the mixed forests in the south, and high negative scores delimit the closed coniferous forest region. *Betula,* Cyperaceae, *Lycopodium,* and Ericaceae all have high positive correlations with component two, whereas *Pinus* and, to a lesser extent, *Larix* and *Picea* are negatively correl-

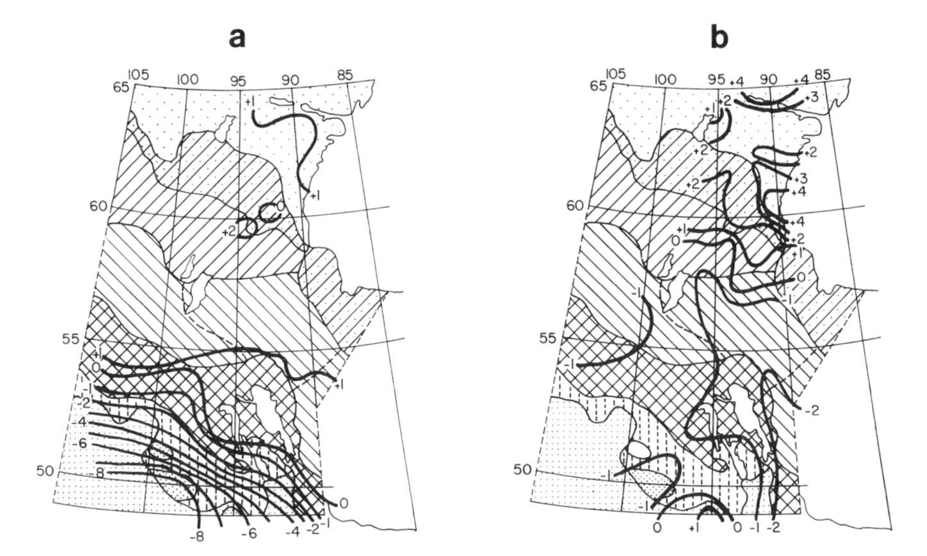

Figure 5.11 Maps of (a) the first principal component scores and (b) the second principal component scores plotted on a base map showing the vegetation–landform units of Figure 53. (After H. J. B. Birks, Webb, and Berti, 1975.)

ated with the second component. The principal components results are thus similar to the canonical variates results (Figure 5.6). Further discussion of these principal components analysis results and of subsequent components is given by H. J. B. Birks, Webb, and Berti (1975).

A broadly similar picture emerges from a principal coordinates analysis of the same data (Figure 5.8). The first two coordinate axes account for 72.6% of the trace of the transformed dissimilarity matrix (Gower, 1966b, 1967a), indicating that these axes provide an efficient low-dimensional representation of the original dissimilarities between samples. As in the results of the principal components and canonical variates analyses (Figures 5.6 and 5.7), the first principal coordinate axis effectively separates southwestern grassland, aspen parkland, and deciduous forest samples (with the exception of one aspen parkland sample, no. 104 of Lichti-Federovich and Ritchie, 1968) from the rest. Samples from the mixed forest units occupy an intermediate position on the first axis. The second principal coordinate axis contrasts, in part, the gradient in pollen composition between the tundra and forest–tundra samples (high positive values) and closed coniferous forest samples, particularly from region A (negative values). Samples from the open coniferous forest unit and regions B and C of the closed coniferous forest unit occupy an intermediate position on the second axis. In addition, samples from the deciduous forest unit are separated from samples from the aspen parkland and grassland units (high negative values) on the second axis. The principal coordinates analysis results are thus broadly similar to the principal components analysis results but differ in minor details.

In discussing the configurations of points shown in Figures 5.7 and 5.8, it was convenient to discuss trends in the data in terms of positions on two orthogonal axes. Although this can be a useful way of summarising patterns in data, there is no reason why trends should be linear (see Subsection 4.6.2 and I. C. Prentice, 1977). Further, while the axes in principal components analysis and principal coordinates analysis satisfy certain optimality criteria, the axes in non-linear mapping and non-metric multidimensional scaling have no such justification; only the relative positions of point are informative.

The results of the non-linear mapping analysis are shown in Figure 5.9. The final mapping error is 16.1%. This is comparatively low for a 19-dimensional data set, and suggests that the two-dimensional representation shown in Figure 5.9 is a fairly accurate summarisation of the data. It can be seen that the samples from the southwest appear in the lower left-hand corner of Figure 5.9, the deciduous forest samples being separated from the aspen parkland and grassland samples. The broad band of points in the upper part of Figure 5.9 shows the same kind of trend as is displayed along the second axes in Figures 5.7 and 5.8, from tundra to closed coniferous forest region A, but samples from the mixed forest units are more intermingled with other samples than is the case in Figures 5.7 and 5.8. Overall, although the non-linear mapping displays some of the structure in the data, the results show a less clear grouping by vegetation–landform unit than those obtained from the principal components and principal coordinates analyses. H. J. B. Birks (unpublished data) has observed similar results in using non-linear mapping in the comparative analysis of other modern pollen data sets.

The two-dimensional representation of the 131 Canadian surface samples obtained by non-metric multidimensional scaling is shown in Figure 5.10. The stress function, measuring the goodness-of-fit of this representation to the original dissimilarities, takes the value 0.123, which can be regarded as indicating a reasonably good fit. The results presented in Figure 5.10 are very similar to those shown in Figures 5.7 and 5.8, particularly the latter.

The results of the four scaling methods suggest that the major contrasts in pollen composition within the modern pollen data from the western interior of Canada are (1) between the pollen assemblages from the aspen parkland, grassland, and deciduous forest units and the rest; (2) between the tundra and forest–tundra assemblages and the closed coniferous forest spectra, especially region A; (3) between spectra from the open and closed coniferous forest units and the mixed forest areas; and (4) between the deciduous forest assemblages and the aspen parkland and grassland spectra. If just one method had been used, it is possible that the data could be distorted by the particular method. However, the similarities between the results of the four methods (Figures 5.7–5.10) indicate that these plots are not markedly distorting the structure in the data; the four methods of analysis make different assumptions about how one should measure the dissimilarity between pairs of surface samples, and about how the distance between the corresponding pair of points in the representation is related to this dissimilarity, yet there is an impressive agreement about what constitutes the main patterns and structure in the data.

The similarity between Figures 5.7–5.10 and the canonical variates analysis plot shown in Figure 5.6 indicates that these patterns are broadly the same whether or not the samples are first grouped by vegetation–landform unit; imposing this grouping on the data has done little to alter the basic patterns displayed in the data. However, significant overlap between many groups of samples is consistently present, implying that a numerical partitioning of the pollen data might not recognise some of the sample groups based on vegetation–landform units. Before investigating this, however, we present a brief discussion of the relative merits of the different scaling methods illustrated above.

Ideally, one would like to have theoretical reasons for specifying a single most appropriate method of analysis. Difficulties in specifying relevant measures of dissimilarity have already been discussed in Section 2.6 (see also I. C. Prentice, 1980), and we greatly doubt that a consensus view will ever emerge. In the absence of such a general theory, one has to fall back on less absolute criteria. The simplicity and computational advantages of principal coordinates analysis and, in particular, principal components analysis have already been discussed. Moreover, it commonly happens, as in our example, that different methods of analysis yield broadly similar results (e.g., Gower, 1966b, 1972; I. C. Prentice, 1978; Gauch, Whittaker, and Singer, 1981). However, such general agreement is not always obtained (e.g., Fasham, 1977; I. C. Prentice, 1977; Noy-Meir and Whittaker, 1978). If there are many total dissimilarities in the original data, that is, samples with no taxa in common, metric methods of analysis such as principal components analysis attempt to equate the corresponding distances between totally dissimilar samples, causing the so-called horseshoe effect (D. G. Kendall, 1975). This effect is unlikely to be critical in the analysis of most, if not all, pollen percentage data, but palaeoecological studies of presence/absence data (e.g., plant macrofossil data) could be affected in this manner. Non-metric scaling can avoid this problem if ties in the dissimilarities are disregarded (D. G. Kendall, 1971a, 1971b; Sibson, 1972a, 1972b; I. C. Prentice, 1977). Further, the less stringent monotonicity criteria of non-metric scaling (compared to metric scaling procedures) can reduce the number of dimensions required to convey the same information. Non-metric scaling is usually most useful when metric methods perform poorly in terms of the variability explained by the first few axes or produce marked ‘arching’ or ‘horseshoe effects’ (I. C. Prentice, 1980). The choice of an appropriate scaling method will thus be governed by several factors, such as the type of data to be analysed and the complexity of the expected structure. No overall recommendation can be made as to a ‘best’ method to be used with all data; it is useful to employ several different methods of analysis, as has been done with the Canadian surface-sample data.

As we stressed in Section 2.5, it can be very helpful to combine scaling methods with partitioning, or cluster analysis, techniques, and this strategy is adopted

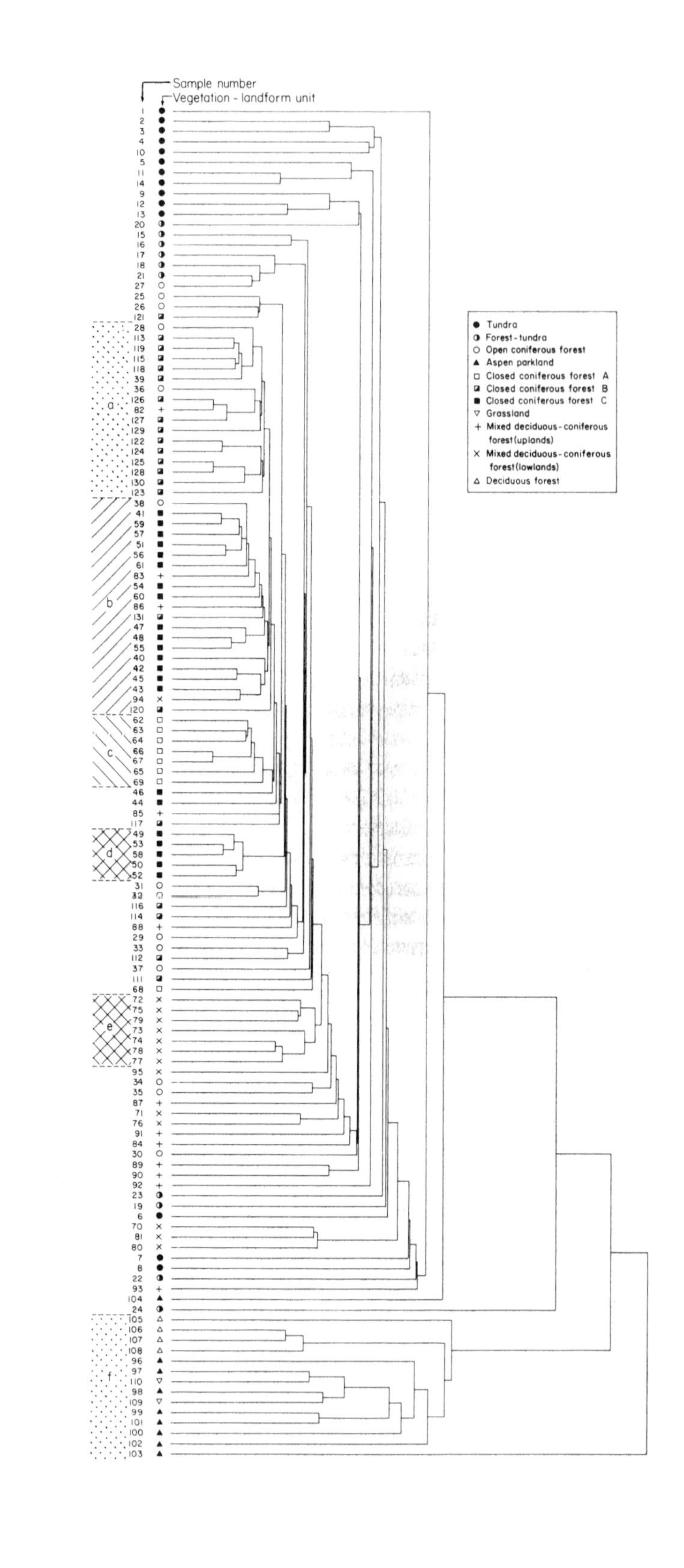

Sample number
Vegetation - landform unit
Tundra
Forest-tundra
Open coniferous forest
Aspen parkland
Closed coniferous forest A
Closed coniferous forest B
Closed coniferous forest C
Grassland
Mixed deciduous-coniferous forest (uplands)
Mixed deciduous-coniferous forest (lowlands)
Deciduous forest
a
b
c
d
e
f

in the investigation of the Canadian surface-sample data. The choice of which clustering criteria to employ requires careful thought. The geometrical representations given in Figures 5.7–5.10 indicate a much less clear grouping of the data than is evident in the studies described in Sections 4.4 and 4.6. We have already remarked on the fact that different clustering criteria are predisposed to finding particular kinds of structure in the data, and may considerably distort the results towards their 'ideal'. A knowledge of the structure might enable one to suggest an appropriate method of analysis which would display the structure more clearly; but, once stated, the circularity is evident: the structure in the data is not known *a priori,* but is being investigated with the assistance of clustering procedures. This is a problem which has exercised many researchers, and a critical discussion of suggested solutions is presented by Gordon (1981, Chapter 6).

A variety of clustering procedures has been used in the analysis of modern pollen data, including single link (H. J. B. Birks, 1973a; Caseldine and Gordon, 1978), group average link (Adam, 1970; Andrews and Nichols, 1981; Elliott-Fisk *et al.,* 1982; Dodson, 1983), information analysis (Kershaw, 1973), centroid (Markgraf *et al.,* 1981), and minimum sum-of-squares (O'Sullivan and Riley, 1974; H. J. B. Birks, Webb, and Berti, 1975; W. L. Strong, 1977; Caseldine and Gordon, 1978; Bonny, 1978, 1980; Andrews and Nichols, 1981; Lamb, 1982, 1984). Other classification procedures employed include Lefkovitch's (1976) polythetic divisive method (I. C. Prentice, 1983a); non-hierarchical plexus techniques (Dabrowski, 1975), in which dissimilarities between pairs of samples are arranged in a network or plexus (see Section 6.5); and direct presentation of matrices of dissimilarity or similarity coefficients between individual pairs or groups of surface samples (Lichti-Federovich and Ritchie, 1965; Ritchie, 1974; Dabrowski, 1975; Kershaw and Hyland, 1975; Brush and DeFries, 1981).

We present here a clustering study of the Lichti-Federovich and Ritchie data, using two different clustering criteria: the minimum sum-of-squares criterion and the single link criterion using Euclidean distance as the measure of dissimilarity between pairs of samples. The sum-of-squares criterion is described in Section 4.4. It tends to produce compact, 'spherical shaped' groups of objects. As this could misrepresent the structure in the data, the single link method has also been used. This method finds clusters of any 'shape' provided these clusters are well separated from one another. The continuous nature of the variation in the surface pollen data would lead us to expect a fair amount of 'chaining' in the single link dendrogram, but it will be a useful check on the tendencies of the sum-of-squares criterion to see what groupings are suggested in the lower parts of the single link dendrogram before the onset of chaining.

The dendrogram for the single link analysis is shown in Figure 5.12. These results indicate that several small groups of samples are amalgamated together at a

Figure 5.12 Single link dendrogram for the 131 individual surface-pollen samples. The vegetation–landform units from which the samples were collected are indicated. Clusters a–f are delimited on the figure and are discussed in the text. Sample numbers follow Lichti-Federovich and Ritchie (1968).

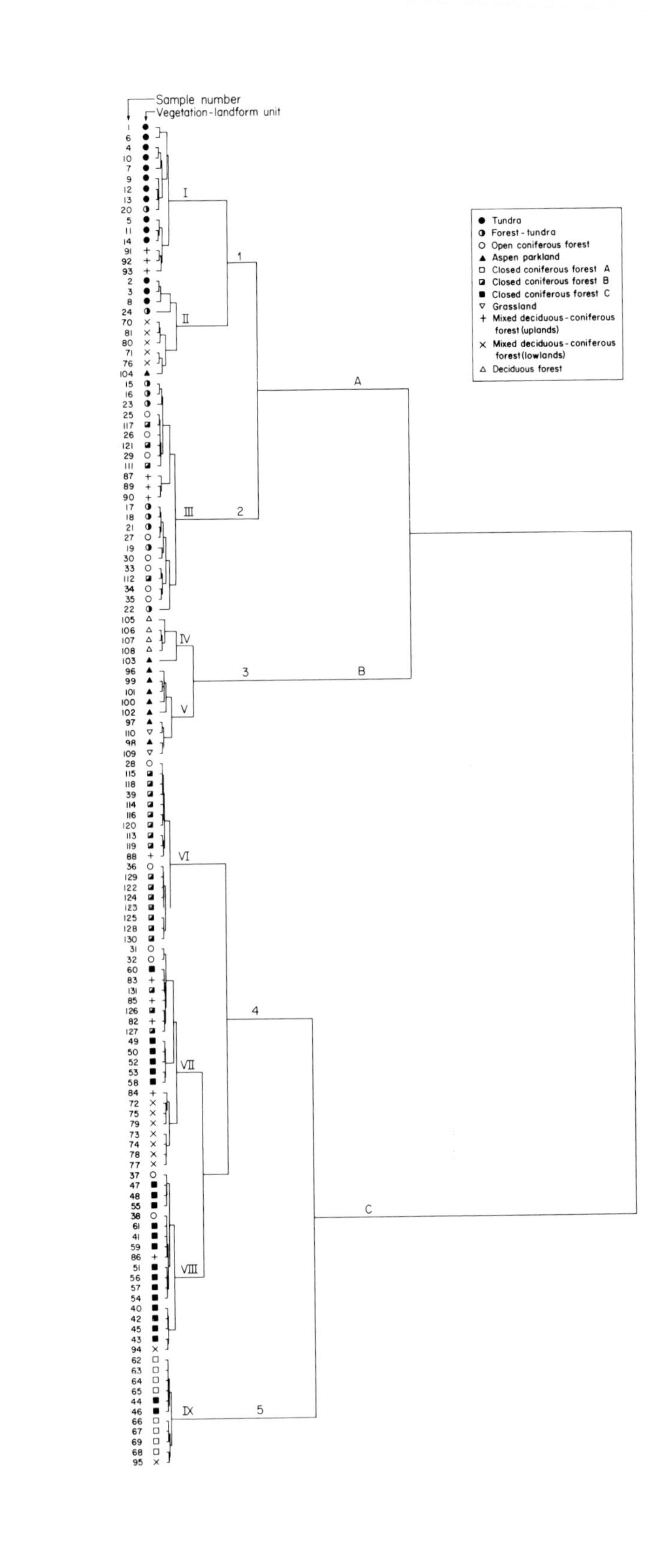

Sample number
Vegetation-landform unit
1 6 4 10 7 9 12 13 20 5 11 14 91 92 93 2 3 8 24 70 81 80 71 76 104 15 16 23 25 117 26 121 29 111 87 89 90 17 18 21 27 19 30 33 112 34 35 22 105 106 107 108 103 96 99 101 100 102 97 110 98 109 28 115 118 39 114 116 120 113 119 88 36 129 122 124 123 125 128 130 31 32 60 83 131 85 126 82 127 49 50 52 53 58 84 72 75 79 73 74 78 77 37 47 48 55 38 61 41 59 86 51 56 57 54 40 42 45 43 94 62 63 64 65 44 46 66 67 69 68 95
I
II
III
IV
V
VI
VII
VIII
IX
1
2
3
4
5
A
B
C
Tundra
Forest - tundra
Open coniferous forest
Aspen parkland
Closed coniferous forest A
Closed coniferous forest B
Closed coniferous forest C
Grassland
Mixed deciduous - coniferous forest (uplands)
Mixed deciduous - coniferous forest (lowlands)
Deciduous forest

low level (groups a–e in Figure 5.12). These groups consist primarily of samples collected from particular vegetation–landform units. Group a contains 14 samples from the closed coniferous forest region B along with two samples from the open coniferous forest and one sample from the mixed forest upland unit; group b contains 15 samples from the closed coniferous forest region C along with one sample from the open coniferous forest, two from closed coniferous forest region B, and three from the mixed forest region; group c contains seven samples, all dominated by *Pinus* pollen from the closed coniferous forest region A; group d consists entirely of five samples from the closed coniferous forest region C; and group e contains eight samples that are all from the mixed forest lowland unit. In addition, there is a rather variable group (group f on Figure 5.12) that contains all but one of the samples from the southwestern aspen parkland (96–103), deciduous forest (105–108), and grassland (109, 110) units; at a lower level, the deciduous forest samples are distinguished from the other samples in this group. There is also some grouping of samples from the tundra (2–5, 9–14) and forest–tundra (15–18, 21) units. There is, however, a large amount of chaining, and it is difficult to delimit any substantial groups other than those marked on Figure 5.12.

The dendrogram for the agglomerative minimum sum-of-squares clustering is shown in Figure 5.13. Groups of samples have been delimited for the three-group, five-group, and nine-group levels on the basis of the dendrogram. These clusters, A–C, 1–5, and I–IX, are labelled on the dendrogram. The numbers of samples from each of the 11 vegetation–landform units in each of the three-, five-, and nine-group clusterings are shown in Tables 5.1, 5.2, and 5.3, respectively. These tables provide a comparison between the numerical partitioning of the samples based solely on their pollen and spore composition and the classification based on the vegetation–landform units from which the samples were collected.

At the three-group level (Figure 5.13 and Table 5.1), cluster B consists entirely of samples from the deciduous forest, aspen parkland, and grassland units, and cluster C contains only samples from the closed coniferous forest, mixed forest, or open coniferous forest units. Cluster A contains all the samples from the tundra and forest–tundra units, as well as some samples from the open coniferous forest, closed coniferous forest region B, mixed forest upland and lowland units, and one sample from the aspen parkland. When the five-group partition is considered (Figure 5.13 and Table 5.2), cluster 1 contains all the samples from the tundra, along with some samples from the forest–tundra and the mixed forest regions and one sample from the aspen parkland. Cluster 2 consists of forest–tundra, open coniferous forest, closed coniferous forest region B, and mixed forest upland samples. All the deciduous forest and grassland samples are grouped together in cluster 3 along with eight of the nine aspen parkland samples. Cluster 4 consists

Figure 5.13 Minimum sum-of-squares dendrogram for the 131 individual surface-pollen samples. The vegetation–landform units from which the samples were collected are indicated. Clusters A–C, 1–5, and I–IX are marked and are discussed in the text. Sample numbers follow Lichti-Federovich and Ritchie (1968).

Table 5.1
Comparison of the Three-Group Clustering of Samples Suggested by the Agglomerative Minimum Sum-of-Squares Algorithm

	Vegetation–landform unit[a]										
Cluster	T	F-T	OCF	CCF-A	CCF-B	CCF-C	MFU	MFL	DF	AP	G
A	14	10	8	—	4	—	6	5	—	1	—
B	—	—	—	—	—	—	—	—	4	8	2
C	—	—	6	8	18	22	6	9	—	—	—
Total no. of samples	14	10	14	8	22	22	12	14	4	9	2

[a] The vegetation–landform units are as follows: tundra (T), forest–tundra (F-T), open coniferous forest (OCF), closed coniferous forest region A (CCF-A), closed coniferous forest region B (CCF-B), closed coniferous forest region C (CCF-C), mixed coniferous–deciduous forest (uplands) (MFU), mixed coniferous–deciduous forest (lowlands) (MFL), deciduous forest (DF), aspen parkland (AP), grassland (G).

predominantly of samples from the closed coniferous forest regions B and C, but with some open coniferous and mixed forest samples. Cluster 5 contains all the samples from the closed coniferous forest region A, and two region C samples and one from the mixed forest lowland unit.

None of the clusters at the nine-group level (Figure 5.13 and Table 5.3) consists entirely of samples collected from one vegetation–landform unit. In each of clusters I, IV, V, VI, VIII, and IX, 70% or more of the samples are, however, from the same vegetation–landform unit. Cluster I largely contains samples from the tundra, cluster IV consists almost entirely of samples from the deciduous

Table 5.2
Comparison of the Five-Group Clustering of Samples Suggested by the Agglomerative Minimum Sum-of-Squares Algorithm

	Vegetation–landform unit[a]										
Cluster	T	F-T	OCF	CCF-A	CCF-B	CCF-C	MFU	MFL	DF	AP	G
1	14	2	—	—	—	—	3	5	—	1	—
2	—	8	8	—	4	—	3	—	—	—	—
3	—	—	—	—	—	—	—	—	4	8	2
4	—	—	6	—	18	20	6	8	—	—	—
5	—	—	—	8	—	2	—	1	—	—	—
Total no. of samples	14	10	14	8	22	22	12	14	4	9	2

[a] The interpretation of the codes for the vegetation–landform units is given in the footnote to Table 5.1.

Table 5.3

Comparison of the Nine-Group Clustering of Samples Suggested by the Agglomerative Minimum Sum-of-Squares Algorithm[a]

	Vegetation–landform unit[a]										
Cluster	T	F-T	OCF	CCF-A	CCF-B	CCF-C	MFU	MFL	DF	AP	G
I	11	1	—	—	—	—	3	—	—	—	—
II	3	1	—	—	—	—	—	5	—	1	—
III	—	8	8	—	4	—	3	—	—	—	—
IV	—	—	—	—	—	—	—	—	4	1	—
V	—	—	—	—	—	—	—	—	—	7	2
VI	—	—	2	—	15	—	1	—	—	—	—
VII	—	—	2	—	3	6	4	7	—	—	—
VIII	—	—	2	—	—	14	1	1	—	—	—
IX	—	—	—	8	—	2	—	1	—	—	—
Total no. of samples	14	10	14	8	22	22	12	14	4	9	2

[a] The interpretation of the codes for the vegetation–landform units is given in the footnote to Table 5.1.

forest, cluster V largely contains samples from the aspen parkland, cluster VI contains samples from the closed coniferous forest region B, cluster VIII consists mainly of samples from the closed coniferous forest region C, and cluster IX contains all the samples from closed coniferous forest region A. Neither the samples from the forest–tundra nor those from the open coniferous forest form discrete clusters (Table 5.3), but taken together they dominate cluster III. This suggests palynological similarity between these two vegetation–landform units. Recall that there was considerable overlap among samples from these two units on the scaling plots as well (Figures 5.6–5.10). The group of samples from the mixed forest unit, both upland and lowland, fail to dominate a single cluster (Table 5.3), and are widely scattered among clusters. This results from the considerable variability in pollen composition of the samples from the mixed forest region (see Figs. 11 and 12 in Lichti-Federovich and Ritchie, 1968). A similar scatter of mixed-forest samples is apparent in the scaling results (see Figures 5.6–5.10).

The partition of the surface-sample data into g groups (for $g = 3, 5, 9$), which has been obtained by sectioning the dendrogram in Figure 5.13, is not guaranteed to be the minimum sum-of-squares partition for each value of g; the agglomerative algorithm merely approximates this by amalgamating at each stage the pair of groups which leads to a minimum increase in the sum-of-squares (Gordon, 1981, Chapter 3). As a check, the data were also analysed using another approximating algorithm, the hybrid algorithm described by Gordon and Henderson (1977), which for a specified value of g seeks to minimise the total within-group sum-of-squares. Partitions into three, five, and nine groups were obtained, these values being selected to allow comparisons with the groupings A–C, 1–5, and I–IX derived from the agglomerative algorithm (see Figure 5.13). In each case, the partition produced by the hybrid algorithm had a lower total within-group sum-of-squares than the corresponding partition obtained from the agglomerative algorithm. Of course, one cannot assert that these hybrid partitions have the lowest possible sum-of-squares for the specified values of g; moreover, there is no guarantee that the sum-of-squares criterion is the most appropriate clustering criterion to employ with these data.

Broadly similar results are suggested by the two algorithms at the three- and five-group levels, with 90% and 89%, respectively, of the samples being similarly grouped. There are larger differences between the two partitions into nine groups, and these results are described more fully below. The partition into nine groups produced by the hybrid algorithm is given in Table 5.4, and Table 5.5 compares this partition with the vegetation–landform units from which the samples were collected. A cross-classification table of the agglomerative and hybrid algorithm results at this level is presented in Table 5.6. The coding of the nine hybrid groups has been chosen so as to produce as high a matching as possible between the corresponding pair of groups labelled by the same number in Roman and Arabic scripts. This maximises the sum of the diagonal entries in Table 5.6, 110 out of 131 samples being displayed on the diagonal.

The main difference between the two partitions is that the agglomerative algorithm suggests a group (cluster IV in Figure 5.13) containing all four samples from the deciduous forest unit and one sample from the aspen parkland (Table 5.3) that is not suggested by the hybrid algorithm results. These five samples (103, 105–108) are grouped instead with the other samples from the aspen parkland and grassland units (Table 5.5). Other, more minor, differences between the two results concern

Table 5.4

Composition of the Minimum Sum-of-Squares Partition into Nine Groups Produced by the Hybrid Algorithm

Cluster code	Members of cluster[a]
1	{1, 5, 9, 11–14, 20, 89, 91–93}
2	{70, 71, 73, 76, 80, 81, 104}
3	{15–19, 21–23, 25–27, 29, 30, 33, 34, 87, 90, 111, 112, 117, 121}
4	{2–4, 6–8, 10, 24}
5	{96–103, 105–110}
6	{28, 36, 39, 82, 88, 113–116, 118–120, 122–125, 127–130}
7	{31, 32, 35, 49, 50, 52–54, 58, 60, 72, 74, 75, 77–79, 84, 85, 126}
8	{37, 38, 40–43, 45, 47, 48, 51, 55–57, 59, 61, 83, 86, 94, 131}
9	{44, 46, 62–69, 95}

[a] Sample numbers follow Lichti-Federovich and Ritchie (1968).

Table 5.5
Comparison of the Nine-Group Clustering of Samples Suggested by the Hybrid Algorithm

Cluster	Vegetation–landform unit[a]										
	T	F-T	OCF	CCF-A	CCF-B	CCF-C	MFU	MFL	DF	AP	G
1	7	1	—	—	—	—	4	—	—	—	—
2	—	—	—	—	—	—	—	6	—	1	—
3	—	8	7	—	4	—	2	—	—	—	—
4	7	1	—	—	—	—	—	—	—	—	—
5	—	—	—	—	—	—	—	—	4	8	2
6	—	—	2	—	16	—	2	—	—	—	—
7	—	—	3	—	1	7	2	6	—	—	—
8	—	—	2	—	1	13	2	1	—	—	—
9	—	—	—	8	—	2	—	1	—	—	—
Total no. of samples	14	10	14	8	22	22	12	14	4	9	2

[a] The interpretation of the codes for the vegetation–landform units is given in the footnote to Table 5.1.

the assignment of pollen samples collected at or near the boundaries of particular vegetation–landform units, particularly in the tundra, forest–tundra, open coniferous forest, and closed coniferous forest B regions (e.g., samples 7 and 10). Numerical methods for comparing the various partitions are discussed in the next subsection.

Synthesis of the numerical partitioning results suggests, as do the scaling results, that samples from the deciduous forest, aspen parkland, and grassland

Table 5.6
Cross-Classification Table Showing the Number of Samples Similarly Grouped by the Agglomerative and Hybrid Minimum Sum-of-Squares Algorithms in the Nine-Group Partitions

Agglomerative partition	Hybrid partition								
	1	2	3	4	5	6	7	8	9
I	11	—	—	4	—	—	—	—	—
II	—	6	—	4	—	—	—	—	—
III	1	—	21	—	—	—	1	—	—
IV	—	—	—	—	5	—	—	—	—
V	—	—	—	—	9	—	—	—	—
VI	—	—	—	—	—	18	—	—	—
VII	—	1	—	—	—	2	17	2	—
VIII	—	—	—	—	—	—	1	17	—
IX	—	—	—	—	—	—	—	—	11

units are distinct from the other units in their pollen composition, and that within these samples there are some differences in pollen composition between the deciduous forest samples and the aspen parkland and grassland samples. The numerical partitions also suggest that spectra from the tundra are relatively distinct palynologically, whereas samples from the forest–tundra and open coniferous forest units are consistently grouped together, usually with some samples from region B of the closed coniferous forest. The large numbers of samples from the vast area of closed coniferous forest are partitioned into four groups, one dominated by samples from region A, one by samples from region C, and one by samples from region B. The fourth group contains samples from regions B and C and several samples from the mixed forest regions. Most of the samples from the lowlands of the mixed coniferous–deciduous forest region are placed in one or other of two groups but many samples from the uplands of the mixed forest region are not separated from samples from the tundra, forest–tundra, open coniferous forest, or closed coniferous forest units (particularly regions B and C).

These numerical partitioning results generally support the grouping of the pollen samples into the vegetation–landform units from which the samples were collected. The two major exceptions are that samples from the forest–tundra and open coniferous forest are not clearly distinguishable from one another, and that samples from the uplands within the mixed forest region do not form a distinct group. In addition, the results indicate that samples from three other units (tundra, open coniferous forest, and region C of the closed coniferous forest region) can possibly be subdivided (see Tables 5.3 and 5.5). The geographical extent of the numerical partitions can be examined by superimposing the partitions onto a map of the vegetation–landform units (see Figure 5.1). If there is any geographical continuity in the extent of these groupings, areas of broadly similar pollen composition could be usefully defined (I. C. Prentice, 1983a). The differences between these areas can then be studied by means of canonical variates analysis using the new areas as *a priori* groupings (Elliott-Fisk *et al.*, 1982).

This partitioning of the surface-pollen samples into groups or clusters should not, however, obscure the basic continuity within the data (H. J. B. Birks, Webb, and Berti, 1975), as illustrated by the results of single link cluster analysis (Figure 5.12) and of the various scaling methods (Figures 5.6–5.10). Overlap tends to occur among samples from adjacent vegetation–landform units in each of the geometrical representations. Despite the fact that minimum sum-of-squares partitioning imposes a grouping on the data, the continuity among samples is still apparent in the partitioning results as samples from adjacent units often appear together in the same group (see Tables 5.3 and 5.5). Whether this overlap results from transport of pollen across boundaries of vegetation–landform units (cf. Janssen, 1970, 1973), from the nature and continuity of the vegetation (see Ritchie, 1959, 1960; Rowe, 1966; Carleton and Maycock, 1978, 1980; E. A. Johnson, 1981), or from inherent variability within the pollen data cannot be determined without quantitative vegetational data. Results from Michigan (T. Webb, 1974a, 1974b), Wisconsin (Webb, 1974c), and Finland (I. C. Prentice, 1978) indicate that at this scale the continuous changes in pollen assemblages may largely reflect the continuous variation in the modern vegetation.

Numerical groupings of modern surface-sample data such as of Lichti-Federovich and Ritchie's (1968) data should thus be viewed as convenient 'reference points' that describe and characterise patterns of pollen composition recognisable within a large and more-or-less continuous multidimensional field of variation. The groupings are equivalent to 'noda' (*sensu* Poore, 1955) in descriptive plant ecology and to 'floristic elements' (*sensu* Jardine, 1972; H. J. B. Birks, 1976a) in descriptive phytogeography. We discussed in Section 3.1 the desirability (or otherwise) of grouping stratigraphical pollen samples into pollen zones, and we emphasised that the value of any classification should be judged in relation to the purposes for which it is required. For the purposes of displaying, structuring, and summarising large amounts of complex, multivariate surface-sample data, numerical partitions provide a convenient synthesis and an effective way of presenting the mass of information contained in the original data. In addition, the partitions can be used to make comparisons with the vegetation groupings from which the samples were collected, and hence provide a means of establishing pollen–vegetation relationships or 'corresponding patterns' (see Figure 5.1). Methods for quantifying such comparisons are described in the next subsection.

The same numerical classificatory methods discussed above have been used to analyse other types of Quaternary surface-sample data such as modern assemblages of cladocera (e.g., Whiteside, 1970; Beales, 1976; Hofmann, 1978; Synerholm, 1979; Norton *et al.*, 1981; Binford, 1982), seeds and fruits (e.g., H. H. Birks, 1973; Spicer, 1981), diatoms (e.g., Bruno and Lowe, 1980; Norton *et al.*, 1981), foraminifers (e.g., Buzas, 1967; Mello and Buzas, 1968; Vilks, Anthony, and Williams, 1970; Hecht, 1973; Malmgren and Kennett, 1973), ostracods (e.g., Kaesler, 1966; Maddocks, 1966), and trace fossils (e.g., Kitchell and Clark, 1979). These methods are of wide potential applicability in Quaternary palaeoecology.

5.3.5 Comparisons of Modern Pollen and Vegetation Patterns

In the previous four subsections we have described various methods of presenting and analysing modern pollen data. This subsection considers ways of assessing the extent to which patterns of variation displayed in the pollen data are a faithful reflection of variations in the vegetation or general ecological setting from which the modern samples originate (see Figure 5.1). A range of methods of comparison is available, depending on how fully the information about the contemporary vegetation has been quantified.

Quantitative vegetational data may be available from all the sites from which surface samples have been collected (e.g., Andersen, 1970; T. Webb, 1974a; P. A. Delcourt *et al.*, 1983). Such vegetation data can take the form of tree counts; measurements of basal area, canopy area, or tree volume; or estimates of plant cover, density, or frequency in plots around the sample sites. The type and detail of the vegetation data collected will depend on the geographical scale and the aims of the investigation. The vegetation data can be analysed numerically in similar ways to pollen data using the methods described earlier in this chapter. Some thought has to be given to relevant ways of conducting such analyses. For example, in classification studies the appropriate measures of dissimilarity between vegetation samples may differ from the appropriate dissimilarity coefficients for relative pollen data; considerable care should thus be given to the selection of relevant measures of dissimilarity and to the questions of standardisation, normalisation, and transformation of vegetational data (see Austin and Grieg-Smith, 1968; Orloci, 1972; Noy-Meir, 1973; Noy-Meir, Walker, and Williams, 1975; Noy-Meir and Whittaker, 1977, 1978; Greig-Smith, 1980, 1983). Orloci (1978), Whittaker (1978a, 1978b), Gauch (1982), and Greig-Smith (1983) provide up-to-date reviews of the use of numerical methods for partitioning and scaling vegetational data.

In studies in which detailed vegetational data are available, the pollen data and the vegetational data can be classified independently, and the two classifications compared. For example, Kershaw (1973) used an agglomerative information theory algorithm to partition qualitative vegetational data from 14 plots in the northern Queensland rain forest. He collected surface pollen samples from the same 14 plots and partitioned the modern pollen data using various numerical methods. The partitions of the vegetational and pollen data were then compared by inspection. Markgraf *et al.* (1981) adopted a similar approach in their analysis of modern pollen and vegetational data from Argentina. They classified modern pollen spectra and associated quantitative vegetation data independently using centroid cluster analysis and principal components analysis. Comparisons of the pollen and vegetation partitions were made visually, and corresponding vegetation–pollen patterns were established.

In many instances, however, detailed vegetation data are not available in surface-pollen studies. Instead, broad, often rather generalised vegetational characterisations of the surface-pollen sites are frequently available, for example, the vegetation–landform units used by Lichti-Federovich and Ritchie (1968) and Janssen (1981a), the vegetation regions used by Wright *et al.* (1967), and the vegetation types of H. J. B. Birks (1973a) and A. M. Davis (1980). In these cases, the samples can be partitioned on the basis of the vegetation type from which the surface samples were collected and can be independently partitioned on the basis of their pollen composition. The questions then arise about how similar the two partitions are, and thus whether there is a satisfactory pollen–vegetation relationship at the particular scale of the study. In the previous subsection we presented cross-classification tables (Tables 5.1–5.3, 5.5) in which the various numerical partitions of the Lichti-Federovich and Ritchie (1968) data on the basis of pollen composition were compared with a partition of the samples on the basis of the vegetation–landform units from which the samples were collected.

Many numerical methods for comparing partitions have been proposed; some of these are discussed by Anderberg (1973), Sneath and Sokal (1973), Rohlf (1974), and Gordon (1980a). The coefficient of Rand (1971) examines all pairs of individuals and evaluates the proportion of these pairs for which

either (1) both individuals are placed together in a group in both partitions;
or (2) both individuals are placed in separate groups in both partitions.

This proportion can be evaluated straightforwardly from cross-classification tables such as Tables 5.1–5.3 and 5.5. If n denotes the total number of objects under investigation, and n_{ij} of these are placed in the ith group of the first partition and the jth group of the second partition, Rand's coefficient is defined by

$$c = 1 - \left[\frac{1}{2}\left\{\sum_i\left(\sum_j n_{ij}\right)^2 + \sum_j\left(\sum_i n_{ij}\right)^2\right\} - \sum_i\sum_j (n_{ij}^2)\right] \Big/ \left[\frac{1}{2}n(n-1)\right].$$

Rand's c ranges from near 0 (dissimilar classifications) to 1 (identical classifications). P. E. Green and Rao (1969) described a coefficient proposed by S. C. Johnson that reduces to the complement of Rand's c.

For the purpose of comparing the various numerical partitions of the surface samples from the western interior of Canada based on their pollen composition with the classification of the samples based on the vegetation–landform units from which the samples were collected, the original 11 vegetation–landform units defined by Lichti-Federovich and Ritchie (1968) were used. In addition, floristically and ecologically related units were grouped together to form seven broader vegetation units (tundra; forest–tundra and open coniferous forest; closed coniferous forest region A; closed coniferous forest region B; closed coniferous forest region C; mixed forest uplands and mixed forest lowlands; deciduous forest, aspen parkland, and grassland) and three very broad units (tundra, forest–tundra, and open coniferous forest; closed coniferous forest and mixed forest; deciduous forest, aspen parkland, and grassland). The partitions of the 131 samples into these 11, 7, and 3 vegetational groupings were compared with the partitions into 3, 5, and 9 groups suggested by the agglomerative algorithm (Figure 5.13 and Tables 5.1–5.3) and the partitions into 9 groups (Table 5.5) and 11 groups suggested by the hybrid algorithm. The values of Rand's c for these comparisons are presented in Table 5.7.

There are high values of Rand's c between the original grouping of the samples into 11 vegetation–landform units and the partition of the samples into 9 groups by the agglomerative algorithm and into 9 or 11 groups by the hybrid algorithm. These values confirm the conclusion that there is a strong relationship between contemporary vegetation types and modern pollen assemblages at the scale at which Lichti-Federovich and Ritchie (1968) sampled. The categorisations of the vegetation into smaller numbers of more broadly defined units also match reasonably well some of the partitions of the pollen data suggested by the algorithms.

We turn now to an assessment of the methods of analysis described in Subsections 5.3.1 and 5.3.2. In the absence of quantitative information about vegetation, an informal assessment can be made of the extent to which the vegetational features are reflected in the displays and maps of pollen data described in Subsec-

Table 5.7

Matrix of Rand's (1971) Coefficients between Partitions of the Lichti-Federovich and Ritchie (1968) Data Based on Vegetation–Landform Units and Partitions Suggested by Several Numerical Classifications of the Surface-Pollen Data

Numerical pollen classification	Vegetation–landform classification		
	3 groups	7 groups	11 groups
Agglomerative (3 groups)	0.76	0.65	0.64
Agglomerative (5 groups)	0.69	0.76	0.77
Agglomerative (9 groups)	0.61	0.86	0.87
Hybrid (9 groups)	0.61	0.86	0.87
Hybrid (11 groups)	0.59	0.85	0.88

tions 5.3.1 and 5.3.2. If quantitative vegetational data are available, these can also be mapped and analysed in the manner described earlier: each taxon can be mapped separately (e.g., T. Webb, 1974a; Bernabo and Webb, 1977; I. C. Prentice, 1978; P. A. Delcourt, *et al.*, 1983) or the two data sets can be subjected to individual principal components analyses and the sample scores on the principal axes mapped (e.g., T. Webb, 1974a) or compared visually (e.g., Markgraf *et al.*, 1981). T. Webb (1973) and Brubaker (1975) implemented the approach of mapping sample scores in their comparison of modern pollen spectra with pollen assemblages deposited just before the onset of European settlement in Michigan (about 1830 A.D.).

The correspondence between pairs of maps, either of component scores or of corresponding pollen and vegetation values, can be assessed by eye (e.g., T. Webb, 1973, 1974a; I. C. Prentice, 1978). Alternatively, one can seek to quantify the agreements, and identify areas of high or low agreement. Numerical methods of comparing two spatially recorded variables can lead to a single measure of the agreement between the variables. Alternatively (Robinson, 1962), the map can be divided into a large number of subregions, and some measure evaluated of the agreement between the two variables within each subregion, for example, their correlation coefficient. The measures of agreement in the subregions can then be mapped to display areas of high agreement. However, the meaning of the correlation between two variables when the observations obtained on each variable are themselves correlated can be questioned.

Recent studies have used a single correlation coefficient to compare two pollen maps (T. Webb, Laseski, and Bernabo, 1978; T. Webb, Yeracaris, and Richard, 1978) and to compare pollen maps with vegetation maps (T. Webb, Laseski, and Bernabo, 1978; P. A. Delcourt *et al.*, 1983). Particularly in this latter application, we have reservations about the appropriateness of comparing percentage data by the correlation coefficient, which gives a measure of the strength of the *linear* relationship between two variables. Because of the constraint that percentages must sum to 100, a perfect relationship between pollen and vegetation is very likely to be reflected in a more complicated functional form (see Subsection 5.4.1), and a rank correlation coefficient (M. G. Kendall, 1970) might prove a more informative measure of agreement.

The methodology described above allows the comparison of the behaviour of a single variable in the two maps; the variable could be the amount of a single taxon (either pollen or vegetation), or the sample scores on a single axis in a geometrical representation. This section is concluded with a brief discussion of canonical correlation analysis which allows the simultaneous comparison of several variables.

Given two sets of variables, canonical correlation analysis (Morrison, 1976, Chapter 7) obtains pairs of linear combinations of variables, one from each set, which are maximally correlated. Thus, if p_k and v_k denote the proportion of the pollen spectrum and the vegetation, respectively, which constitutes the kth taxon, one has a set of pollen variables $(p_1, \ldots, p_t)$ and a set of vegetation variables $(v_1, \ldots, v_t)$. We seek the linear combinations

$$g_1 = \sum_{k=1}^{t} a_k p_k \qquad \text{and} \qquad h_1 = \sum_{k=1}^{t} b_k v_k$$

for which the correlation between g_1 and h_1 is as large as possible. The pair (g_1, h_1) is called the first pair of canonical variables; one can seek further pairs which are uncorrelated with earlier pairs but, subject to this constraint, maximise the correlation between g_j and h_j $(j = 2, 3, \ldots)$. The strength of the relationship between the two sets of variables is given by the set of canonical correlations. The interpretation of the canonical variables can be quite challenging, although Meredith (1964), Monmonier and Finn (1973), and Gittins (1979) report that interpretation can often be helped by evaluating the correlation between each canonical variable and each of the original variables (see also Reyment, 1972). The methodology can also be applied to the case in which the variables in each set are scores on the axes of a geometrical representation (Kershaw, 1973), but the possibly complicated relationship between such scores and the original variables can make interpretation even more difficult.

Canonical correlation analysis of modern pollen and associated vegetation data (14 taxa) from 64 sites in Lower Michigan collected by T. Webb (1974a) yields high canonical correlations (0.93, 0.88, 0.83, 0.64, 0.59). Redundancy analysis (see Section 6.6 and Cooley and Lohnes, 1971) indicates that 42% of the variance within the pollen data is covered by the first five canonical variables of the vegetation data, and that 50% of the vegetation variance is covered by the corresponding canonical variables of the pollen data. Although currently available significance tests for canonical correlations assume that the variables are multivariate normally distributed, the correlations obtained suggest that pollen–vegetation 'corresponding patterns' exist within Lower Michigan. T. Webb (1974a) reached a similar conclusion by comparing visually individual pollen and vegetation maps, and maps of principal component scores for the pollen and vegetation data sets.

The conclusions reached from the comparisons of pollen and vegetation data such as those described thus far in the chapter confirm the existence of a link between the two, and indicate how one might measure the strength of the relationship in some instances. They also, however, show that the relationship is not straightforward, and emphasise the importance of studies which attempt to resolve some of the difficulties which arise, for example, in specifying an appropriate catchment area for a pollen sample (see T. Webb, Laseski, and Bernabo, 1978; T. Webb, Yeracaris, and Richard, 1978; I. C. Prentice, 1982a; Bradshaw and Webb, 1985). Such conclusions will not surprise experienced palynologists. However, studies of the kind described in this section enable one to have more confidence in proceeding to the next stage of an investigation: attempting to model quantitatively the relationship between modern pollen and contemporary vegetation (see Figure 5.1).

5.4 Modelling Modern Pollen–Vegetation Relationships

In this section we review the second major approach to the analysis of modern pollen data (see Figure 5.1), namely, the modelling of pollen–vegetation relationships and the estimation of contemporary quantitative pollen-representation factors or so-called correction factors, such as $\hat{R}_k$ in Figure 1.1. Representation factors quantify the relationship between modern pollen values of the different taxa at one or more sites and the abundance of the taxa in the surrounding vegetation, as measured by, for example, basal area, crown area, or growing-stock volume. By assuming that representation factors are invariant in space and time, one can apply them in reverse to transform fossil pollen spectra into estimates of the former abundance and population size of the taxa in the past vegetation (e.g., $\hat{f}_k$ in Figure 1.1). General introductions to the derivation and use of representation factors include M. B. Davis (1969b), Livingstone (1969), A. M. Solomon and Harrington (1979), H. J. B. Birks and H. H. Birks (1980), and I. C. Prentice (1982a).

It has long been known (e.g., von Post, 1918, 1967; Pohl, 1937; Erdtman, 1943) that plants differ in their pollen production and dispersal, and hence in their pollen representation. Von Post (1918, 1967) suggested estimating representation indices from modern pollen and vegetation data as an aid to interpreting pollen stratigraphical data (see also Fagerlind, 1952), and early estimates of representation factors were made by, for example, Müller (1937), Steinberg (1944), Iversen (1947, 1952–1953), Jonassen (1950), Tsukada (1957, 1958), and Curtis (1959). The first mathematical model of pollen–vegetation relationships was formalised by Fagerlind (1952) and developed by M. B. Davis (1963) as the *R*-value model, in which a taxon's *R* value is defined (p. 898) as the ratio 'between pollen percentage and vegetational percentage'. The *R*-value model is described and discussed more fully in Subsection 5.4.1. In particular, it is shown that there is an important indeterminacy in *R* values, as only the ratios of pairs of *R* values for different taxa at a site are important.

Quite apart from this indeterminacy, it has often been found that a taxon's *R* value, suitably standardised, may differ markedly at different sites. It is thus important to compare sets of *R* values from several sites, in order to establish if there are groups of sites with similar sets of *R* values and to detect 'outliers' or anomalous sites which would merit further investigation. Such comparisons must take account of the property of *R* values mentioned above, namely, that only ratios of *R* values are informative. Some relevant methods of comparison are described in Subsection 5.4.2.

It is clearly preferable to obtain *R*-value estimates based on data from several sites, because such estimates are likely to be more reliable than estimates obtained from a single site (Wright, 1967; Livingstone, 1968, 1969). The set of sites used in *R*-value estimation would usually comprise sites that had been shown to have broadly similar *R* values using the methods of Subsection 5.4.2. Various ways have been proposed for combining modern pollen and vegetation data from several sites so as to estimate a single set of *R* values. These methods are critically reviewed in Subsection 5.4.3; the recommended procedure of estimation involves fitting the *R*-value model by means of the statistical method of maximum likelihood.

When the *R*-value model has been fitted to different data sets, serious inadequacies have often emerged, and more general models have recently been proposed in an attempt to model pollen–vegetation relationships more realistically. One of the major limitations of the *R*-value model is that it assumes that *all* the pollen deposited at a site is derived from a specific vegetational area around the site. The size of this area will vary depending on local site characteristics and topography. It is typically several thousand square metres for small basins (20–40 m in diameter) within forests, and up to several thousand square kilometres for medium- or large-sized lakes or bogs (>250 m in diameter) (see Tauber, 1965; T. Webb, Laseski, and Bernabo, 1978; Jacobson and Bradshaw, 1981). However, pollen grains may be transported over very great distances, and some of the pollen deposited at a site may well be derived from outside the vegetation area sampled and the presumed pollen-source area. This has led to the development of more elaborate but more realistic models which incorporate a background component of pollen originating from outside the primary vegetational source area. These models are described and discussed in Subsection 5.4.4.

Finally, we discuss in Subsection 5.4.5 the potential usefulness of the general approach of modelling quantitatively pollen–vegetation relationships.

Interested readers can find further details of the underlying theory of pollen–vegetation models in Fagerlind (1952), M. B. Davis (1963), Andersen (1970), Parsons and Prentice (1981), and I. C. Prentice and Parsons (1983). This section draws extensively on these sources.

5.4.1 The R-Value Model

Implicit in the R-value model of Fagerlind (1952) and M. B. Davis (1963) is the assumption that y_{ik}, the absolute accumulation of pollen grains of taxon k at the ith sampling location, is related to x_{ik}, the absolute abundance of taxon k in the vegetation near the ith site, by the formula

$$y_{ik} = \alpha_k x_{ik}, \tag{5.4.1}$$

where α_k is a constant productivity or representation factor for taxon k (Mosimann and Greenstreet, 1971). Usually α_k cannot be estimated directly because only percentage data are available. In the notation of Figure 1.1, the underlying proportion in the sediment of pollen grains of taxon k is

$$u_{ik} \equiv y_{ik}/y_{i\cdot}, \qquad \text{where} \quad y_{i\cdot} \equiv \sum_{l=1}^{t} y_{il}, \tag{5.4.2}$$

and an estimate of u_{ik} is given by p_{ik}, the proportion of the pollen sum at the ith site which comprises pollen of taxon k. The proportion of the vegetation at the ith site which belongs to the kth species is clearly

$$x_{ik}/x_{i\cdot}, \qquad \text{where} \quad x_{i\cdot} \equiv \sum_{l=1}^{t} x_{il}.$$

M. B. Davis (1963) sought to estimate

$$R_{ik} \equiv u_{ik}/(x_{ik}/x_{i\cdot}), \tag{5.4.3}$$

namely, the ratio of the pollen proportion to vegetation proportion of the kth taxon at the ith site. If v_{ik} denotes an estimate of the vegetation proportion $(x_{ik}/x_{i\cdot})$, an estimate of R_{ik} is

$$\hat{R}_{ik} \equiv p_{ik}/v_{ik}. \tag{5.4.4}$$

Initially, R values were estimated from a single site, with pollen proportions p_{01}, p_{02}, ..., p_{0t}, and in the notation of Figure 1.1,

$$\hat{R}_k \equiv p_{0k}/v_k \qquad k = 1, 2, \ldots, t.$$

M. B. Davis (1963) emphasised that an R value by itself does not convey any precise information, since R values for a taxon at different sites may differ because of different combinations and abundances of taxa (see also Andersen, 1970, 1980c). However, the *ratios* of R values are informative and should, in theory, be constant from one site to another. This is because, from Equations (5.4.1)–(5.4.3)

$$\begin{aligned} R_{ik}/R_{il} &= (u_{ik}x_{i\cdot}/x_{ik})/(u_{il}x_{i\cdot}/x_{il}) \\ &= (y_{ik}/x_{ik})/(y_{il}/x_{il}) \\ &= \alpha_k/\alpha_l. \end{aligned} \tag{5.4.5}$$

To illustrate the theoretical constancy of R-value ratios, consider the hypothetical example given in Table 5.8, which is based on one presented by M. B.

Table 5.8

Hypothetical Example Comparing Vegetation and Pollen Proportions and R values for Three Taxa (a, b, c) at Three Sites[a]

	Vegetation		Pollen		R value	Relative R value
Taxon	Absolute abundance x	Proportion v	Absolute abundance y	Proportion p	R	R/R_c
			Site 1			
a	4000	0.4	80,000	0.8	2	10
b	1000	0.1	10,000	0.1	1	5
c	5000	0.5	10,000	0.1	0.2	1
			Site 2			
a	1600	0.2	32,000	0.5	2.5	10
b	2400	0.3	24,000	0.375	1.25	5
c	4000	0.5	8,000	0.125	0.25	1
			Site 3			
a	3400	0.68	68,000	0.85	1.25	10
b	1100	0.22	11,000	0.1375	0.625	5
c	500	0.10	1,000	0.0125	0.125	1

[a] Based on an example presented by M. B. Davis (1963).

Davis (1963). In this example, it is assumed that the estimates p and v perfectly estimate the underlying pollen and vegetation proportions; in practice, sampling and measurement errors will also be present but these are ignored for the time being.

At each of the sites in Table 5.8,

$$\alpha_a = 20, \qquad \alpha_b = 10, \qquad \alpha_c = 2,$$

but the R values at the three sites, given in the second-to-last column of Table 5.8, differ. However, the *ratios* of the R values are constant, and equal to the ratios of the α's. Although M. B. Davis (1963) clearly pointed out this important property of R values (see also Andersen, 1970, 1980c), several workers (e.g., Janssen, 1967a; Comanor 1968; West, 1971; Miller, 1973; Faegri and Iversen, 1975; A. M. Solomon *et al.*, 1980; Crowder and Starling, 1980; Starling and Crowder, 1981) have presented sets of unstandardised R values as evidence for varying pollen representations in different ecological settings or geographical areas.

One way of removing this indeterminacy in R values is always to express them relative to a standard taxon, as so-called relative R values *sensu* Andersen (1970, 1973), that is,

$$\hat{R}_{(\mathrm{rel})ik} = \hat{R}_{ik}/\hat{R}_{i1}, \qquad (5.4.6)$$

where $\hat{R}_{i1}$ is the estimated R value at site i of the standard taxon 1. The R values in the hypothetical example considered in Table 5.8 are standardised in this manner with respect to taxon c. Comparisons using this approach can be eased by selecting the reference taxon 1 to be one whose R value can be estimated reliably; this often suggests selection of a common taxon whose R value is not ranked near the top or bottom of the set. An alternative method of standardising R values was proposed by Livingstone (1968) (see the Appendix in Parsons and Prentice, 1981, for details).

The R-value model is applicable at different spatial scales, but careful thought must be given to the choice of an appropriate 'source area' whose vegetation is considered to supply all the pollen grains at the site under study (I. C. Prentice, 1982a). At a local scale, pollen deposited in small basins (20–40 m in diameter) (e.g., Andersen, 1973, 1975, 1978a, 1980c) and mor humus profiles within forests (e.g., Iversen, 1964, 1969)—as reflected by modern pollen spectra obtained from moss cushions or surface litter on the forest floor—is largely derived from vegetation within a 20–30-m radius of the site (Andersen, 1967, 1970, 1973). On the other hand, reconstructions of regional vegetational history invariably depend on pollen analyses of lakes or bogs at least 250 m in diameter. Pollen spectra from such basins are derived from vegetation within a much larger area compared with spectra deposited in small basins (Tauber, 1965; T. Webb, Laseski, and Bernabo, 1978; Jacobson and Bradshaw, 1981) because there is proportionally less local (*sensu* Janssen, 1966, 1973) pollen input to medium or large basins (Tauber, 1965).

The R-value model assumes that all pollen deposited at a site is derived from the sampled primary vegetational source area and that the individual plants within the vegetation are 'evenly distributed throughout a uniform area of infinite size' (M. B. Davis, 1963, p. 898). The selection of the size of vegetational area to be sampled is very critical and inevitably somewhat problematical (Lichti-Federovich and Ritchie, 1965; Berglund, 1973; Parsons and Prentice, 1981; Bradshaw and Webb, 1985). Vegetational source areas of several thousand square kilometres are rarely uniform, due to differences in, for example, slope, aspect, soils, topography, and drainage. Plants are rarely, if ever, distributed evenly within such an area. As pollen deposition decreases gradually rather than abruptly with distance (Turner, 1964; Tauber, 1965; Janssen, 1966, 1973); as different pollen types vary in size, shape, and weight; and as the rate of decrease in pollen deposition with distance is very different for different pollen types (Tauber, 1965; Janssen, 1966), taxa differ in their pollen dispersal capabilities and depositional characteristics. Well-dispersed pollen grains will tend to be better represented in large rather than in small basins (Tauber, 1965). Basin size may thus affect the relative representation of different pollen types. In addition, there is a complex but largely unexplored relationship between pollen production and dispersal and plant distribution within a source area, particularly at the broad spatial scales represented by pollen spectra deposited in medium- or large-sized basins (Livingstone, 1969; Oldfield, 1970; A. M. Solomon and Harrington, 1979; Parsons and Prentice, 1981). Such a relationship is likely to be different for different taxa. For example, poorly dispersed pollen of regionally rare taxa that only grow in fringing fen woodlands around a lake site may be well represented in pollen assemblages despite the low relative abundance of the taxa within the *total* vegetation sampled. Such taxa will acquire high *R* values. In contrast, poorly dispersed pollen of taxa equally rare in the vegetation but that do not grow near the site will be poorly represented in pollen spectra and will acquire very low *R* values. Small amounts of pollen from outside the presumed source area can markedly increase *R*-value estimates (Faegri, 1966; Janssen, 1967a; Livingstone, 1968, 1969; Parsons and Prentice, 1981), particularly at sites close to the range limits of the taxon or where the taxon is rare in the surrounding vegetation. Infinite *R* values occur for taxa absent in the sampled vegetation but whose pollen is present at a site.

The likely primary source area for a given size and type of site is usually selected by experience (e.g. Berglund, 1973; T. Webb, Laseski, and Bernabo, 1978); for example, T. Webb (1974a) used a 22-km radius (1500 km^2 area) around medium-sized lakes in Michigan, Parsons, Prentice, and Saarnisto (1980) used 16-km (800 km^2) and 25-km (2000 km^2) radii in Finland, and T. Webb *et al.* (1981) used an inverse-square weighting by distance of plant abundance within a 30-km radius (2800 km^2) in Wisconsin and Michigan. Kabailiene (1969) attempted to predict from theoretical pollen-dispersal models the likely source areas for different pollen types. The main limitation of her approach is that it relies on theoretical dispersal curves that may not accurately model pollen dispersal in a forested environment (cf. Tauber, 1965, 1977; Andersen, 1973). Kabailiene (1969) also considered the effects of differential pollen dispersal on the estimation of pollen-representation factors. Bradshaw and Webb (1985) have demonstrated the effects of varying the size of vegetational area surveyed around a site and the size of basin on the quantitative derivation of pollen–vegetation relationships. They show empirically that different pollen types have different source areas and that estimates of pollen-representation factors differ for sites of different size (see also Brush and DeFries, 1981; T. Webb *et al.*, 1981; Heide and Bradshaw, 1982). The selection of any primary source area is inevitably therefore a compromise (Parsons and Prentice, 1981). Size and type of basin should be standardised as far as possible in any attempt to estimate *R* values, and should be similar to the sites from which fossil stratigraphical data of interest are available. It is this problem of defining realistic pollen-source areas that has perhaps resulted in the *R*-value model being discarded by many palynologists; for example, A. M. Solomon and Harrington (1979, p. 345) conclude that 'while the model seems to deal adequately with pollen analysis data from a theoretical standpoint, it has little validity from a practical one, for its assumptions are far from reality'.

Thus far, attention has been restricted to the comparison of pollen and vegetation *proportions*. This has certain implications, which may not be immediately obvious. Firstly, an increase in the pollen proportion of a taxon need not imply an increase in the corresponding vegetation proportion (Fagerlind, 1952; M. B. Davis, 1963); consider, for example, taxon c at sites 1 and 2 in Table 5.8. Secondly, the linear relationship between pollen and plant abundances stated in

Equation (5.4.1) does not imply a linear relationship between the proportions of pollen and vegetation (Andersen, 1970; Faegri and Iversen, 1975; T. Webb *et al.*, 1981). This will be illustrated under the simplifying assumption that the vegetation proportions are known exactly. We concentrate here on a single taxon k, which contributes a proportion v_k to the surrounding vegetation. We assume that the pollen-representation factor of taxon k *relative to all other taxa in the vegetation* is β. The factor β is related to the set of representation factors $\alpha_1, \ldots, \alpha_t$ and the set of vegetation proportions $v_1, \ldots, v_t$. To be precise,

$$\beta = \alpha_k/\bar{\alpha}_k, \qquad \text{where} \quad \bar{\alpha}_k \equiv \sum_{l \neq k} \alpha_l v_l \Big/ \sum_{l \neq k} v_l;$$

$\bar{\alpha}_k$ is a weighted average of the representation factors of all the taxa except the kth.

Under this assumption, u_k, the proportion of pollen in the underlying sediment which belongs to the kth taxon, is given by

$$u_k = \beta v_k/(\beta v_k + 1 - v_k). \tag{5.4.7}$$

Plots of u against v for several values of β are given in Figure 5.14 (see also Fagerlind, 1952; Livingstone, 1968; Andersen, 1970; Faegri and Iversen, 1975, Figure 25; T. Webb *et al.*, 1981). It can be seen that the plot of pollen proportion against vegetation proportion is linear only if $\beta = 1$. However, the tangent to the curve at the origin has slope β, and so if attention were restricted to small values of v and/or values of β near 1, the relationship would be close to linear (T. Webb *et al.*, 1981). This result will be discussed further in Subsection 5.4.4.

These properties indicate that it would clearly be preferable to work throughout with absolute pollen and vegetation data, thus avoiding difficulties associated with the fact that proportions must sum to 1 and hence that the proportions of two species will be interdependent. Thus, given suitable data, one could obtain an estimate $\hat{\alpha}_k$ for the representation factor for taxon k in Equation (5.4.1), and under the assumption that these representation factors are constant in time and space, transform y_{ik}, the observed absolute abundance in fossil sample i of pollen of taxon k, to obtain an estimate of the absolute abundance of taxon k in the surrounding vegetation at time i from $y_{ik}/\hat{\alpha}_k$ (e.g., Mosimann and Greenstreet, 1971; Bradshaw, 1978; Ritchie, 1982).

Andersen (1967, 1970, 1973) investigated a more general linear model relating modern pollen-accumulation rates to absolute tree abundances in his work on local pollen deposition within Danish forests. In our notation, his model is

$$y_{ik} = \alpha_k x_{ik} + y_{0k}, \tag{5.4.8}$$

where α_k and y_{0k} are taxon-specific constants, α_k being the productivity, representation, or 'slope' factor, and y_{0k} being a constant background term, the amount of pollen derived from outside the vegetation plot. Andersen obtained x_{ik} by estimating the crown areas of each tree species in 20- and 30-m radius plots. From the centre of each plot he counted the pollen in moss cushions to determine y_{ik}. To eliminate percentage constraints, Andersen related all his pollen counts to the

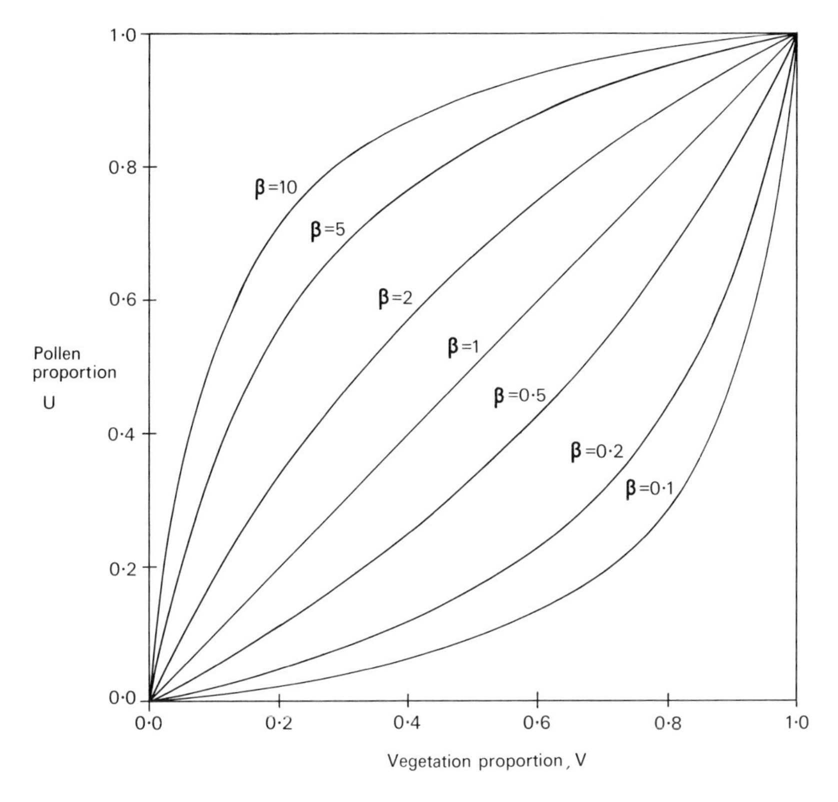

Figure 5.14 The relationship between pollen and vegetation proportions of a taxon implied by the linear model of Equation (5.4.1) relating absolute amounts of pollen to absolute amounts of vegetation; β is the pollen-representation factor for the taxon relative to all other taxa in the vegetation.

exotic pollen rain, namely, pollen of taxa derived from outside the forest (e.g., pollen of agricultural weeds and crops). He demonstrated that this exotic rain was proportionally constant throughout the forest (see also Andersen, 1974); his y_{ik} can thus be regarded as approximations to pollen-accumulation rates. Andersen (1970) showed that data from 40 to 50 sites had high correlations between y_{ik} and x_{ik}; he used least-squares linear regression to estimate the slope α_k and the intercept y_{0k}, representing the pollen-production factor and the amount of pollen derived from outside the plot, respectively. (The use of this method of analysis assumes that x_{ik} can be measured without error). Andersen then normalised the α_k's relative to a reference taxon, *Fagus sylvatica,* to derive relative pollen-production factors for two forests. Bradshaw (1978, 1981a) used a similar model in his work on pollen deposition within British woodlands. The assumption of linearity implicit in the model is clearly supported by Andersen's (1967, 1970) and Bradshaw's (1978, 1981a) data, with their high correlations between y_{ik} and x_{ik}. If

y_{0k} is set to 0 for all taxa, the resulting model is the original R-value model of M. B. Davis (1963), since Equation (5.4.8) reduces to Equation (5.4.1).

As discussed by T. Webb, Laseski, and Bernabo (1978), T. Webb *et al.* (1981), and Parsons and Prentice (1981), absolute data are more equivocal for other spatial scales and from other depositional environments, in particular lakes and mires (bogs and fens).

Pollen-accumulation rates from lake sediments are very variable, as a result of the methods of estimation (see Section 1.2) and site-dependent processes within and between lakes (see Subsection 6.2.4). Pollen recruitment to lakes is complex, consisting of an aerial component C_a and a waterborne component C_w carried by streams and surface run-off (see H. J. B. Birks and H. H. Birks, 1980). The waterborne component is frequently, if not invariably, larger than the aerial component (Peck, 1973; Bonny, 1976). It is not known whether C_a and C_w have different pollen-source areas, but Bonny (1978) presents evidence that suggests different source areas for these components. In addition, the ratio C_a/C_w varies through time within and between lakes (Pennington, 1979). Recent pollen-accumulation rates can, with care and some difficulty, be obtained from surficial lake sediments (e.g., M. B. Davis *et al.*, 1973). However, observed modern and fossil pollen-accumulation rates commonly vary two- to five-fold within and between lakes despite relatively constant sediment-accumulation rates (M. B. Davis *et al.*, 1973). This high variability in time and space results from complex and, as yet, poorly understood site-dependent changes in pollen recruitment (e.g., changing C_a/C_w ratios), sedimentation, and redeposition (e.g., Pennington, 1973, 1979; M. B. Davis *et al.*, 1973; H. J. B. Birks, 1976b), and sediment focussing (e.g., M. B. Davis and Ford, 1982; Bennett, 1983b). Attempts to model relationships between recent (*ca.* 150 years) pollen-accumulation rates and absolute tree abundances in 1500-km^2 areas around 29 lakes in Michigan were unsuccessful (M. B. Davis *et al.*, 1973), largely because of the high inherent variability of the absolute pollen data. As T. Webb, Laseski, and Bernabo (1978, p. 1159) comment, 'influx data merely replace one source of uncertainties in a data set by another source'.

In contrast, bogs and fens have a simpler pollen recruitment, because there is an insignificant waterborne component. Pollen in peats should, in theory, represent the absolute aerial component only. However, the estimation of this is difficult. Modern pollen-accumulation rates are, in practice, extremely difficult to estimate from undecomposed surface peats, and consistent and reliable quantitative sampling of deeper, fibrous peats in varying states of decomposition is also very difficult. Moreover, peat-accumulation rates commonly show large and often erratic changes over short time intervals (e.g., Aaby and Tauber, 1975). Important fine-scale variations in local vegetational cover and microtopography of the mire surface (e.g., Aaby, Jacobsen, and Jacobsen, 1979; Oldfield, Brown, and Thompson, 1979) can also occur. These local sources of variability in peat- and pollen-accumulation rates and processes result in the highly variable and erratic pollen-accumulation rates through time that are so characteristic of peats (e.g., Donner, Alhonen, Eronen, Jungner, and Vuorela, 1978; Beckett, 1979; Ovenden, 1982).

This inherently high variation in present and past pollen-accumulation rates from peats renders them unsuitable for deriving modern estimates of α_k in Equation (5.4.1) or for reconstructing past plant abundances.

Percentage pollen data from lakes and mires have proportionately smaller error components (Maher, 1972a) and are consistently less variable and less site-dependent than their absolute counterparts. Although, in theory, it is preferable to use absolute data, thereby avoiding problems of proportional constraints, relative data are, in practice, more appropriate for reconstructing past plant abundances at the regional scale (*ca.* 2000 km^2) by means of modern pollen-representation factors (T. Webb, Laseski, and Bernabo, 1978; T. Webb *et al.*, 1981; Parsons and Prentice, 1981). The rest of this chapter, therefore, only considers methods for the estimation and comparison of pollen-representation factors from relative pollen percentage data, namely, the *R*-value model, and models that incorporate a 'background' percentage component analogous to the y_{0k} term in Andersen's (1970) model for absolute data.

5.4.2 Comparing *R*-Value Estimates from Different Sites

If *R* values have been estimated at several sites within a vegetational formation, vegetation–landform unit, or forest, it is informative to compare these estimates. The model described in the previous subsection assumed that a taxon's α value and relative *R* value were constant, and it is important to investigate the extent to which this assumption is valid, and the nature of any variation in *R* values. Thus, it is relevant to detect whether there are any systematic site-to-site differences in *R* values that might reflect local ecological conditions. If such variation occurs today, presumably such variation also occurred in the past, thereby limiting the potential applicability of *R* values for transforming fossil pollen spectra into estimates of former plant abundance. Further, it is important to detect groups of sites with comparable *R*-value estimates, so that more reliable *R* values can be estimated on the basis of these site groups rather than on the basis of single sites (see Subsection 5.4.3 and Parsons and Prentice, 1981).

We emphasised earlier that, in the *R*-value model, only the *ratios* of *R* values for different taxa are important; no change is made to the estimates of past abundance of taxa if all the *R* values at a site are multiplied by the same positive constant (see M. B. Davis, 1963 and Subsection 6.3.1). Comparison of sets of *R*-value estimates from several sites thus requires a 'size-free' method of assessing the resemblance between sets of *R* values. The *R* value spectrum for *t* taxa estimated at the *i*th site, $\hat{\mathbf{R}}_i \equiv \hat{R}_{i1}, \hat{R}_{i2}, \ldots, \hat{R}_{it}$, may be represented as a vector in *t*-dimensional space. The vectors representing two identical *R*-value spectra would point in the same *direction* in this space, the precise lengths of the vectors being unimportant (Parsons and Prentice, 1981).

Parsons *et al.* (1980) and Parsons and Prentice (1981) present classification studies of sets of *R* vectors. Both clustering and scaling studies were undertaken.

The clustering study employed the single link method of analysis, using as a measure of the similarity between two vectors, $\hat{\mathbf{R}}_i$ and $\hat{\mathbf{R}}_j$, the cosine of the angle between them:

$$\cos \theta(\hat{\mathbf{R}}_i, \hat{\mathbf{R}}_j) = \frac{\sum_{k=1}^{t} \hat{R}_{ik}\hat{R}_{jk}}{|\hat{\mathbf{R}}_i|\,|\hat{\mathbf{R}}_j|},$$

where

$$|\hat{\mathbf{R}}_i| \equiv \left(\sum_{k=1}^{t} \hat{R}_{ik}^2\right)^{1/2},$$

the length of the vector $\hat{\mathbf{R}}_i$.

In the scaling study, the 'size-effect' was eliminated by normalising each vector so that it was of unit length, by defining

$$\begin{aligned} \mathbf{R}_i^* &= \hat{\mathbf{R}}_i/|\hat{\mathbf{R}}_i| \\ &= (\hat{R}_{i1}/|\hat{\mathbf{R}}_i|, \hat{R}_{i2}/|\hat{\mathbf{R}}_i|, \ldots, \hat{R}_{it}/|\hat{\mathbf{R}}_i|). \end{aligned}$$

The normalised vectors $\mathbf{R}_i^*$ ($i = 1, \ldots, n$) were then subjected to a principal components analysis (of the covariance matrix; thus, no further standardisation of the vectors was implemented). The results were presented as scatter plots showing the positions of the sites on the first two principal component axes (Parsons and Prentice, 1981). This low-dimensional representation adequately portrayed the main features of the data, sites with similar R-value spectra being positioned close to each other, and sites with dissimilar R values being located far apart (see Parsons and Prentice, 1981, for further details). The component scores of the sites were also mapped (Parsons *et al.*, 1980) to put the results into a geographical context.

One problem is that an infinite estimate for an R value is obtained if pollen of a taxon is present but the taxon is absent from the vegetation in the presumed pollen-source area; this situation commonly occurs close to the limits of range of a taxon (e.g., M. B. Davis *et al.*, 1973); for example, *Pinus* pollen is commonly blown long distances beyond the geographical distribution of pine trees (Aario, 1940; H. J. B. Birks, 1973a; I. C. Prentice, 1978). When the spectrum of R values has been normalised, it will be dominated by the R values of such taxa, all other information having been lost. Two alternative methods can be used to avoid this difficulty (Parsons and Prentice, 1981). Firstly, one can compare the reciprocals of R values, provided that none of these is infinite (i.e., whenever a taxon is present in the vegetation surrounding a site, its pollen is also recorded at the site). Secondly, one can partition the data set into groups, such that all sites within a group have the same taxa present in the vegetation; only pollen from these taxa would be considered in that part of the analysis (Parsons *et al.*, 1980; Parsons and Prentice, 1981).

Although either of these expedients avoids difficulties associated with infinite R values, neither is wholly satisfactory, in that one is ignoring information that indicates that the R-value model is inadequate: it does not allow for the possibility of pollen from outside the presumed source area reaching the sampling site. A preferable course of action is the development of more realistic quantitative models of the pollen–vegetation relationship. Some more general models are described in Subsection 5.4.4.

Parsons *et al.* (1980) and Parsons and Prentice (1981) have used the two classification methods described above to compare estimates of R values from subsets of 51 sites in Finland (I. C. Prentice, 1978), and from 22 sites in northeastern North America (Livingstone, 1968), respectively. The two methods gave consistent results and detected important and unsuspected differences between sites on the basis of their R-value estimates that could, in part, be related to local anomalies in the representation of certain taxa and, in part, to systematic geographical patterns in pollen representation. The methods are simple and robust, and provide a powerful means of comparing sites quantitatively on the basis of their R-value spectra.

5.4.3 Estimating R Values from Several Sites

Given data from n separate sites on the sampled pollen proportions $\{p_{ik}\ (i = 1, \ldots, n)\}$ and vegetation proportions $\{v_{ik}\ (i = 1, \ldots, n)\}$ of the kth taxon, it is relevant to consider ways of combining the information from the different sites so as to obtain a single estimate of the R value for the kth taxon. Several methods of estimation are as follows:

(i) $\hat{R}_k$ is defined to be the sum of the pollen proportions divided by the sum of the vegetation proportions (e.g., Müller, 1937; Steinberg, 1944; Janssen, 1967a; Whitehead and Tan, 1969; Andersen, 1970; Donner, 1972; Bradshaw, 1978, 1981a; Kershaw, 1979; Grabandt, 1980; Huttunen, 1980; Ritchie, 1982),

$$\hat{R}_k \equiv \sum_{i=1}^{n} p_{ik} \Big/ \sum_{i=1}^{n} v_{ik}.$$

(ii) $\hat{R}_k$ is defined to be the arithmetic mean of the R-value estimates from the n sites (e.g., Tsukada, 1958; Comanor, 1968; Livingstone, 1968; Miller, 1973; A. M. Solomon *et al.*, 1980),

$$\hat{R}_k \equiv \frac{1}{n} \sum_{i=1}^{n} \hat{R}_{ik} = \frac{1}{n} \sum_{i=1}^{n} (p_{ik}/v_{ik}).$$

(iii) $\hat{R}_k$ is defined to be the geometric mean of the R-value estimates from the n sites (Kabailiene, 1969),

$$\hat{R}_k \equiv \prod_{i=1}^{n} (\hat{R}_{ik})^{1/n} = \prod_{i=1}^{n} (p_{ik}/v_{ik})^{1/n}.$$

The first method of estimation cannot be recommended, since it does not preserve the ratio of R values. Thus, for the data presented in Table 5.8, this method of estimation would yield estimates

$$\hat{R}_{\mathrm{a}} = \frac{2.15}{1.28} = 1.680,$$

$$\hat{R}_{\mathrm{b}} = \frac{0.6125}{0.62} = 0.988,$$

and

$$\hat{R}_{\mathrm{c}} = \frac{0.2375}{1.1} = 0.216.$$

Standardising these estimates relative to taxon c, we have

$$\hat{R}_{\mathrm{a}}/\hat{R}_{\mathrm{c}} = 7.78 \qquad \text{and} \qquad \hat{R}_{\mathrm{a}}/\hat{R}_{\mathrm{c}} = 4.58,$$

instead of the correct values of 10 and 5.

For the artificial data in Table 5.8, in which the R values are estimated without error, the second and third methods of estimation both preserve the ratios of R values. In practice, R-value estimates will contain some error, and one therefore requires reliable and robust methods of estimation. The second and third methods of estimation can be markedly influenced by extreme values from one or more sites; in particular, the second method can lead to a large overestimate and the third method to a large underestimate, although the dangers of this happening will be reduced by a preliminary analysis of the data along the lines suggested in the previous subsection.

Parsons and Prentice (1981) obtained R-value estimates from data recorded at several sites using the statistical method of maximum likelihood estimation (A. W. F. Edwards, 1972). The R-value model assumes (see Subsection 5.4.1) that y_{ik}, the absolute pollen deposition of taxon k at site i, is related to x_{ik}, the absolute abundance of taxon k in the vegetation around site i, by the Equation (5.4.1)

$$y_{ik} = \alpha_k x_{ik},$$

where α_k is a representation factor for taxon k that is constant from site to site. Summing over k for both sides of this equation, we obtain

$$y_{i\cdot} \equiv \sum_{k=1}^{t} y_{ik} = \sum_{k=1}^{t} \alpha_k x_{ik} \qquad (5.4.9)$$

From Equations (5.4.1) and (5.4.9), the underlying proportion of pollen, u_{ik}, in the sediment at the ith site which comprises the kth taxon is

$$u_{ik} \equiv y_{ik}/y_{i\cdot} = \alpha_k x_{ik} \Big/ \sum_{l=1}^{t} \alpha_l x_{il}. \qquad (5.4.10)$$

Under the assumption that

$$v_{ik} = x_{ik}/x_{i\cdot} \qquad (i = 1, \ldots, n;\ k = 1, \ldots, t),$$

that is, that the vegetation proportions surrounding each site are known exactly, Equation (5.4.10) can be written as

$$u_{ik} = \alpha_k v_{ik} \Big/ \sum_{l=1}^{t} \alpha_l v_{il}. \tag{5.4.11}$$

The method of maximum likelihood uses the fact that the pollen counts are multinomially distributed (see Section 2.2) with parameters u_{ik} specified by Equation (5.4.11) to find those values of $\{\alpha_k\ (k = 1, \ldots, t)\}$ that are most consonant with the observed data. This involves maximising the expression

$$\sum_{i=1}^{n} \sum_{k=1}^{t} m_{ik} \log(u_{ik}), \tag{5.4.12}$$

where u_{ik} is defined by Equation (5.4.11) and m_{ik} is the number of pollen grains of taxon k counted at the ith site.

To avoid difficulties associated with infinite R values ($m_{ik} > 0$ but $v_{ik} = 0$), the data can be divided into groups, within each of which the same taxa are always present in the vegetation; each group of sites provides a separate estimate of $\boldsymbol{\alpha} \equiv (\alpha_1, \alpha_2, \ldots, \alpha_t)$.

It can be seen from Equation (5.4.11) that the vector $\boldsymbol{\alpha}$ contains an indeterminacy, in that multiplying each α_k by the same constant would not alter the value of u_{ik}; in other words, only the ratios of the α_k's are important. If the indeterminacy is removed by presetting one of the α_k's (say, α_1) equal to 1, then since

$$\alpha_k/\alpha_1 = R_k/R_1,$$

all the other α estimates will be $R_{(\mathrm{rel})}$ values *sensu* Andersen (1970).

Standard deviations associated with these R-value estimates are provided by likelihood theory (Parsons and Prentice, 1981). The standard deviations indicate the amount of variability present due to the fact that the pollen counts are random variables. The assumption is made that the vegetation proportions v_{ik} are measured without error; this means that the standard deviations specified will, in fact, over-state the precision with which the representation factors are estimated.

This approach assumes that the R values are constant within the group of sites investigated; no account is taken of any systematic site-to-site variations in R values. It is thus important to conduct preliminary investigations, such as those described in the previous subsection, to detect groups of sites with broadly similar R values.

Parsons and Prentice (1981) used this maximum likelihood procedure to estimate R values from Livingstone's (1968) data for several sites in northeastern North America. They conducted four separate analyses, categorising the sites by the amount of oak in the vegetation. Their R-value estimates generally correspond to Livingstone's estimates, based on averaging R-value estimates from individual sites, for taxa such as *Abies, Acer, Fagus,* and *Picea* (when expressed relative to

Betula). Parsons and Prentice's estimates differ somewhat for *Pinus* and *Tsuga* (0.85–1.4 ± 0.1 and 0.17–1.4 ± 0.04–0.2 compared with 1.4 and 0.96, respectively). The largest differences arise in the estimation of R values for *Quercus*. Maximum likelihood estimates for sites where *Quercus* is greater than 2.5% of the vegetation are 0.49 ± 0.08, whereas the estimate is 11 ± 1.7 for sites where oak is less than 0.5% in the vegetation. This difference suggests that the high R value may be due to long-distance transport of oak pollen from outside the area sampled for tree composition. Such transport would result in high *Quercus* pollen percentages at sites where oak trees are very rare (Livingstone, 1968). In such instances, a model of pollen–vegetation relationships that includes a background component is clearly more appropriate. We will discuss such models in the next subsection. Livingstone (1968) averaged his R-value estimates to derive a value for *Quercus* of 30 based on all sites.

Parsons *et al.* (1980) derived maximum likelihood estimates of R values for sites in southern Finland using the modern pollen data of I. C. Prentice (1978) and vegetation data for 16-km and 25-km radii around each site. Their estimates for the two radii are similar and generally agree with Donner's (1972) independent estimates based on averaging pollen and vegetation data from seven sites (see Table 5.9). Parsons *et al.* (1980) also estimated R values for northern Finland; these estimates are reasonably consistent with those from southern Finland (Table 5.9). They also derived maximum likelihood estimates of R values for subsets of sites in southern and northern Finland that were suggested by numerical classifications of the R-value spectra at individual sites.

Although the maximum likelihood estimates of R values may often correspond to the R-value estimates derived by some simple averaging of the data from several sites, the maximum likelihood estimates have several advantages. The

Table 5.9

Maximum Likelihood Estimates of R Values for Southern and Northern Finland Expressed Relative to *Pinus*[a]

	Pinus	*Betula*	*Picea*	*Alnus*
Southern Finland (21 sites)[b]	1.0 (1.0)	1.8 (2.1) ±0.1 (±0.1)	0.42 (0.43) ±0.01 (±0.01)	3.4 (4.0) ±0.1 (±0.2)
Northern Finland (12 sites)[b]	1.0 (1.0)	2.6 (1.6) ±0.08 (±0.04)	0.38 (0.38) ±0.02 (±0.02)	—
Donner (1972) estimates				
Southern Finland[c]	1.0	3.0	0.30	3.3
Southern Finland[b]	1.0	2.8	0.19	5.5

[a] The maximum likelihood estimates and associated standard deviations are based on a 16-km radius. Results for 25-km radius are given in parentheses. (From Parsons *et al.*, 1980.)

[b] Forest inventory data of 1951 to 1953.

[c] Forest inventory data of 1921 to 1924.

method of maximum likelihood explicitly models the pollen counts by a multinomial distribution, thus effectively weighting the contribution of each site in the estimation by the amount of data recorded at (and hence precision of estimation possible from) that site. In addition, each estimate has an associated standard deviation, giving an indication of the reliability of that estimate. This can act as a check against unwarranted conclusions being drawn from the available data.

5.4.4 Pollen-Representation Models That Incorporate a Background Component

It has been seen earlier that the R-value model can be a seriously inadequate representation of the pollen-vegetation relationship if much of the pollen deposited at a site originates from outside the presumed source area within which the vegetation proportions are estimated. This has led various researchers to consider more general models incorporating a 'background' component of pollen. One such model has already been described in Subsection 5.4.1, the model of Andersen (1967, 1970, 1973) relating y_{ik}, the modern pollen-accumulation rate of taxon k at site i, to x_{ik}, the absolute abundance of species k at site i, by Equation (5.4.8):

$$y_{ik} = \alpha_k x_{ik} + y_{0k}.$$

In this equation, y_{0k} denotes a background amount of pollen of taxon k, originating from outside the vegetation plot.

Bradshaw (1978, 1981a), T. Webb *et al.* (1981), Heide and Bradshaw (1982), P. A. Delcourt *et al.* (1983), and Bradshaw and Webb (1985) have considered an analogous model (see also Leopold, 1964; H. J. B. Birks, 1973b) relating u_{ik}, the underlying proportion of pollen of taxon k at site i, to v_{ik}, the proportion of taxon k in the vegetation surrounding site i, by

$$u_{ik} = r_k v_{ik} + u_{0k} \tag{5.4.13}$$

where r_k is a constant representation factor for taxon k, and u_{0k} is a constant proportion of background pollen of taxon k.

The model relating pollen and vegetation proportions when the absolute amounts of pollen and vegetation are linearly related with no background term (Equation (5.4.1); alternatively, Equation (5.4.8) with $y_{0k} = 0$) was outlined near the end of Subsection 5.4.1 (see Equation (5.4.7)). It was noted that the equation was not a straight line (Figure 5.14), although it would be close to linear for small values of v_k and/or values of $\beta \equiv \alpha_k/\bar{\alpha}_k$ close to 1. T. Webb *et al.* (1981) (see also Heide and Bradshaw, 1982) conducted a similar study of conditions under which the more general model of Equation (5.4.8) might be reasonably approximated by a linear relationship of the form of Equation (5.4.13), and illustrated their approach by the use of geometric-mean regression (Riggs, Guarnieri, and Addelman, 1978).

The visual and computational simplicity of this linear model provides a strong incentive to use this method of analysis, and it is certainly valuable to prepare scatter diagrams of pollen and vegetation proportions in the manner of Leopold (1964), H. J. B. Birks (1973b), T. Webb *et al.* (1981), Heide and Bradshaw (1982), P. A. Delcourt *et al.* (1983), and Bradshaw and Webb (1985); these can illustrate data sets for which the model is wholly inappropriate, and highlight anomalous sites which require further study (I. C. Prentice, 1982a). However, we have a preference for more precisely formulated models, in which assumptions are explicitly stated. Two such models of pollen–vegetation relationships based on percentage data and incorporating a background component have been proposed by I. C. Prentice and Parsons (1983), and these are described below.

In I. C. Prentice and Parsons' (1983) first model (equivalent to Parsons *et al.*'s (1980) and Parsons and Prentice's (1981) 'background model'), Equation (5.4.8) is rewritten as

$$y_{ik} - Y_{0k} = \alpha_k x_{ik}\,.$$

Some manipulation of this equation yields

$$\frac{u_{ik} - y_{0k}/y_{i\cdot}}{1 - \left(\sum_{l=1}^{t} y_{0l}\right) \Big/ y_{i\cdot}} = \frac{\alpha_k v_{ik}}{\sum_{l=1}^{t} \alpha_l v_{il}}\,.$$

If one defines $z_k = y_{0k}/y_{i\cdot}$, this can be rewritten as

$$\frac{u_{ik} - z_k}{1 - \sum_{l=1}^{t} z_l} = \frac{\alpha_k v_{ik}}{\sum_{l=1}^{t} \alpha_l v_{il}}\,,$$

or

$$u_{ik} = \frac{\alpha_k v_{ik}}{\sum_{l=1}^{t} \alpha_l v_{il}} \left(1 - \sum_{l=1}^{t} z_l\right) + z_k\,. \tag{5.4.14}$$

In I. C. Prentice and Parsons' (1983) second model, both sides of Equation (5.4.8) are summed, to yield

$$y_{i\cdot} = \sum_{l=1}^{t} \alpha_l x_{il} + \sum_{l=1}^{t} y_{0l}\,. \tag{5.4.15}$$

From Equations (5.4.8) and (5.4.15), we obtain

$$\begin{aligned} u_{ik} &= \frac{\alpha_k x_{ik} + y_{0k}}{\sum_{l=1}^{t} (\alpha_l x_{il} + y_{0l})} \\ &= \frac{\alpha_k v_{ik} + y_{0k}/x_{i\cdot}}{\sum_{l=1}^{t} (\alpha_l v_{il} + y_{0l}/x_{i\cdot})} \end{aligned}$$

by dividing the numerator and denominator of the right-hand side by $x_{i\cdot}$. If one defines $z_k = y_{0k}/x_{i\cdot}$, this can be rewritten as

$$u_{ik} = \frac{\alpha_k v_{ik} + z_k}{\sum_{l=1}^{t} (\alpha_l v_{il} + z_l)} = \frac{\alpha_k(v_{ik} + z_k/\alpha_k)}{\sum_{l=1}^{t} \alpha_l(v_{il} + z_l/\alpha_l)}. \tag{5.4.16}$$

In each of the two models, z_k is assumed to be independent of any properties of the ith site, but the parameters $(z_1, \ldots, z_t)$ have different interpretations in the two models.

In the first model, it is assumed that there is a constant background pollen proportion $z_k \equiv y_{0k}/y_{i\cdot}$ for each taxon. For this assumption to be valid, it is necessary that $y_{i\cdot}$ is a constant, that is, that the total absolute amount of pollen deposited at each site is the same. This is very unlikely to hold for most data sets; however, if y_{0k} is always considerably less than $y_{i\cdot}$, it may be a reasonable approximation to reality (Parsons and Prentice, 1981).

In the second model, it is assumed that $z_k(\equiv y_{0k}/x_{i\cdot})$ does not depend on i; for some data sets, the assumption that $x_{i\cdot}$, the total absolute amount of vegetation surrounding the ith site, is a constant may be a reasonable approximation to reality. In this model, one can interpret $(z_k/\alpha_k)x_{i\cdot}$ as the 'abundance of extra vegetation' contributing pollen to the ith site.

The total absolute pollen accumulation at site i, $y_{i\cdot}$, will depend on the very different pollen productivities and abundances of different taxa (see Pohl, 1937; Andersen, 1970, 1974) as well as on site-dependent processes of pollen recruitment and sedimentation (M. B. Davis *et al.*, 1973). This may vary from site to site more than $x_{i\cdot}$, the absolute total amount of vegetation. Thus, in theory, I. C. Prentice and Parsons's (1983) model 2 may be more realistic than model 1; however, I. C. Prentice and Parsons (1983) report no consistent difference in the performances of the two models when applied to different data sets.

Parsons *et al.* (1980) discuss the interpretation of z values. Clearly if a taxon has a z value of 0, the simple R-value model is appropriate for that taxon. In contrast, taxa with large z values and low α values may have such a large background pollen component that any relationships between pollen and vegetational abundance are obscured. This can occur, for example, at the northern range limits of trees such as *Picea, Pinus,* and *Betula,* where long-distance, extra-regional pollen from areas further south may predominate (e.g., Aario, 1940; I. C. Prentice, 1978). High z values and low α values can also occur, however, where pollen is almost exclusively local in origin, for example, from fen woods around a lake. Such uncertainties in the interpretation of z values illustrate the main limitation of current models of pollen–vegetation relationships, in that no account is taken of the spatial distributions of individual taxa within the presumed source area used for estimating the proportions of the taxa in the vegetation. This problem is discussed further in Subsection 6.3.3.

Parsons and Prentice (1981) and I. C. Prentice and Parsons (1983) outline the computational problems of estimating $\{(\alpha_k, z_k)\ (k = 1, \ldots, t)\}$ using pollen and vegetation percentage data. For each model, maximum likelihood estimates are found by maximising the expression (5.4.12), with u_{ik} specified by (5.4.14) for the first model and by (5.4.16) for the second model. The estimates are obtained numerically, with the assistance of an iterative function-maximisation computer program. In each case, indeterminacy in the estimation of the α_k's is resolved by setting $\alpha_1 = 1$, and the z_k's are restricted to be non-negative (cf. the model implicit in T. Webb *et al.*, 1981, Heide and Bradshaw, 1982, and P. A. Delcourt *et al.*, 1983). In the first model, the z_k's are proportions and the condition

$$\sum_{k=1}^{t} z_k < 1$$

is also imposed. The standard deviations associated with the estimates in each model can also be calculated (I.C. Prentice and Parsons, 1983).

The adequacy of a model for the pollen–vegetation relationship can be examined with the assistance of plots and maps of residuals (see Parsons *et al.*, 1980). The residual Δ_{ik} is defined to be the difference between p_{ik}, the observed pollen proportion, and the pollen proportion predicted by the model. For example, if $\hat{\alpha}_k$ and $\hat{z}_k$ denote the estimates obtained under the first model,

$$\Delta_{ik} \equiv p_{ik} - \left\{ \frac{\hat{\alpha}_k v_{ik}}{\sum_{l=1}^{t} \hat{\alpha}_l v_{il}} \left(1 - \sum_{l=1}^{t} \hat{z}_l\right) + \hat{z}_k \right\}.$$

If the model were adequate, one would expect small positive and negative values of Δ_{ik}, with no geographical trend present. However, in their analysis of data from Finland, Parsons *et al.* (1980) observed regional patterns and also anomalies in the residuals (Δ_{ik}), highlighting inadequacies in the model which merit further study. Possible reasons for the anomalies are complex, and include local pollen input from lakeside and streamside vegetation, long-distance pollen transport, and different patterns of flowering intensity of the same taxon in different parts of Finland related to differences in, for example, climate, land-use history, and silviculture practices. The advantage of these background models is that they permit not only the detection of anomalies in pollen representation but also the quantification of such anomalies.

Applications of these models are presented by Parsons *et al.* (1980), Parsons and Prentice (1981), I. C. Prentice (1982a), and I. C. Prentice and Parsons (1983). Parsons *et al.* (1980) applied the first model to data from Finland. Estimates of α_k (relative to *Pinus*) and of z_k are given in Table 5.10 for 51 sites, for a subset of 21 sites in southern Finland, and for a subset of 16 sites in northern Finland, using forest data from areas with 16-km and 25-km radii around each site. These estimates include very high z values for *Pinus* and *Betula,* and variable z values for the other taxa. The estimates of z for *Pinus* (high) and *Picea* (low) are not unexpected, but the estimates for *Betula* (0.27–0.28) and *Alnus* (0.024–0.077) are

Table 5.10
Estimates of α Values and z Values for Finland, Southern Finland, and Northern Finland[a]

	α[b]				100z (%)			
	Pinus	*Betula*	*Picea*	*Alnus*	*Pinus*	*Betula*	*Picea*	*Alnus*
All Finland (51 sites)								
16-km radius	1.0	0.42	2.9	7.1	47	27	1.5	3.0
25-km radius	1.0	0.47	2.3	13	45	27	1.5	2.4
Southern Finland (21 sites)								
16-km radius	1.0	0.0	5.2	0.0	47	27	0.0	7.7
25-km radius	1.0	0.0	7.0	0.0	48	27	0.0	7.7
Northern Finland (16 sites)								
16-km radius	1.0	23[c]	1.5	—	48	0.0[c]	4.3[c]	3.0
25-km radius	1.0	0.23	0.94	—	40	28	1.8	2.4

[a] Table based on I. C. Prentice and Parsons' (1983) model 1 and Parsons *et al.* (1980).
[b] The α values are expressed relative to *Pinus*.
[c] Anomalous figures, due possibly to this group of sites being too small and/or vegetationally uniform to give reliable estimates (Parsons *et al.*, 1980).

higher than expected. There are also differences in the ranking of the α estimates compared with the ranking of the R-value estimates given in Table 5.9. Parsons *et al.* (1980) also present maps of the residuals (Δ_{ik}), which suggest that the model may have overcompensated for long-distance dispersed pollen in the extreme north; for example, the residual values for pine pollen are usually positive in pine-forest areas and negative in birch-forest areas.

Parsons and Prentice (1981) applied the first model to Livingstone's (1968) data from northeastern North America. With α for *Betula* specified to be 1, the estimates of all the other α_k, except that for *Fagus,* were smaller than the corresponding R-value estimates. All species other than *Betula* and *Fagus* had positive z values in compensation. *Quercus's* R-value estimates (30, Livingstone, 1968; 0.49–11, Parsons and Prentice, 1981) were replaced by estimates for α of 0.06 and z of 0.038, suggesting that the reduction in *Quercus's* representation factor can mostly be explained by transport of oak pollen from outside the presumed source area.

I. C. Prentice and Parsons (1983) applied both background models to data of T. Webb *et al.* (1981) from Wisconsin and Upper Michigan, and obtained broadly consistent results from the two analyses. This type of investigation is particularly valuable, since if comparable conclusions are reached from analyses based on two different models, one can have more confidence in the validity of the main conclusions; in particular, the assumptions that z_k (the background component) does not depend on i (a given site) in both models does not seem to be critical for these data. However, the interaction between theory and applications of modelling pollen–vegetation relationships is still at an early stage, and it would be valuable if further comparative studies were conducted, particularly replication studies in which several data sets were collected from the same area and analysed separately before comparison (cf. T. Webb *et al.*, 1981; Heide and Bradshaw, 1982). However as T. Webb *et al.* (1981) comment, there is, at present a critical lack of suitable and extensive modern vegetation and pollen data sets with which to evaluate different models of pollen–vegetation relationships.

5.4.5 Discussion and Conclusions

The idea of estimating modern pollen-representation factors and using them to aid interpretation of pollen stratigraphical data is long established, and was generally accepted by pollen analysts prior to 1963 (e.g., Iversen, 1947, 1952–1953; Tsukada, 1957, 1958). The *R*-value model, as formalised by Fagerlind (1952) and presented by M. B. Davis (1963), was the first quantitative *model* of modern pollen–vegetation relationships. Davis went on to apply *R*-value estimates to reconstruct past plant abundances from fossil pollen spectra. The failure of this reconstruction resulted in much adverse criticism of the *R*-value model (e.g., Faegri, 1966; Janssen, 1970; A. M. Solomon and Harrington, 1979). The approach was neglected until Andersen (1967, 1970) and, more recently, Parsons and Prentice (1981), T. Webb *et al.* (1981), and I. C. Prentice and Parsons (1983) developed new models of modern pollen–vegetation relationships as means of deriving representation factors to transform pollen stratigraphical data into estimates of past plant abundances (see Section 6.3).

We consider briefly why M. B. Davis's (1963) approach was unsuccessful before discussing whether recent developments provide realistic models of modern pollen–vegetation relationships. Our discussion of M. B. Davis's (1963) study follows Parsons and Prentice (1981). We feel that their conclusions require reiteration in order to evaluate the advances that have been made in this field since 1963 and to provide a perspective for assessing the present state of the art in modelling pollen–vegetation relationships.

M. B. Davis's (1963) pioneering study failed for two main reasons (see Parsons and Prentice, 1981): (1) the small size of her modern and fossil data sets resulted in estimated past vegetation proportions with standard deviations of the same order of magnitude (see Section 2.4 and Parsons *et al.*, 1983 for details), and (2) the vegetational area she used for estimating modern vegetational proportions was far too small (7 km^2), considering the large amount of background pollen in the modern sample from trees outside the area, particularly *Pinus* and *Quercus* (see M. B. Davis and Goodlett, 1960 and Livingstone, 1968 for details). As Parsons and Prentice (1981) and Parsons *et al.* (1983) discuss, the first problem can be overcome by estimating pollen-representation factors from modern data at several sites (see Subsection 5.4.3). The problem of background pollen is more difficult to circumvent, because the vegetational area chosen must be sufficiently large to ensure that 0 or only very small amounts of pollen originate from outside it and hence that the *R*-value model is acceptable. The newly developed models (see Subsection 5.4.4) are useful here, as they provide a means of quantifying the background component for different taxa, and thus of evaluating the reliability of a given set of *R*-value estimates (see I. C. Prentice, 1982a). Parsons and Prentice (1981, p. 147) conclude that 'Davis' problems with *R*-values are attributable, not to the *R*-value model, but to the way it was deployed.' In the event, the idea of pollen-representation factors was generally abandoned. However, they can, if estimated carefully and used critically, be valuable tools for interpreting pollen stratigraphical data (see Section 6.3). We do not agree with Janssen's (1981b) conclusion that their use is 'a doubtful enterprise' or Faegri and Iversen's (1975, p. 158) view that 'it is impossible to establish *R* values of general usefulness.' Representation factors are as potentially important now as they were when von Post (1918) first suggested deriving representation indices as aids in interpretation.

Reliable representation factors require realistic models of modern pollen–vegetation relationships. It is thus pertinent to ask how reliable existing models are, and how well they correspond to our knowledge of pollen-recruitment processes. The basic *R*-value model (Equations (5.4.4) and (5.4.1)) assumes that there is no background pollen, *all* pollen deposited at a site being derived from the sampled vegetational area, and that plants are distributed evenly within such an area (*ca.* 1000–2000 km^2 in regional-scale studies). There is, at present, no reliable theoretical means of calculating appropriate source areas. They are chosen by experience; their suitability (or lack of it) for *R*-value estimation can be evaluated by the background-component models. In practice, there is always some background pollen, and, as discussed in Subsection 5.4.1, small amounts of pollen from outside the presumed source area can cause havoc with *R*-value estimates. The *R*-value model is thus clearly a poor representation of modern pollen–vegetation relationships at a regional scale. Andersen's (1970) model (Equation (5.4.8)) explicitly considers background pollen at the local scale (1000–3000 m^2), and closely fits observed patterns of local absolute pollen deposition and tree abundances within forests. The background models specified by Equations (5.4.14) and (5.4.16) represent regional-scale pollen deposition at a site as two components, one derived from plants within the sampled source area and one for plants outside this area. These models appear more realistic for mires than for lakes: bogs and fens only receive an aerial pollen component from local, extra-local, regional, and extra-regional sources (*sensu* Janssen, 1973), whereas pollen recruitment to lakes is more complex, with local, extra-local, regional, and extra-regional aerial components and also local, extra-local, and possibly regional waterborne components. These components may be derived from different source areas, and their relative and absolute magnitudes may vary in space and time (see Bonny, 1978; Pennington, 1979; Jacobson and Bradshaw, 1981).

The *R*-value model ignores all information about spatial aspects of the local and regional vegetation, and background models only distinguish between vegetation inside and outside a specific area. As discussed in Subsection 6.3.3, a multitude of plant abundances and vegetational patterns can occur within source areas of 1000 to 2000 km^2. Existing models of pollen–vegetation relationships are thus inevitably crude, but we suspect that they may be the forerunners of more realistic models that explicitly consider vegetational patterns and different pollen components at various spatial scales. The state of existing palynological theory and methodology and the shortage of suitable data sets are the major limitations to deriving more sophisticated quantitative models. Of the existing models of *modern* regional pollen–vegetation relationships, the background models are clearly preferable to the *R*-value model. However, modelling modern pollen–vegetation relationships is not an end in itself. It is a means to an end, the end being robust representation factors that are invariant in time and space, and that can be used with confidence to transform pollen stratigraphical data into reliable estimates of past plant abundance. Problems of constancy in time and space of the parameters of the regional background models (α_k, z_k) immediately arise when they are used to transform stratigraphical data. Pollen analysts using representation factors thus face a critical dilemma concerning *R*-value and background models, discussion of which is deferred to Subsection 6.3.3.

Further advances in modelling pollen–vegetation relationships require a greater understanding than we presently have of the likely source areas and dispersal distances of different pollen types and components for a range of site sizes and types. Such an understanding is likely to come from an interaction between theoretical (e.g., Tauber, 1965; Kabailiene, 1969; A. M. Solomon and Harrington, 1979) and empirical studies on modern pollen deposition in relation to vegetation at different spatial scales (e.g., T. Webb *et al.*, 1981; Bradshaw and Webb, 1985). Estimation of reliable representation factors requires accurate and reliable models of pollen–vegetation relationships at the spatial scales of relevance to the palaeoecologist (Livingstone, 1968, 1969; Oldfield, 1970). Despite the importance of representation factors in quantitative palynology, we are still some way from such models except at the local scale (e.g., Andersen, 1970; Janssen, 1981b).

In this chapter we have reviewed the wide range of numerical techniques that are currently available to analyse modern pollen data. We will turn in the next chapter to the problem of the quantitative interpretation of fossil pollen data in terms of past populations, communities, and environments. Many of these reconstructions utilise modern pollen data, and we consider the application of pollen-representation factors and surface pollen spectra to the interpretation of pollen stratigraphical data.

CHAPTER 6

The Interpretation of Pollen Stratigraphical Data

6.1 Quantitative Approaches to Interpretation

The interpretation of Quaternary pollen stratigraphical data is a complex, difficult, and exacting task. It requires a knowledge of the physiological, population, and community ecology of the plants represented in the fossil pollen assemblages, as well as information concerning the relationships, both qualitative and quantitative, between modern pollen assemblages and the contemporary vegetation and environment from which the assemblages are derived. In addition, information on stratigraphical changes and trends in the pollen values of taxa considered individually and collectively is frequently required in detailed ecological interpretations of pollen stratigraphical data, particularly when the emphasis of the study is on population changes, community dynamics, and fine-scale vegetational differentiation.

In general, interpretation can follow the logical sequence of reconstruction of past floras, reconstruction of past plant populations, reconstruction of past vegetation, and reconstruction of past environments. As outlined in Chapter 1, at least six basic questions can be asked in the interpretation of pollen stratigraphical data (H. J. B. Birks and H. H. Birks, 1980):

1. What taxa were present in the past flora?
2. What were the relative abundances or population sizes of the taxa present in the past flora?
3. What plant communities or vegetation types were present in the past?
4. Where did these communities occur in the past?

5. When did these communities occur in the past?
6. What was the environment in which these plant communities existed in the past?

Numerical methods can assist the Quaternary pollen analyst in providing answers to questions 2, 3, and 6, namely, in the reconstruction of past plant populations, past plant communities, and past environments. It is important, however, to emphasise at the outset that we view all the numerical methods discussed in this chapter as nothing more than tools, sometimes very crude tools, that can, in some instances, assist in the interpretation of pollen stratigraphical data. Numerical methods cannot replace the sound ecological knowledge and experience that are so essential for reliable and critical interpretations of Quaternary pollen analytical data. The methods can, however, help in detecting features of the data that may otherwise be overlooked (e.g., D. G. Green, 1982), in suggesting new lines of enquiry (e.g., Ritchie and Yarranton, 1978b), and in generating new hypotheses that may be subsequently falsified by further independent observations and data (e.g., Bernabo, 1981). In addition, numerical methods can summarise complex patterns in both fossil and modern data in an efficient and unambiguous way that can aid interpretation (e.g., H. J. B. Birks, 1976b; Ritchie, 1977). Numerical procedures can also expose limitations in our current interpretative procedures, and can prompt us to re-consider certain assumptions about the nature of vegetation and the processes, rates, and directions of population change and community dynamics (e.g., Ritchie and Yarranton, 1978b; D. G. Green, 1981, 1982; D. Walker and Pittelkow, 1981; D. Walker, 1982a, 1982b; Bennett, 1983c). We thus recommend a combination and critical interaction of qualitative and quantitative appoaches in the detailed interpretation of pollen-analytical data. As D. Walker and Pittelkow (1981, p. 50) emphasise, 'the most sophisticated mechanical manipulation of data is but a preliminary to the judicious exercise of ecological judgement'.

Mathematical methods can be useful in detecting, quantifying, and modelling stratigraphical and hence temporal changes and trends in stratigraphical data for individual pollen taxa. Such methods of sequence-splitting, curve-fitting, and time series analysis involve separate numerical analyses of each taxon and, given reliable data, can detect subtle stratigraphical patterns within and between individual pollen curves (D. Walker and Wilson, 1978; D. G. Green, 1981). When the results of analyses of several individual taxa are compared, unsuspected patterns of stratigraphical change among taxa may emerge (e.g., D. Walker and Pittelkow, 1981; D. G. Green, 1982). The results of such analyses can thus, in some instances, provide insights into patterns of temporal change within and between plant populations in the past. These methods are discussed in Section 6.2 and are illustrated with reference to data on pollen-accumulation rates from Abernethy Forest.

Quantitative procedures can also be used to reconstruct past plant populations by transforming fossil pollen values into estimates of past plant abundances. These methods involve the use of modern pollen-representation factors, with or without the incorporation of a background component, and provide a means of reconstructing the former abundance of taxa and past population sizes within the presumed pollen source-area of the site being investigated. The relevant numerical methods are considered in Section 6.3.

Different numerical methods can help in the comparison of modern and fossil pollen spectra, and hence in the reconstruction of past plant communities and vegetation types in terms of modern analogues, particularly at the scale of vegetation formations and vegetation–landform units. This is the so-called comparative approach to the reconstruction of past vegetation (H. J. B. Birks and H. H. Birks, 1980). The appropriate numerical methods form the basis of Section 6.4.

Quantitative methods can also assist in the detection of groups of fossil pollen taxa that are significantly correlated or associated together, so-called recurrent groups (H. J. B. Birks and H. H. Birks, 1980). Whilst the numerical delimitation of such groups is relatively straightforward, the ecological interpretation of recurrent groups is often considerably more difficult. Recurrent groups are discussed in Section 6.5.

Finally, numerical methods can provide a means of reconstructing past environments directly from pollen stratigraphical data by using modern transfer or calibration functions that quantitatively relate contemporary pollen assemblages to present-day environmental factors (usually climate). These modern transfer functions, like pollen-representation factors, can be applied to fossil pollen assemblages to derive quantitative estimates of the past environment (e.g., H. S. Cole, 1969; T. Webb, 1971; T. Webb and Bryson, 1972). These methods are considered in Section 6.6.

The various approaches to the interpretation of pollen stratigraphical data discussed in this chapter are summarised as a series of stages in Figure 6.1. Many of the approaches are applicable in theory to both pollen percentages and accumulation rates, whereas some are only possible with reliable accumulation rates. In practice, pollen percentages are used almost exclusively for the reconstruction of past populations, vegetation types, and environments.

The interpretative approaches reviewed here are almost all descriptive in character, being concerned primarily with the reconstruction of past populations, communities, and environments (see Figure 6.1). We are thus largely concerned with descriptive palaeoecology (*sensu* H. J. B. Birks and H. H. Birks, 1980). We do not consider in detail the interpretation and use of pollen stratigraphical data along the lines recently proposed by Colinvaux (1983) 'to test contemporary ecological hypotheses' and by D. Walker (1982b) for 'the solution of major problems in ecological theory', such as the widely discussed topics of diversity, stability, and complexity, of community integration, and of resilience and stability of ecological systems. D. Walker (1982b, pp. 420–421), in a stimulating and provocative essay, suggests that 'vegetation history has an important and timely contribution to make', that 'the time is ripe', and that the palaeoecological 'baby has not disappeared with the bathwater, but has been a bit slow to make ecologically interesting conversation'.

We concur with Faegri (1974, p. 64) that it is essential in Quaternary pollen

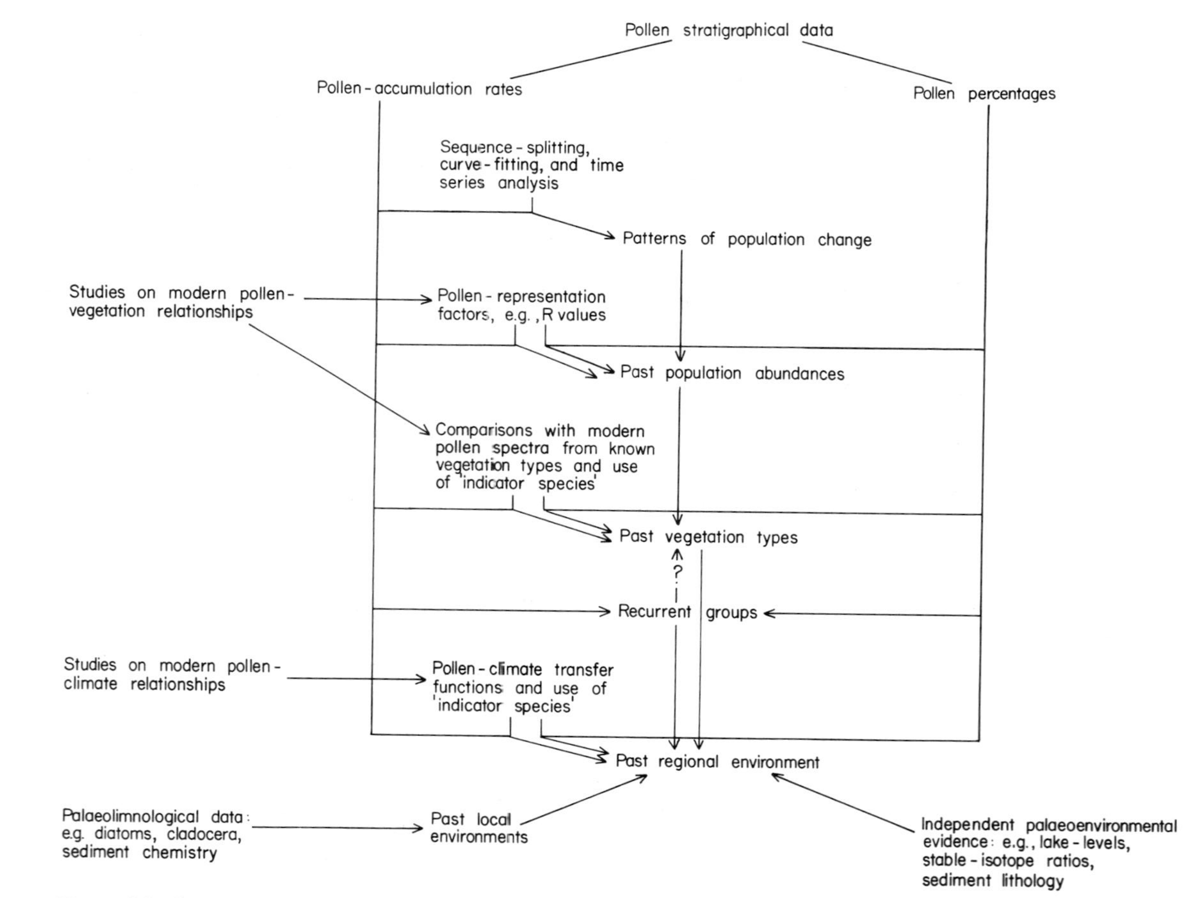

Figure 6.1 Stages in the interpretation of pollen stratigraphical data, with particular reference to the use of numerical methods in the reconstruction of past plant populations, vegetation types, and environments.

analysis to complete phase I first, namely, 'establishing the facts, the sequence of vegetational events as they could be read out of the diagrams'. Only then can Faegri's (1974, p. 67) later phases be tackled, in the hope of leading to 'a detailed phytosociological reconstruction, and through that to a detailed ecological analysis' and hence to 'the most exciting future uses of pollen analysis: as a tool of palaeoecological analysis'. The attainment of these phases remains a major challenge to many, if not all, Quaternary pollen analysts and is an essential stage before attempting to solve 'major problems in ecological theory' with pollen stratigraphical data.

Some ecologists (e.g., Connell and Sousa, 1983) have questioned whether pollen analysis can contribute to problems of ecological stability and resilience in natural communities (cf. D. Walker, 1982a), for two reasons. Firstly, much of the pollen analytical data has generally broad spatial (about 10^8–10^{12} m^2) and temporal (5000–15000 years) scales and limited spatial (10^8 m^2) and temporal (5–50 years) precision (see T. Webb, Laseski, and Bernabo, 1978; T. Webb *et al.*, 1981; T. Webb, 1981, 1982 for discussions of scale and precision in pollen analysis). Secondly, there are difficulties in distinguishing vegetational changes that have occurred as a direct response to environmental change (T. Webb, 1981; I. C. Prentice, 1983b) from vegetational changes resulting from biological processes under conditions of environmental stability (D. G. Green, 1981; D. Walker, 1982a).

Moreover, there is currently a considerable amount of controversy among ecologists as to the status, validity, and potential falsifiability of many widely accepted ecological theories. There has been frequent confusion between falsifiable hypotheses and unfalsifiable but useful concepts in theoretical ecology (see, e.g., Dayton, 1979; McIntosh, 1980; Peters, 1980; D. R. Strong, 1980). It remains to be seen whether some of the problems relating to diversity, stability, complexity, and resilience (see, e.g., D. Goodman, 1975; Abele and Walters, 1979a, 1979b; Pimm, 1984) and to niche availability and competition (see, e.g., Wiens, 1977; Vuilleumier, 1979; Simberloff, 1982) require solutions or indeed whether they are even potentially falsifiable or soluble. Similarly, attempts to interpret patterns in pollen stratigraphical data in terms of r- and K- life-history phenomena (e.g., M. B. Davis, 1976; Flenley, 1982; Tsukada and Sugita, 1982) require evaluation of the underlying assumptions and concepts of modern life-history traits (see, e.g., Stearns, 1976; Caswell, 1982) before these ideas can usefully contribute to our understanding of past population dynamics.

Many of these ecological concepts were developed by theoretical ecologists, and assume that populations and communities are in equilibrium with their environment, that communities are saturated, and that the environment is constant (Schaffer and Leigh, 1976). Environment varies continuously not only in space but also in time (T. Webb, 1981). For example, climate has changed over a variety of temporal scales in the historical past. Presumably, climate has been equally variable in prehistoric times. Population, community, and environmental reconstructions should therefore be attempted first before considering ecological concepts and theories that implicitly assume equilibrium, saturation, and constant environment. Without a careful completion of Faegri's (1974) phases, some consideration of the spatial and temporal resolution and inherent scales of pollen stratigraphical data, and a critical evaluation of current ecological concepts and theories, there is a danger that D. Walker's (1982b) palaeoecological baby may be forced to run before it can walk. It is surely better that it make slow but ecologically relevant, rather than precociously irrelevant, conversations!

Many approaches are potentially useful and important in providing insights into the diverse patterns observable in the late-Quaternary pollen stratigraphical record and their underlying causal processes. The reconstruction of past populations, communities, and environments remains, however, for most investigators, the major aim in the interpretation of pollen stratigraphical data.

6.2 Sequence-Splitting, Curve-Fitting, and Time Series Analysis

Watts (1973) has suggested that for certain palaeoecological problems, pollen analysts have been preoccupied with stratigraphical sub-divisions, zonations, and correlations as a means of summarising and analysing pollen stratigraphical data prior to their interpretation (see also D. Walker, 1972). This concern with zonation and correlation may reflect the primarily geological origins of pollen analysis and its early use as a technique for relative dating (see reviews by H. J. B. Birks, 1982a and D. Walker, 1982b). Whilst there is little doubt that pollen zones are useful for the description, discussion, comparison, and correlation of pollen data and as a framework for broad-scale reconstructions of past vegetation (see Chapter 3), the interpretation of pollen stratigraphical data in terms of plant populations changing through time can benefit from alternative methods of data analysis and summarisation (D. Walker and Wilson, 1978). These alternative methods assume that pollen curves are records, admittedly complex and imperfect records, of plant populations that may increase, remain stable, oscillate, decline, or become extinct with time (Watts, 1973). The methods are particularly useful in situations in which past vegetation and its constituent plant populations have varied continuously in space and continually in time, as proposed by Gleason's (1939) individualistic concept of vegetation; and as a result, the past vegetation has no convincing modern analogue (see, e.g., Livingstone, 1969; Watts, 1973; H. J. B. Birks, 1976b, 1981a; M. B. Davis, 1976, 1981; Ritchie, 1977; D. Walker, 1982b for discussions of 'no-analogue' vegetation types).

By analysing quantitatively the stratigraphical record of individual pollen taxa and by fitting a time scale to this record, pollen data provide records of population changes of taxa (at least of their haploid generations!) over long periods of time (see Figure 6.1), time series far longer than is possible by direct observation. In many instances, particularly when the vegetation of the past has no satisfactory modern analogue, it is valuable to consider initially the history of individual taxa, viewed as populations, and to detect and to quantify stratigraphical changes and trends within the pollen values of individual taxa (see Figure 6.1), prior to assembling these histories into a reconstruction of the vegetation of which the taxa formed a part (e.g., Ritchie, 1977, 1981, 1982).

The methods described in this section can only be used to analyse pollen-accumulation rates (grains cm^{-2} $year^{-1}$). The many sources of error and bias present in such data are outlined in Section 1.2. In assuming that the estimated pollen-accumulation rates accurately reflect the population size of the taxa in the pollen catchment area, one is making stringent palynological assumptions about the data that may not be fully justified (see Bennett, 1983b). A discussion of these assumptions, and of the extent to which one might expect to obtain data which reasonably satisfy the assumptions, is postponed until Subsection 6.2.4.

Using pollen-accumulation rates, it is possible to display and to study the independent variations in pollen deposition of each separate taxon through time (e.g., M. B. Davis, 1969a, 1976, 1981; Donner, 1972; Ritchie, 1977, 1982; H. H. Birks and Mathewes, 1978). For each taxon separately, D. Walker and Flenley (1979) plotted for every sample the percentage deviation from the taxon's overall mean value, in an attempt to portray the mean 'performance' of a taxon through time, and the variations about that mean. The results of this type of representation can be strongly influenced by a few extreme values that may result from independent sedimentary phenomena within the basin (e.g., H. J. B. Birks, 1976b) and hence not be related to the population behaviour of the taxon of interest.

D. Walker and Wilson (1978) presented a more elaborate method of analysing pollen stratigraphical data of individual taxa. The data, or *sequence,* for each taxon are first split into *sections* on the basis of presence/absence data only, in an attempt to distinguish intervals in which the taxon is essentially absent from those in which the taxon is effectively present. Each of the second type of section is then split into sections of distinct but homogeneous mean and standard deviation. This latter method of analysis is very similar to the zonation of pollen stratigraphical data described in Chapter 3, except that it is based on a single taxon at a time. A second difference is that D. Walker and Wilson (1978) present a procedure for deciding on the number of divisions to be implemented based on more formal statistical significance tests.

The means, standard deviations, and mean/standard deviation ratios (D. Walker and Pittelkow, 1981) of the data in each section are used to characterise the sections. Finally, some sections are analysed further by modelling the pollen-accumulation rates as linear, quadratic, or more general functions of the time variable. These methods of analysis are described in Subsection 6.2.1, and applications of them are reviewed in Subsection 6.2.2.

More elaborate methods of analysing the behaviour of a variable over time, and investigating the relationship of two such variables, are available under the portmanteau heading of 'time series analysis'. Introductions to the mathematics of time series analysis are given by J. C. Davis (1973), Ord (1979), and Chatfield (1980). Usher (1973), Platt and Denman (1975), M. Williamson (1975), and Stephenson (1978) discuss the ecological interpretation and applications of time series techniques.

Data that can be subjected to time series analysis are of the following kind: the values taken by several variables are known at selected times in the past; for example, $y_{1k}, y_{2k}, \ldots, y_{nk}$ could denote the pollen-accumulation rates of the kth taxon at n different times (the subscript k will sometimes be dropped, as investigation is often restricted to a single variable at a time). In the great majority of time series methodology, it is assumed that the observations are made at equal intervals of time. There are clearly difficulties in meeting this requirement with pollen stratigraphical data; discussion of the implications of this is presented in Subsections 6.2.3 and 6.2.4.

Two main modes of analysing time series are referred to as analysis in the *time domain* and analysis in the *frequency domain*. The main tool used in the time domain is the autocorrelation coefficient: the autocorrelation coefficient at lag l is simply the correlation coefficient among readings in the same sequence which are l time intervals apart, for example, among $(y_1, y_{l+1}), (y_2, y_{l+2}), \ldots, (y_{n-l}, y_n)$. This gives a measure of the similarity of readings which are separated by l time intervals. A plot of the sample autocorrelation coefficient for a range of values of l (a correlogram) can assist in the assessment of the behaviour of the variable over time; for example, one might be able to detect periodicities in its behaviour. One can also compare two different variables by use of a similar function, the cross-correlation coefficient, to detect patterns of temporal variation between variables.

In the frequency domain, the power spectrum of a time series gives an indication of the different frequencies of variation which account for most of the variability in the data, and can thus help one to detect periodicities within the data.

The use of these time series procedures in the analysis of pollen stratigraphical data is discussed in Subsection 6.2.3.

6.2.1 Sequence-Splitting and Curve-Fitting

In the first method of analysis described by Walker and Wilson (1978), the stratigraphical sequence for a pollen taxon consisting of n levels is represented as n presence/absence variables $A_1, A_2, \ldots, A_n$, where

$$A_i = 1 \qquad \text{if} \quad y_i > 0,$$

that is, if the taxon is present at the ith level; and

$$A_i = 0 \qquad \text{if} \quad y_i = 0,$$

that is, if the taxon is absent from the ith level. It is assumed that the A_i are independent random variables, and that their probability distribution, under the null hypothesis that the frequencies of presence and absence have not changed over time, is given by

$$\text{prob}(A_i = 1) = \theta \qquad (i = 1, 2, \ldots, n).$$

The alternative hypothesis is that after the tth observation, there is a change in the probability distribution of the presence/absence variables, that is,

$$\text{prob}(A_i = 1) = \theta_1 \qquad (i = 1, 2, \ldots, t),$$
$$\text{prob}(A_i = 1) = \theta_2 \qquad (i = t + 1, t + 2, \ldots, n).$$

To determine whether a significant change occurred after the tth observation, the likelihood ratio Λ_t is calculated, where

$$\Lambda_t = \frac{\hat{\theta}_1^{n_1}(1 - \hat{\theta}_1)^{t-n_1}\hat{\theta}_2^{n_2}(1 - \hat{\theta}_2)^{n-t-n_2}}{\hat{\theta}^{n_1+n_2}(1 - \hat{\theta})^{n-n_1-n_2}} \tag{6.2.1}$$

and

$$\hat{\theta} = n_1/t, \qquad \hat{\theta}_2 = n_2/(n - t), \qquad \hat{\theta} = (n_1 + n_2)/n,$$

$$n_1 = \sum_{i=1}^{t} a_i, \qquad n_2 = \sum_{i=t+1}^{n} a_i,$$

and $(a_1, \ldots, a_n)$ is the sequence of observed values.

If the suspected 'change point' t is specified before the data are examined, one can compare the value of 2 log Λ_t with a chi-squared distribution on one degree of freedom, large values of the test statistic leading to rejection of the null hypothesis of no change in θ, in favour of the alternative hypothesis that θ changed between the tth and $(t + 1)$th observations. Since t is not known, the maximum value of 2 log Λ_t $(t = 1, 2, \ldots, n - 1)$ is compared with the χ_1^2 distribution at a more stringent significance level (i.e., probability of rejecting the null hypothesis when it is in fact true); D. Walker and Wilson (1978) use a significance level of $0.05/n$.

If a statistically significant split is found, the sections are examined in turn for further presence/absence splits. In these tests, the value of n in expression (6.2.1) now refers to the length of the section in which the test is being conducted.

D. Walker and Wilson (1978) then examine the quantitative data (y_i) within each section in which the taxon is effectively present, conducting a similar test to the one described above. In this case, however, the null hypothesis is that the pollen-accumulation rates in a section are independent observations from a normal $N(\mu, \sigma^2)$ distribution, with mean μ and variance σ^2, namely, that their probability density function is

$$f(y) = \frac{1}{\sqrt{2\pi}\sigma} \exp -[(y - \mu)^2/(2\sigma^2)].$$

The alternative hypothesis is now that the sequence changes after t observations, and that readings up to the tth are independent observations from a normal $N(\mu_1, \sigma_1^2)$ distribution, whereas readings after the tth are independent observations from a normal $N(\mu_2, \sigma_2^2)$ distribution. As before, one can readily test the null hypothesis for a pre-specified value of t. D. Walker and Wilson (1978) conducted a simulation study to suggest values with which the maximum value of the likelihood ratio test statistic should be compared when t is not specified beforehand but is indicated by the data.

If a statistically significant split is found, sections containing four or more levels are in turn analysed for further significant splits. The process is continued until no more significant splits are found. In both methods of analysis, the divisions result in a hierarchy of splits, referred to as first-order splits, second-order splits, and so forth. Although the resulting sections have the same relative position within the hierarchy of splits, there is no correspondence between sections resulting from splits of the same order (see D. Walker and Pittelkow, 1981).

When the sequences for all the pollen taxa have been analysed, the total number of significant presence/absence and quantitative splits between given levels can be used as an index of the amount of population change occurring at about that time (D. G. Green, 1982).

The mean, standard deviation, and mean/standard deviation ratio are then calculated for each section in which the taxon is effectively present. D. Walker and Pittelkow (1981) discuss the use of the mean/standard deviation ratio as a measure of the amount of variation in pollen values within and between sections. They also suggest possible ecological interpretation of these ratios in terms of environmental factors, population dynamics, and biological interactions.

The final stage in D. Walker and Wilson's (1978) approach (see also D. Walker and Pittelkow, 1981) is to fit the data points within sections by linear, quadratic, or cubic functions of time using least-squares regression procedures, in an attempt to characterise and quantify any temporal trends within sections. However, for many data sets the number of samples or data points in each section may be too small to justify the fitting of any polynomial equations.

An alternative method of modelling trends within stratigraphical data is by piecewise regression models (McGee and Carleton, 1970; Hawkins, 1976b; Gordon, 1982b). In this approach, the division of the stratigraphical sequence into sections, and the modelling within each section is conducted in a single analysis. This method, like D. Walker and Wilson's (1978) curve-fitting, will, in general, lead to discontinuities in the curves at the boundaries between sections. There are also more elaborate curve-fitting procedures which lead to continuity and a specified amount of smoothness at the boundaries, for example, spline functions (Hayes, 1974).

By contrast, Watts (1973) suggested that specific population growth models could be fitted to detailed stratigraphical pollen-accumulation data for the time interval (generally 250–1000 years) of tree arrival and subsequent population expansion to asymptotic pollen (and assumed tree population) values. This requires reliably dated and closely spaced pollen counts, such as can be obtained from annually laminated lake sediments. Tsukada (1981, 1982a, 1982b, 1982c, 1983), Tsukada and Sugita (1982), and Bennett (1982, 1983c) have adopted this approach, and fitted exponential and logistic population growth models to a variety of pollen stratigraphical data.

If y denotes the population size (measured as pollen grains cm^{-2} $year^{-1}$), the exponential and logistic models can be derived from simple assumptions about the rate of increase of y with time, dy/dt. In the exponential model, it is assumed that dy/dt is proportional to y, that is,

$$\frac{dy}{dt} = ry,$$

where r is the intrinsic rate of population growth per unit time. This can be integrated, to give

$$\log y = rt + a, \qquad \text{or} \qquad y = \exp(rt + a),$$

where a is a constant of integration. Bennett (1983c) plotted log y against t (obtained from radiocarbon dating) for several tree taxa, estimating r from the slope of the best-fitting straight line (see also Tsukada, 1982c).

One would not expect y to be able to increase exponentially without limit; an upper carrying capacity K for the environment can be incorporated into the logistic model, in which

$$\frac{dy}{dt} = by(K - y),$$

where b is a constant. This can be integrated, to give

$$\log[(K - y)/y] = c - bKt,$$

where c is a constant of integration. If an estimate of K is available, one can obtain estimates of b and c by plotting $\log[(K - y)/y]$ against t. However, the assumption that K and b are constant through time is a considerable one (see Tsukada, 1982a, 1982b; Tsukada and Sugita, 1982).

Tsukada (1981, 1982a, 1982b, 1982c, 1983), Tsukada and Sugita (1982), and Bennett (1982, 1983c) have fitted these models to pollen stratigraphical data for a variety of taxa in Japan, North America, and eastern England. Tsukada and Sugita (1982) and Tsukada (1983) have also fitted a logistic model to phases of declining pollen values in an attempt to compare rates of decline with intrinsic rates of expansion.

Further discussion of the applicability of the methods of analysing pollen stratigraphical data described in this subsection is postponed to Subsection 6.2.4.

6.2.2 Applications of Sequence-Splitting

As an example of sequence-splitting, D. Walker and Wilson's (1978) algorithm has been applied to the pollen-accumulation rates of the nine major pollen taxa in the Abernethy Forest 1974 data set (see Section 1.4 and Figure 1.2). The pollen-accumulation rates of these taxa are plotted in Figure 6.2 against estimated sample age in radiocarbon years B.P. (from H. H. Birks and Mathewes, 1978). These plots are similar in design to stratigraphical pollen diagrams except that sample depths and ages are plotted along the horizontal axis and pollen-accumulation rates along the vertical axis (see M. B. Davis, 1981 and Ritchie, 1982, for examples of comparable plots). This representation is used here in an attempt to portray the pollen-stratigraphical data in a form similar to that used by population ecologists to present time series of populations. Significant presence/absence splits are sug-

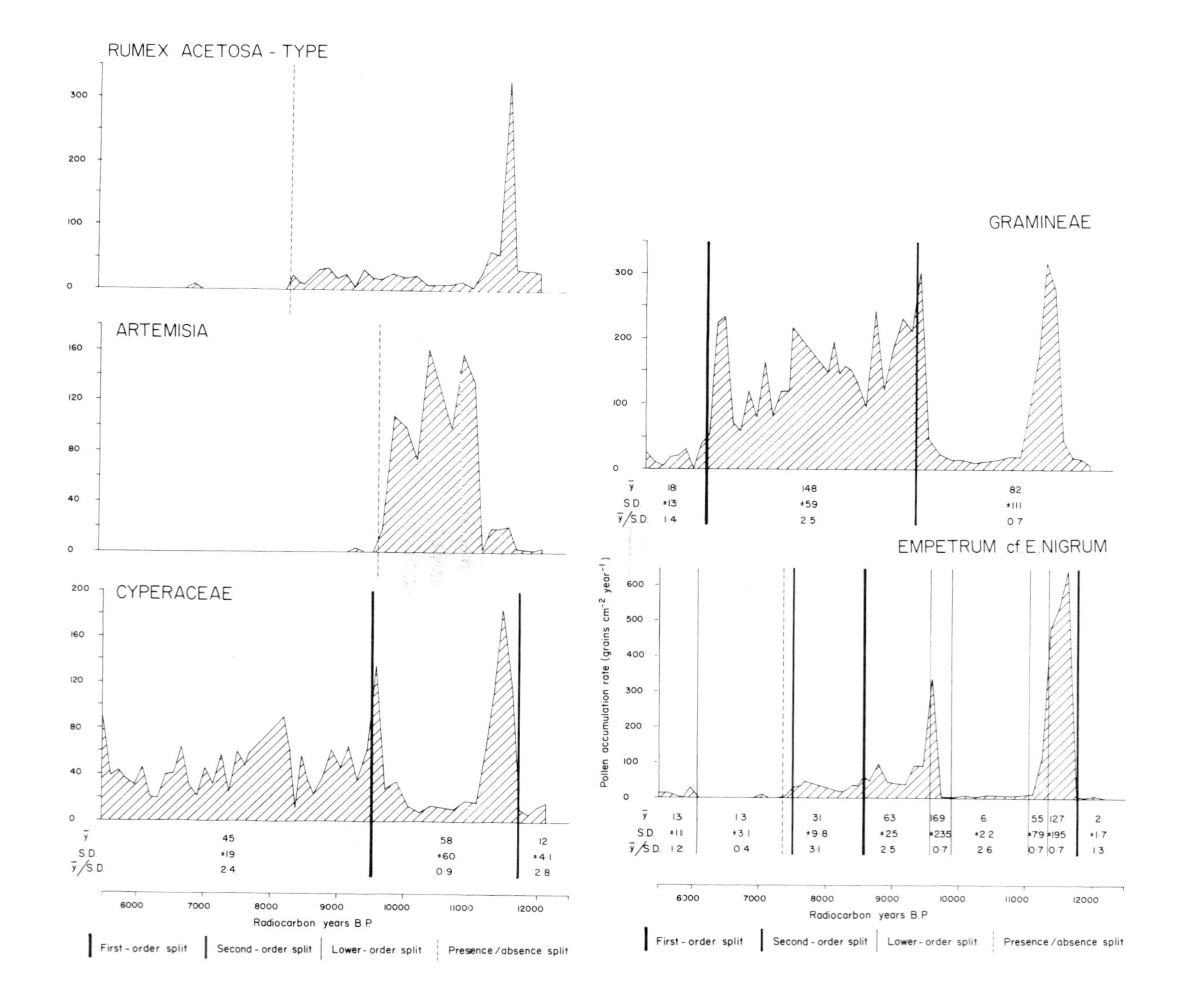
RUMEX ACETOSA - TYPE
ARTEMISIA
CYPERACEAE
GRAMINEAE
EMPETRUM cf. E. NIGRUM
Pollen accumulation rate (grains cm⁻² year⁻¹)
Radiocarbon years B.P.
First - order split
Second - order split
Lower - order split
Presence / absence split

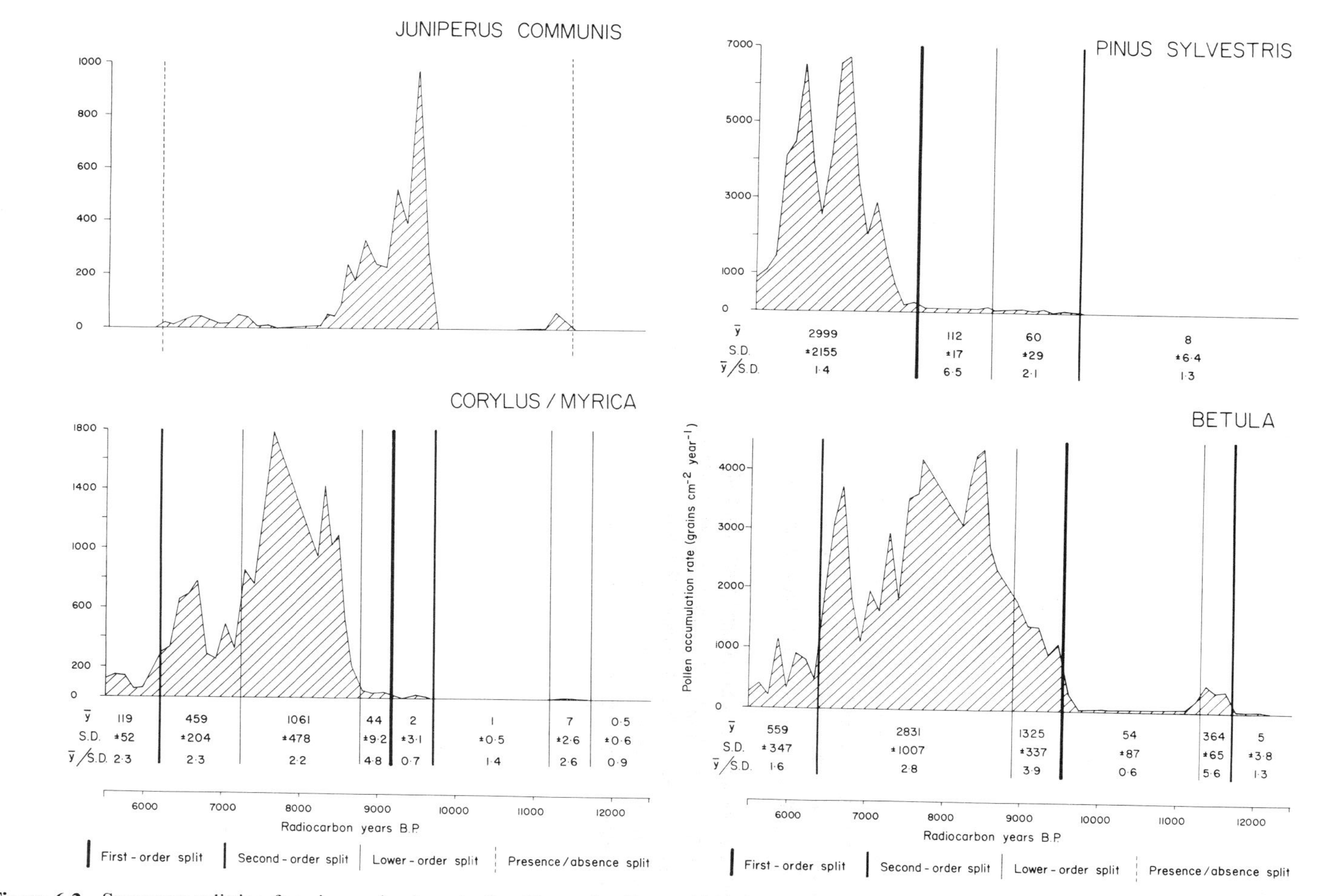

Figure 6.2 Sequence-splitting for nine major taxa in the Abernethy Forest 1974 data set. The curves are of pollen-accumulation rates plotted against radiocarbon years B.P. The first-order, second-order, lower-order, and presence/absence splits are indicated. For each section the mean ($\bar{y}$), standard deviation (S.D.), and mean/standard deviation ratio are shown.

gested for the sequences of *Juniperus*, *Empetrum* cf. *E. nigrum*, *Rumex acetosa*-type, and *Artemisia* pollen. These splits are shown as vertical dashed lines on Figure 6.2. Several statistically significant quantitative splits are suggested by the results of the sequence-splitting algorithm. The first-order, second-order, and lower-order splits are shown on the relevant pollen plot. The mean, standard deviation, and mean/standard deviation ratio for each of the quantitative sections are also shown on Figure 6.2. The sequences for *Betula*, *Pinus*, *Corylus/Myrica*, and *Empetrum* cf. *E. nigrum* pollen are split into several sections with distinct means and standard deviations, whereas the Gramineae and Cyperaceae sequences are only split into three sections.

The sequence-splitting results for individual taxa such as *Betula* clearly reflect major quantitative changes within the time series of birch pollen. There is a phase of low pollen values between 11,200 and 9500 B.P., increased but variable values as tree birches expanded into the area between 9500 and 8900 B.P., high but variable values when birch was abundant or even dominant in the pollen catchment of the Abernethy Forest site between 8900 and 6400 B.P., and rapidly declining values as other tree taxa became established and expanded locally, and as the birch population collapsed.

The stratigraphical and hence temporal patterns of these sequence-splitting results are summarised in Figure 6.3 for the nine taxa shown in Figure 6.2. In addition, the total number of statistically significant presence/absence and quantitative splits between levels is shown for all 63 pollen types included by H. H. Birks and Mathewes (1978) in their basic pollen sum. The optimal sum-of-squares partition for six groups of the pollen percentage data (Section 3.7 and Figure 3.8) is also shown, this partition being based on analysis of the proportions of the same nine taxa shown in Figure 6.2. The sequence-splitting results for all taxa indicate marked concentrations of significant splits, for example, between 9750 and 9600 B.P. (levels 34–33), 11,380 and 11,250 B.P. (levels 43–42), and 11,750 and 11,600 B.P. (levels 46–45) (see Figure 6.3). If the total number of splits between levels is an indication of the amount of population change occurring within the pollen-source area of the Abernethy Forest site (D. G. Green, 1982), these concentrations or clusterings of splits suggest that populations of several different taxa all underwent changes in pollen production, or in population size, performance, or distribution at these times. These temporarily linked responses suggest that the populations of different taxa may have all been limited by strong, overriding environmental controls and that major environmental changes at certain times may have resulted in synchronous population changes (cf. D. Walker, 1982b). This interpretation is consistent with independent palaeoecological records of rapid and large climatic and associated environmental change during the Devensian late-glacial (10,000–13,000 B.P.) in Scotland (Sissons, 1979a, 1979b).

Besides linked responses, there are also periods at Abernethy Forest in which splittings are not clustered to the same extent (see Figure 6.3), particularly between 9500 and 6000 B.P. These suggest that during this time different taxa had a predominantly individualistic behaviour, with populations exhibiting significant

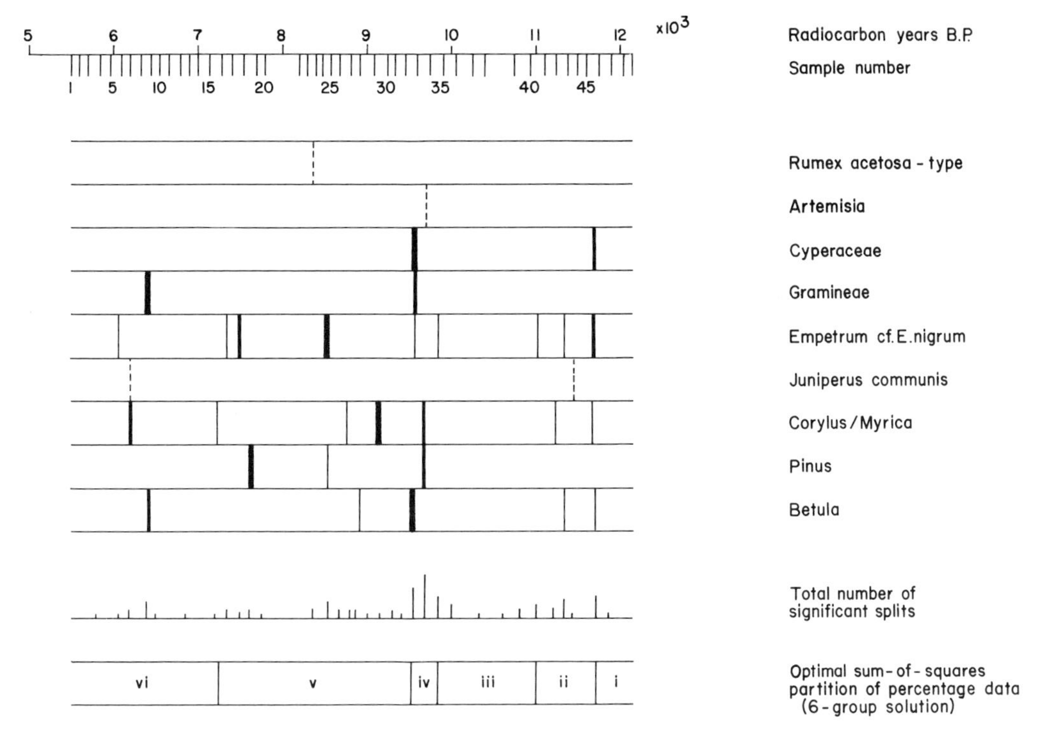

Figure 6.3 Positions of splits for nine major taxa in the Abernethy Forest 1974 data set: first-order split (▌), second-order split (▍), lower-order split (|), presence/absence split (⁞). The total number of significant splits for all taxa and the optimal sum-of-squares partition of the percentage data (six-group solution) are also shown.

expansion or decline independently of each other. The interpretation of these and other sequence-splitting results in terms of linked and individualistic behaviour of populations in the late-glacial and post-glacial is explored more fully by H. J. B. Birks and J. M. Line (in preparation).

There is some correspondence between the numerical zonation based on pollen percentage data of the nine major taxa and the total number of significant splits within the pollen-accumulation rates of all the dry-land pollen taxa (see Figure 6.3). This correspondence is particularly well marked in the lower part of the sequence (9500–12,150 B.P.). The correspondence is less good in the early post-glacial, where there are many marked but short-lived changes in pollen percentages and accumulation rates of certain taxa (see Figures 1.2 and 6.2).

There are, as yet, comparatively few published applications of the use of sequence-splitting as an aid to the interpretation of pollen stratigraphical data. D. Walker and Wilson (1978) illustrated their methods with data from a site in New Guinea. They fitted quadratic and cubic curves to some of the data points within sections, and presented tentative ecological interpretations of these trends.

D. Walker and Pittelkow (1981) analysed three published sets of data using the sequence-splitting approach. The data were from a Finnish post-glacial site studied by Donner (1972), a late-glacial site, Wolf Creek, in central Minnesota (H. J. B. Birks, 1976b), and a post-glacial sequence from Lake Victoria in East Africa (R. L. Kendall, 1969). The time series from the late-glacial at Wolf Creek interestingly shows markedly linked population changes, as in the late-glacial at Abernethy Forest. The post-glacial series from Lake Victoria shows strongly individualistic patterns of splits, whereas the Finnish post-glacial series are intermediate in the temporal patterns of splits within the time series examined.

H. J. B. Birks and J. M. Line (in preparation) have analysed several data sets from late- and post-glacial sequences in Britain and eastern North America. Sequence-splitting results consistently indicate strongly clustered splits in late-glacial sequences, but, with the exception of one post-glacial site in Michigan, all post-glacial sequences exhibit a non-clustered pattern of splitting. D. G. Green (1982) has used sequence-splitting in conjunction with other time series procedures to analyse very detailed, close-interval data from a lake in Nova Scotia. The sequence-splitting results provide insights into the temporal frequency of population change and into questions of long-term vegetational stability and dynamics, and rapid vegetational change following forest fires (see Subsection 6.2.3).

6.2.3 Other Methods of Analysing Time Series

D. G. Green (1981) has pioneered the detailed application of time series procedures as an aid to the interpretation of quantitative pollen stratigraphical data (see also I. Williamson, 1981; D. G. Green, 1982, 1983). D. G. Green (1981) conducted cross-correlation, autocorrelation, and power-spectral analysis of close-interval stratigraphical data on pollen-accumulation rates from a lake in Nova Scotia. Previous applications of these procedures in Quaternary palynology include studies by P. S. Martin and Mosimann (1965), who calculated the autocorrelation of certain pollen percentage curves, in an attempt to characterise the variation to be expected within and between pollen assemblage zones; Kutzbach and Bryson (1974), who computed power spectra from pollen data in an analysis of post-glacial climatic changes; Nichols, Kelly, and Andrews (1978), who calculated power spectra from the values of presumed long-distance dispersed pollen types in post-glacial deposits on Baffin Island as a means of reconstructing changing airflow patterns and investigating the periodicity of these changes over the last 6000 years (cf. Barry, Elliott, and Crane, 1981).

The data that D. G. Green (1981, 1982) analysed were collected by him specifically with the aim of obtaining equal time intervals between samples, an assumption inherent in the use of most time series methods of analysis. Green discusses the various problems involved in attempting to obtain such samples. Small variations in inter-sample time intervals are generally not a problem although they may result in some smudging of peaks in power spectra. Large variations, however, make the results meaningless (cf. I. Williamson, 1981). It can also be crucial to obtain fine-resolution samples. Thus, D. G. Green (1983) obtained simulations of the pollen-accumulation rates under several models of vegetational change, and investigated the effect of altering the sampling intensity and other parameters of the models. To be able to detect certain types of behaviour in the pollen record, close-interval sampling is essential. A second important assumption is that the data are stationary, namely, that the time series contain no trends in their means or variances. Although non-stationarities, such as long-term trends in means or variances, in time series can be removed by detrending (Chatfield, 1980), D. G. Green (1981) concentrated his statistical analyses on one near-stationary section of post-glacial data (4450–2100 years B.P.), for which the sampling interval is about 50 years between samples. A third assumption is that the data are normally distributed about the series mean. Although some such distributional assumption is necessary to enable statistical testing of the significance of values obtained for cross-correlation and autocorrelation functions and power spectra, general analyses in the time and frequency domains can proceed without the necessity for such assumptions (in much the same way that principal components analysis can be deployed as a data exploration technique, without any use of significance testing). In practice, many time series procedures are reasonably robust to small departures from the assumption of normality (Platt and Denman, 1975).

D. G. Green (1981) conducted time domain and frequency domain analyses of individual time series of charcoal, inorganic material, and various pollen types, all expressed as absolute accumulation rates. These analyses indicated a 350-year cycle in frequencies of widespread forest fires. In addition, Green found a 100-year lag between charcoal and inorganic material, suggesting that erosion and in-washing of minerogenic material occurred for up to 100 years after a widespread fire (cf. Swain, 1973; Cwynar, 1978). Trees such as *Larix, Picea, Pinus, Ulmus,* and *Salix* showed significant cross-correlations between their pollen-accumulation rates and charcoal at 50-year lags, whereas *Abies, Acer, Ostrya/Carpinus,* and *Fagus* have significant cross-correlations with charcoal with 350- to 500-year lags. These differences in lag times were interpreted as reflecting the different responses of pioneer and non-pioneer tree taxa following disturbance by fire. From the cross-correlations between individual pollen types and charcoal, and between different pollen types, Green proposed a model of post-fire succession for the pollen catchment of his site, and suggested different frequencies of fire during the mid- and late-post-glacial. These differences in fire frequency may result, in part, from climatic change (e.g., Swain, 1973, 1978; I. C. Prentice, 1983b) and, in part, from changes in forest composition and in the mosaic structure of the forests, related to factors such as aspect, slope, drainage, and soil type (e.g., Swain, 1980; D. Walker, 1982a). The model of post-fire succession that Green proposes for Nova Scotia suggests considerable variability in the possible routes of vegetational succession (see D. Walker, 1970; Horn, 1976; Van Hulst, 1979, 1980; Gittins, 1981, for other examples of variability in successional pathways).

Although there are several, as yet unresolved, problems in the interpretation of time series of charcoal in relation to fire frequencies (see Swain, 1973, 1978,

1980; Cwynar, 1978), and in reconstructing the mosaic structure of forests from a single pollen diagram (see Andersen, 1978a; Jacobson, 1979; Jacobson and Bradshaw, 1981; Ritchie, 1981), D. G. Green's (1981) study is important in indicating the potential for applying, in selected instances, time series methods of analysis to detailed pollen stratigraphical data that have equal time intervals between adjacent samples. With the exception of pollen data from annually banded sediments (e.g., Swain, 1973, 1978; Cwynar, 1978; Tolonen, 1978), such data are very rare and extremely difficult to obtain.

D. G. Green (1982) has also combined sequence-splitting and time series analyses of stratigraphical palynological and charcoal data. He calculated cross-correlation coefficients between the absolute amount of charcoal in a level, and the number of taxa whose sequences are divided after this level by D. Walker and Wilson's (1978) sequence-splitting procedure. Interestingly, high correlations at a lag of 50 or 100 years occur, suggesting that when charcoal influx is high, there are many splits after the stratigraphical levels 50 or 100 years later. When charcoal influx is low, however, there are few significant splits in the overlying levels. These cross-correlations suggest the importance of major fires and associated disturbance in triggering population changes, in permitting the invasion of new species into the area (cf. Watts, 1973; Drury and Nisbet, 1973), and thus in influencing long-term vegetation dynamics (see D. Walker, 1982a). D. G. Green (1982) compared power spectra of different taxa for different sections as a means of tentatively reconstructing the vegetational mosaic and of inferring temporal changes in the mosaic structure of the forests. These comparisons and the cross-correlations between sequence splits and charcoal generate new and important hypotheses about the possible importance and interactions of climatic change, plant migration, competition, and fire on forest dynamics, vegetational change, and resilience over long time periods that can now be tested by detailed pollen analysis of small sites critically positioned in contrasting ecological settings within the present-day vegetational mosaic whose history is of interest (e.g., Jacobson, 1979; Ritchie, 1981; Jacobson and Bradshaw, 1981; I. C. Prentice, 1982a). Given the complex nature of the data, it would have proved difficult, if not impossible, to generate such hypotheses without recourse to numerical methods of analysis.

6.2.4 Discussion

The methods of analysis described in this section make strong assumptions about the data analysed. The two main assumptions, one or both of which are necessary, are as follows: firstly, that statistically reliable and ecologically informative data on pollen-accumulation rates are available; secondly, that the time associated with each datum point can be accurately determined.

Errors inherent in the estimation of pollen-accumulation rates are discussed in Section 1.2. The assumption that observed pollen-accumulation rates are a true reflection of the population size of the taxa in the pollen catchment area is a considerable one. The validity of this assumption will vary not only from site to site (e.g., M. B. Davis *et al.,* 1973; Pennington, 1973, 1979; Hyvärinen, 1975, 1976; Bonny, 1978) but also within a site (e.g., Waddington, 1969; M. B. Davis *et al.,* 1973; Lehman, 1975; Likens and Davis, 1975; Pennington, Cambrey, Eakins, and Harkness, 1976; H. J. B. Birks, 1976b, 1981a; Bonny, 1978; Pennington, 1979; Haworth, 1980; M. B. Davis and Ford, 1982; Bennett, 1983b). At present, there is no entirely satisfactory quantitative way (cf. Lehman, 1975) of distinguishing between observed changes in pollen-accumulation rates that result from sediment focussing and other site-dependent processes from those that result from changes in population size, other than by studying several cores from the same lake (e.g., M. B. Davis and Ford, 1982; Bennett, 1983b). Sequence-splitting is only appropriate, however, if sediment focussing and changes in pollen-recruitment processes and sedimentation have not occurred at the site during the time period of interest.

If these limitations are borne in mind, D. Walker and Wilson's (1978) sequence-splitting procedure can, in some instances, aid interpretation. One advantage is that it retains information that may be lost in conventional zoning of pollen stratigraphical data. Pollen-zone boundaries, delimited either by inspection or by numerical partitioning of quantitative data, tend in practice to be located at or near major stratigraphical changes in the numerically most abundant pollen taxa. These boundaries can overemphasise the changes in the relatively few abundant taxa and underemphasise the changes in the large number of less abundant taxa (D. Walker and Pittelkow, 1981; D. G. Green, 1982), thereby providing a potentially biased summary of the patterns of stratigraphical change within a pollen diagram. Sequence-splitting, on the other hand, summarises all the changes within individual pollen sequences and can show whether the changes among sequences are linked in time or not. It does not assume any order or synchroneity of change among taxa, and thus allows the hypothesis of individualistic behaviour of taxa through time to be examined critically (see Livingstone, 1969). Theophrastus is said to have commented 'There is disorder in the universe and order must be proved, not assumed' (McIntosh, 1980, p. 215). Sequence-splitting provides a means of examining the hypothesis of disorder in vegetational history.

Sequence-splitting can also help in isolating periods of change for certain taxa from periods of little change for other taxa. In contrast, pollen zones are commonly viewed and frequently interpreted as periods of more or less stable vegetational composition (D. Walker, 1966, 1982b; Watts, 1973). The fact that many statistically significant splits frequently occur within pollen zones (see Figure 6.3; D. Walker and Pittelkow, 1981; D. G. Green, 1982) indicates that important and numerically large changes in population size and structure are occurring continually in time, as proposed by Gleason (1939). Sequence-splitting is thus clearly of value in analysing pollen stratigraphical data as long-term records of plant populations through time viewed individualistically and collectively. It can also assist in formulating new hypotheses about ecological processes, vegetation dynamics, and population changes at a variety of temporal and spatial scales. Such hypotheses may be more difficult to formulate if the primary pollen stratigraphical data are only partitioned into pollen zones (cf. the ecological interpretations of the Wolf

Creek data of H. J. B. Birks, 1976b, using pollen zones with the results of D. Walker and Pittelkow, 1981, using sequence-splitting).

It is, however, pertinent to warn against excessive claims being made for the 'statistically tested accuracy' with which sequence-splits have been located. As D. Walker and Wilson (1978) remark, the confidence interval for the 'change point' t can be very wide, particularly if there are few observations in the sequence.

The procedure of splitting a stratigraphical sequence into sections does not make any assumptions about inter-sample time intervals, but a reliable time control is essential for D. Walker and Wilson's (1978) and Watts's (1973) curve-fitting approaches and for the time series applications described in Subsection 6.2.3. If there are inaccuracies in the measurement of the time variable, higher-degree polynomial, exponential, and logistic curves, and spline functions can be considerably affected, particularly if few data points are available for fitting the model. It is unlikely to be appropriate to fit the more elaborate models unless the data are plentiful and reliable.

Strictly speaking, time domain and frequency domain analyses do not require the inter-sample time intervals to be constant throughout the series, as one can obtain data which are regularly spaced in time by interpolation. This is of little practical assistance in the analysis of pollen stratigraphical data, however, as the time resolution is generally relatively imprecise, unless the data are derived from annually laminated sediments. Saarnisto (1979b) describes techniques for detailed sampling of annually laminated lake sediments. As noted earlier, slight inaccuracies in measuring the time variable will tend to obscure features of the data, although marked periodicities may only be blurred rather than totally hidden.

It is important to be aware of the limits of the resolution of one's data (D. G. Green, 1983): the pollen rain from several seasons is spatially and temporally smoothed by the processes of pollen dispersal and sedimentation, and the size of sample required for pollen analysis often smooths the data over a longer time period. D. G. Green (1981) estimated that these factors prevented detection of periodicities of less than 75 years duration in his data. This feature indicates the importance in such studies of very precise sampling procedures such as those developed by Swain (1973) and Cwynar (1978), elaborate and time consuming though these may often be. Schindel (1980) and T. Webb (1982) discuss the problems of detailed sampling in palaeoecology and pollen analysis and outline the possible limits of temporal resolution of biostratigraphical data.

A further problem is the frequent paucity of data points within stationary sections. Chatfield (1980, Section 7.8) discusses the size of data set required to obtain a reliable estimate of the power spectrum. Some authors have recommended that 100 or 200 observations are necessary, but if the data are suitably transformed, Chatfield reports that reasonable estimates can be obtained even with fewer than 100 observations. Less clearly marked features of the data will, of course, require a larger data set for detection than will more evident features.

While it will sometimes be worthwhile expending the very considerable time and effort necessary to obtain reliable and adequate data which can be usefully analysed by time series procedures, our conclusions are that such methods of analysis are of limited applicability to the interpretation of the majority of *currently available* Quaternary pollen stratigraphical data. These remarks are not intended to discourage detailed investigations, such as those conducted by D. G. Green (1981, 1982, 1983), but rather to stress the very great care in sampling design and the immense amount of labour needed to acquire suitable data and to bring such studies to a successful conclusion.

6.3 The Use of Pollen-Representation Factors

In Section 5.4 we discussed the estimation of two types of modern pollen-representation factors: R values estimated from contemporary pollen and vegetation data collected at one or, preferably, several sites, and representation factors that incorporate a background pollen component, estimated from modern pollen and vegetation data at several sites. In this section we consider the use of pollen-representation factors as aids in the interpretation of pollen stratigraphical data, particularly in the estimation and reconstruction of former plant abundances and population sizes (see Figure 6.1). We assume that we have obtained reliable estimates of modern pollen-representation factors based on modern pollen and vegetation data using the methods discussed in Section 5.4, that the factors are invariant in time and space, and thus that the factors can be applied to fossil pollen spectra to estimate the abundance and population size of different taxa in the past (Figure 1.1).

6.3.1 The R-Value Model

We consider first the model in which the modern representation factors are R values (*sensu* M. B. Davis, 1963), estimated from data obtained from one or several sites. Thus, $\hat{R}_k$ is an estimate of the ratio of the pollen proportion to the vegetation proportion for the kth taxon. Given fossil pollen data (p_{ik}), where p_{ik} denotes the proportion of the ith stratigraphical sample which comprises pollen of taxon k, we can transform these data as follows:

$$\hat{f}_{ik} = (p_{ik}/\hat{R}_k) \Big/ \sum_{l=1}^{t} (p_{il}/\hat{R}_l). \qquad (6.3.1)$$

In this equation, $\hat{f}_{ik}$ denotes the transformed proportion of pollen of taxon k in sample i; in other words, the estimated proportion of taxon k in the vegetation at the time represented by sample i within the same fixed area from which the modern vegetation data were collected (see Figure 1.1). An example is given in

Table 6.1

Hypothetical Example of the Use of Modern R Values to Transform Fossil Pollen Proportions

Taxon	Fossil pollen proportion p	R value relative to taxon 6 $\hat{R}$	$p/\hat{R}$	Transformed pollen proportion (estimated vegetation proportion) $\hat{f}$
1	0.35	10	0.0350	0.157
2	0.25	5	0.0500	0.225
3	0.18	4	0.0450	0.202
4	0.10	8	0.0125	0.056
5	0.08	2	0.0400	0.180
6	0.04	1	0.0400	0.180
			0.2225	

Table 6.1. The form of Equation (6.3.1) is such that the same estimated vegetational proportions are obtained if either R values or relative R values are used as $\hat{R}_k$.

If fossil pollen-accumulation rates y_{ik} are available, where y_{ik} is the estimated accumulation rate (grains cm^{-2} year^{-1}) of taxon k in the ith sample, they can be transformed into estimates of the absolute abundance x_{ik} of taxon k at time i within the same area used to estimate α_k, as defined in Equation (5.4.1), by

$$\hat{x}_{ik} = y_{ik}/\hat{\alpha}_k$$

(see Ritchie, 1982, for an example, discussed at the end of this subsection). If, however, one's estimate of representation factors is $\hat{R}_k$, based on relative pollen and vegetation data, then the transformation

$$\tilde{x}_{ik} = y_{ik}/\hat{R}_k$$

provides estimates $\tilde{x}_{ik}$ which are proportional to the absolute abundance of taxon k at time i (see Donner, 1972, for an example).

In applying R-value estimates to pollen stratigraphical data, we are assuming that these representation factors are invariant in time and space. Parsons and Prentice (1981) discuss this assumption in detail. They emphasise that the geographical, and hence the stratigraphical, applicability of a given set of R values must be evaluated empirically. One approach is to apply them to independent modern pollen data and to predict vegetational composition around the sites where the independent data were collected. Parsons *et al.* (1980) discuss other means of studying the geographical applicability of R values. As Parsons and Prentice (1981) and I. C. Prentice (1982a) emphasise, R values should not be applied to fossil pollen spectra that differ markedly in their composition from the modern pollen assemblages used to estimate the R values. It is thus essential to compare fossil pollen assemblages with modern pollen spectra first; only if a satisfactory match can be found should the appropriate R-value estimates, derived from the matching group of modern spectra, be applied to the fossil data. The R values should not be used to transform fossil spectra that lack a modern analogue. We discuss methods of comparing modern and fossil pollen spectra in the next section. This comparative stage prior to the use of R values is particularly important because of the effects of background pollen resulting from pollen transport from outside the area from which the modern vegetation data were collected (cf. M. B. Davis, 1963). If background pollen is present, but a background component is not included in the model, taxa having low local pollen and vegetation percentages will erroneously be assigned relatively large R values (e.g., Janssen, 1967a; Livingstone, 1968; Parsons and Prentice, 1981). In addition, there are probably important geographical variations in pollen representation related to factors such as forest age, structure, and density (M. B. Davis *et al.*, 1973; Parsons *et al.*, 1980; Andersen, 1980c; Bradshaw, 1981a), regional climate (Parsons *et al.*, 1980), local forest conditions (Andersen, 1970), human activity (Parsons *et al.*, 1980; Bradshaw, 1981a), pollen recruitment processes (Jacobson and Bradshaw, 1981), and vegetational composition. The latter is particularly important, because different species within a genus, all with morphologically similar pollen, may have different pollen productivities (e.g., *Pinus;* see Pohl, 1937; M. B. Davis *et al.*, 1973; Brubaker, 1975). Comparison of modern and fossil assemblages prior to the use of R values can help to minimise some of the effects of these sources of variability on R-value estimates and on the resulting reconstructions of past plant abundances and population sizes (Parsons and Prentice, 1981).

Andersen (1980c) has addressed the problem of whether pollen-representation factors are invariant in time. By comparing early post-glacial pollen concentrations with transformed pollen percentages from a small hollow (15 m in diameter) in Denmark, he demonstrated that there is no reason to reject the hypothesis that his (1970) modern representation factors, when applied at the same spatial scale at which they were estimated, are invariant with time.

The R values, estimated in various ways, have been used to transform pollen stratigraphical data in a wide variety of ecological and palaeoecological situations and at various spatial and temporal scales. For example, Iversen (1964, 1969), Andersen (1973, 1975, 1978a, 1980c), Stockmarr (1975), C. A. Baker, Moxey, and Oxford (1978), Bradshaw (1978, 1981a, 1981b), M. E. Edwards (1980), H. J. B. Birks (1982b), and Aaby (1983) have used representation factors to transform relative pollen stratigraphical data from mor humus profiles and small hollows (10–30 m in diameter) within woodlands into relative estimates of tree composition within a 20–30-m radius of the sampling site, usually estimates of basal area, tree volume, or crown area percentages. Iversen (1949, 1952–1953), Mikkelsen (1949, 1963), Tsukada (1957), Smith (1958, 1961), Curtis (1959), Smith and Willis (1961–1962), M. B. Davis (1963, 1965b), Livingstone and Estes (1967), Livingstone (1968), Donner (1972), Vuorinen and Tolonen (1975), Andersen (1978a), Aalto, Taavitsainen, and Vuorela (1980), Huttunen (1980), Ritchie (1982), and

Huntley and Birks (1983) have applied representation factors to transform pollen stratigraphical data from medium- or large-sized lakes or bogs into estimates of former plant abundance within a fixed, but often unspecified, area around the lake or bog, usually estimates of basal area or growing-stock volume percentages. When pollen-accumulation rates are transformed, the resulting estimates of past population size are proportional to absolute tree basal area or volume (Livingstone, 1968; Donner, 1972; Vuorinen and Tolonen, 1975), or to absolute plant cover (Ritchie, 1982).

An example of applying R values to post-glacial relative pollen stratigraphical data is shown in Figure 6.4. The data are from Vakojärvi, a medium-sized lake in

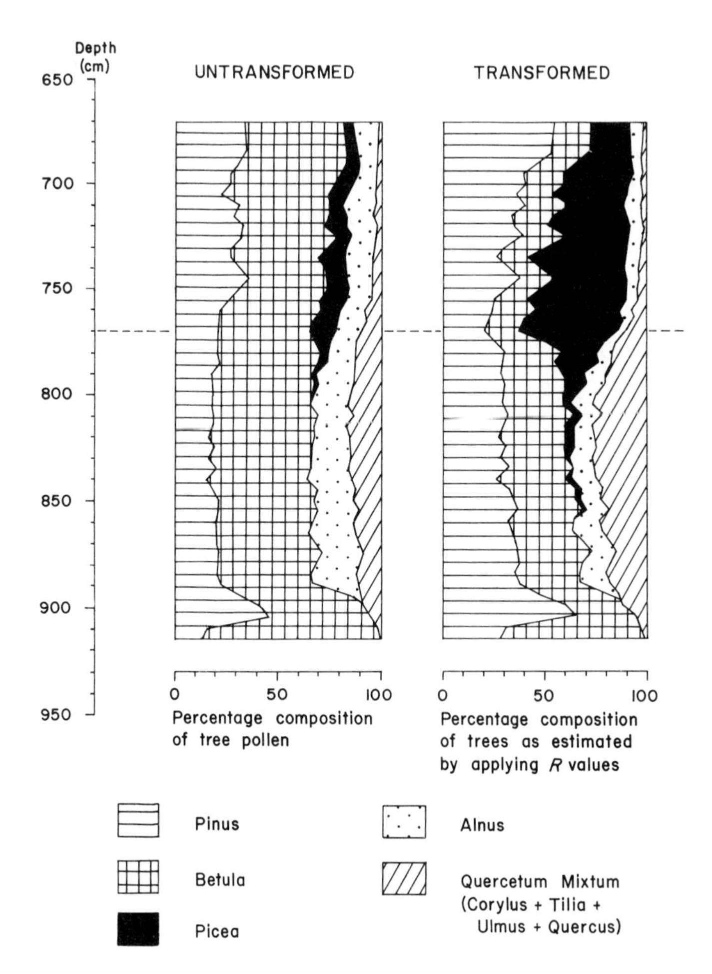

Figure 6.4 The use of pollen-representation factors in transforming pollen stratigraphical data at Lake Vakojärvi, southern Finland into estimates of relative tree composition. (After Donner, 1972.)

southern Finland (Donner, 1972). The relative abundances of the five major tree pollen types are shown. The effect of applying R values is to increase the percentages of *Pinus, Picea,* and Quercetum Mixtum (*Corylus* + *Ulmus* + *Tilia* + *Quercus*). These transformations, if valid, suggest a greater change in plant abundance and forest composition at about 770 cm (*ca.* 3500 B.P.) than is suggested by the untransformed pollen diagram.

In an application of the R-value model to the reconstruction of absolute past population sizes and total vegetational cover around a lake in the northern Yukon, Ritchie (1982) estimated modern representation factors (α_k in Equation (5.4.1)) from recent pollen-accumulation rates (y_{ik}) and the absolute areal coverage (x_{ik}) of the nine main pollen-producing taxa within two representative 2500-ha plots around the lake. These factors were then used to transform pollen-accumulation rates for the last 16,000 years into estimates of absolute cover for the nine taxa within an area of 5000 ha. For many samples, particularly those older than 7500 B.P., the total reconstructed plant cover was significantly less than 5000 ha. Ritchie inferred that the remaining area was bare ground, or was occupied by palynological 'blind spots' (*sensu* M. B. Davis, 1963, 1967a), such as non-pollen producing bryophytes and lichens or by very poor pollen producers such as low-growing herbs. In some samples, particularly those younger than 5000 B.P., the total reconstructed cover slightly exceeded 5000 ha, possibly because of errors in estimating the modern representation factors, or perhaps because of changes in pollen production with time (cf. Andersen, 1980c). Despite these limitations, Ritchie has made the first quantitative reconstruction of the progressive revegetation of a landscape following deglaciation. The use of representation factors to transform pollen-accumulation rates into estimates of population size and total plant cover is the only known way to estimate quantitatively the extent of unvegetated ground, although its existence may be inferred from pollen of 'indicator species' characteristic of bare-ground habitats (see e.g., H. J. B. Birks, 1973b). Ritchie's (1982) approach is of great potential in the detailed quantitative reconstruction of late-glacial vegetation and environments.

6.3.2 Models That Incorporate a Background Component

We now consider the use of pollen-representation factors that incorporate a background pollen component (see Subsection 5.4.4) to transform fossil pollen proportions (p_{ik}). If the parameters α and z have been estimated for I. C. Prentice and Parsons's (1983) model 1 (Equation (5.4.14), Subsection 5.4.4), an estimate of f_{ik}, the proportion of taxon k in the vegetation surrounding the site at the time corresponding to fossil spectrum i, is given by

$$\hat{f}_{ik} = \frac{(p_{ik} - \hat{z}_k)/\hat{\alpha}_k}{\sum_{l=1}^{t} (p_{il} - \hat{z}_l) \Big/ \hat{\alpha}_l}, \tag{6.3.2}$$

where $\hat{\alpha}_k$ is the estimated modern pollen-representation factor for taxon k, and $\hat{z}_k$ is the estimated constant background pollen proportion for taxon k. This transformation is illustrated in Table 6.2, in which the fossil pollen proportions of Table 6.1 are transformed using the estimated parameters of this more general model.

If the parameters α and z are estimated for I. C. Prentice and Parsons's (1983) model 2 (Equation (5.4.16), Subsection 5.4.4), the relevant transformation is

$$\hat{f}_{ik} = \frac{p_{ik}/\hat{\alpha}_k}{\sum_{l=1}^{t} (p_{il}/\hat{\alpha}_l)} \left(1 + \sum_{l=1}^{t} (\hat{z}_l/\hat{\alpha}_l)\right) - \hat{z}_k/\hat{\alpha}_k . \qquad (6.3.3)$$

This transformation is illustrated in Table 6.3 using data from Tables 6.1 and 6.2. These transformations can only be applied to relative pollen stratigraphical data.

In applying these transformations to fossil data, it is assumed that α_k and z_k are both invariant in time and space. These are major assumptions, particularly the latter one that z_k, the background component, has not changed (I. C. Prentice, 1982a). It represents the long-distance pollen component and is presumably some function of the vegetational abundance of taxon k on a broad, regional scale. As such abundances have undoubtedly changed in both time and space since the last glaciation (see, e.g., Bernabo and Webb, 1977; Huntley and Birks, 1983), it seems likely that z_k has also varied over 10,000 years (Parsons and Prentice, 1981). Subtraction of z_k from fossil pollen spectra should *only* be done if the fossil spectra lie within the range of the modern spectra used to estimate α_k and z_k (I. C. Prentice, 1982a). Thus, the fossil spectra must contain all the same taxa in similar proportions as in the modern spectra. Such precise matches are comparatively rare, particularly for fossil spectra older than about 5000 B.P. (I. C. Prentice, 1982a; cf. H. R. Delcourt, P. A. Delcourt, Webb, and Overpeck, 1983).

Table 6.2

The Use of Modern Pollen-Representation Factors Incorporating a Background Pollen Component Estimated for I. C. Prentice and Parsons's (1983) Model 1 to Transform Fossil Pollen Proportions

Taxon	Fossil pollen proportion p	Representation factors $\hat{\alpha}$	$\hat{z}$	$(p - \hat{z})/\hat{\alpha}$	Transformed pollen proportion (estimated vegetation proportion) $\hat{f}$
1	0.35	8	0.15	0.0250	0.115
2	0.25	4	0.05	0.0500	0.230
3	0.18	3	0.03	0.0500	0.230
4	0.10	8	0.0	0.0125	0.057
5	0.08	2	0.0	0.0400	0.184
6	0.04	1	0.0	0.0400	0.184
				0.2175	

Table 6.3

The Use of Modern Pollen-Representation Factors Incorporating a Background Pollen Component Estimated for I. C. Prentice and Parsons's (1983) Model 2 to Transform Fossil Pollen Proportions

Taxon	Fossil pollen proportion p	Representation factors $\hat{\alpha}$	$\hat{z}$	$p/\hat{\alpha}$	$\hat{z}/\hat{\alpha}$	Transformed pollen proportion (estimated vegetation proportion) $\hat{f}$
1	0.35	8	0.15	0.0438	0.0188	0.157
2	0.25	4	0.05	0.0625	0.0125	0.239
3	0.18	3	0.03	0.0600	0.0100	0.231
4	0.10	8	0.0	0.0125	0.0	0.050
5	0.08	2	0.0	0.0400	0.0	0.161
6	0.04	1	0.0	0.0400	0.0	0.161
				0.2588	0.0413	

Bradshaw (1978, 1981a) proposes certain restricted conditions under which these assumptions may be reasonable, and he has applied estimates of modern representation factors with a background component to the reconstruction at the local scale of tree populations around small hollows in southeast England (see also T. Webb *et al.*, 1981). Heide (1981) has adopted a similar approach in her work in Wisconsin, and has assumed constant background pollen percentages at two nearby sites for the last 3000 years. At a very different spatial and temporal scale, H. R. Delcourt and P. A. Delcourt (1984) have presented estimates of the percentage abundance of *Abies* and *Picea* trees during the last 18,000 years in the southeastern United States by applying Equation (6.3.2) to fossil data from 13 sites within an area of 230,000 km^2.

In all these studies, α_k and z_k were estimated by regression analysis rather than by the maximum likelihood estimation procedures of Parsons and Prentice (1981) and I. C. Prentice and Parsons (1983). Unlike the maximum likelihood procedures, regression analysis can yield negative intercepts for z_k. In the H. R. Delcourt and P. A. Delcourt (1984) study, $\hat{z}$ was negative for both *Picea* (−5.28%) and *Abies* (−2.35%), and Equation (6.3.2) was applied to those fossil spectra for which $p_{ik} \geq \hat{z}_k$. We have grave reservations about the advisability of using such transformations when the estimated z values are both negative and of the same order of absolute magnitude as the pollen proportions which are to be transformed.

I. C. Prentice and Parsons (1983) regard pollen–vegetation models that incorporate a background component as tools for investigating and quantifying relationships between modern pollen and contemporary vegetation rather than as robust and proven techniques for reconstructing past plant abundances from pollen stratigraphical data. They emphasise that the 'models are in their infancy'. We strongly support this cautious approach, particularly as factors such as taxon abundance and vegetation density, structure, and composition, all of which have undoubtedly changed with time, may influence the magnitude of z_k. The background component can also vary within a taxon's range, especially at and near range limits (see Parsons *et al.*, 1980; Parsons and Prentice, 1981). Forest management practices over many hundreds of years may also have altered the magnitude of α_k, particularly in northwestern Europe (I. C. Prentice, 1982a).

Moreover, there are often relatively large standard deviations associated with estimates of model parameters (see, e.g., I. C. Prentice and Parsons, 1983). Statistical theory can assist here in indicating the amount of data necessary for reliable estimation of the models' parameters. We noted in Section 2.4 and Subsection 5.4.5 the work of Parsons *et al.* (1983) that provides the approximate standard deviations associated with reconstructed vegetation proportions when R values are estimated from data at a single site. Similar (though computationally more involved) work on obtaining standard deviations associated with $\hat{f}_k$ under more elaborate models would lead to a more realistic assessment of the reliability of such estimates. However, we stress that these estimates and their standard deviations are only valid under the assumption that *the basic model relating pollen to vegetation is correct*. If the model is seriously inadequate, the reconstructions could be misleading, and the asserted precision could over-state the accuracy of estimation.

Clearly, background models should be applied to a wide range of independent modern pollen data before their potential applicability to fossil data can be adequately assessed. Their use as interpretative tools (e.g., Bradshaw, 1981a; Heide, 1981; H. R. Delcourt and P. A. Delcourt, 1984) for transforming pollen stratigraphical data may be premature, and the results obtained could have a misleading precision.

6.3.3 Discussion

We conclude this section by discussing the main problems and limitations in the use of pollen-representation factors for reconstructing past plant abundances, and by suggesting potential uses of representation factors in the light of these limitations.

The major problem of the R-value model (see Subsection 5.4.1) is that it assumes that a pollen assemblage is derived entirely from the vegetation within an arbitrarily selected area around a site and thus that there is no background pollen. When R values are used to transform fossil pollen spectra, the resulting reconstruction is, by definition, the past plant abundances within a similar-sized area around the site of interest. The problems of delimiting appropriate primary pollen-source areas for the estimation of R values are discussed in Subsection 5.4.1. As plants differ greatly in their abundance and spatial distribution within vegetation, as different taxa produce vastly different amounts of pollen, and as different pollen types differ in their settling velocities and dispersal abilities, the area required around a site to reduce the background pollen component to 0 or near 0 will vary from taxon to taxon (see Bradshaw and Webb, 1985). Some taxa produce huge amounts of widely dispersed, light pollen (e.g., *Betula, Pinus*) and thus will require a large area for the *R*-value model to be valid, whereas taxa with a low production of poorly dispersed, heavy pollen (e.g., *Acer, Tilia*) require a small area. Any specified area is inevitably therefore a compromise, and the validity of the *R*-value model will vary from taxon to taxon for a given site and presumed pollen-source area. Models incorporating a background component (Subsection 5.4.4) can suggest if a particular source area is appropriate for *R*-value estimation by providing estimates of the background component for different taxa. Unfortunately, these models do not provide a realistic, alternative procedure for transforming pollen stratigraphical data (Parsons and Prentice, 1981).

When representation factors based on models with a background component are used to transform fossil spectra, it is assumed that both the productivity factor α_k and the background component z_k are constant in time and space. As discussed above, this assumed constancy of z_k does not seem likely over any extended time period or large geographical area, because z_k is a function, in part at least, of the regional abundance of the taxon in question. The assumption of constancy of z values severely limits the applicability of these representation factors as tools for reconstructing past plant populations.

Pollen analysts attempting quantitative reconstructions of past populations thus face a dilemma. Should they use representation factors based on the simple *R*-value model that makes fewer assumptions about parameter constancy in time and space but that may be a poor and unrealistic model of modern pollen–vegetation relationships, particularly at the regional scale; or should they use representation factors based on a background-component model that makes major assumptions about parameter constancy but that is a more realistic model of modern pollen–vegetation relationships? There seems, at present, no obvious solution to this dilemma. Both types of models should be used together wherever possible (I. C. Prentice, 1982a). The validity of a given set of modern *R*-value estimates should be tested by applying the background-component models to the same data to assess the magnitude and importance of background pollen. If the background component is low for all taxa of interest, the *R*-value estimates can then be used to transform fossil assemblages with some degree of confidence.

An additional limitation in the use of *all* representation factors for population reconstruction arises from the inherent spatial heterogeneity of vegetation. If representation factors are used to transform fossil spectra, the resulting reconstructions are plant abundances within an area of similar size to the vegetational area used for estimating the modern representation factors. At the local scale of small hollows (10–30 m) and mor humus profiles, these reconstructions relate simply to plant abundances within a 20–30-m radius (*ca.* 1000–3000 m^2) of the sampling site. Problems of vegetational heterogeneity and differential pollen dispersal at this fine scale are minimal (Andersen, 1970, 1973, 1980c). These problems become more acute, however, when pollen spectra are transformed from medium- or large-sized lakes or bogs with pollen-source areas of at least 1000–2000 km^2 (see T. Webb, Laseski, and Bernabo, 1978; T. Webb, *et al.*, 1981;

Jacobson and Bradshaw, 1981, for discussions of pollen-source areas). Many population sizes, spatial distributions, and vegetational patterns are possible within such a large area around a site, many of which could produce similar pollen spectra (Kabeiliene, 1969; Oldfield, 1970). The *R*-value model takes no account of any spatial information concerning the vegetation surrounding the site, and the background models only distinguish between vegetation inside and outside a specified region around the site. More elaborate models, building on the work of, for example, Turner (1964), Tauber (1965, 1967, 1977), Janssen (1966, 1970, 1973), and Kabailiene (1969), could assist in the investigation of the spatial relationships between modern pollen and vegetation. However, given the multiplicity of different patterns of vegetation that can theoretically produce a given pollen spectrum, there is a limit to the extent of information which one can hope to obtain from pollen stratigraphical data from a single site. In order to be able to reconstruct regional and local vegetation patterns, population abundances, and vegetational differentiation with any confidence, it is essential to have detailed information from a network of small sites (Livingstone, 1969; Kabailiene, 1969; Janssen, 1973; Jacobson, 1975, 1979; Vuorela, 1977, 1980; Andersen, 1978a; Jacobson and Bradshaw, 1981; Ritchie, 1981; T. Webb *et al.*, 1981). Livingstone (1969) and I. C. Prentice (1982a) outline possible, but as yet untried, methods for separating different components (local, regional, etc.) of the pollen rain at two or more sites. It seems likely that more intricate models of the spatial relationships between modern pollen and vegetation will be developed in the future, and that attempts will be made to reconstruct fine-scale patterns of past vegetation using accurately dated fossil pollen from many sites; however, the theoretical and practical difficulties involved in completing successfully such an investigation should not be underestimated.

Reconstructions of past populations will inevitably therefore be rather imprecise, given our current models of pollen–vegetation relationships. However, we agree with Parsons and Prentice's (1981, p. 148) conclusion that 'any imprecision is not so much the fault of the pollen record as ours in failing to accurately model and estimate the pollen–vegetation relationship'.

Despite these limitations, there are at least three broad types of studies where *R* values, if applied carefully and critically to fossil pollen data, can provide new and potentially important insights into past populations, vegetational cover and composition, and population and community dynamics. These are (1) local-scale studies involving small sites (10–30 m in diameter); (2) regional-scale studies involving medium-sized sites (250–1000 m in diameter); and (3) broad-scale studies involving arrays of contemporaneous data from many sites.

At the local scale, pollen stratigraphical data can be transformed into a reliable temporal record of estimated population abundances within a 20–30-m radius of the site (e.g., Andersen, 1973, 1978a; C. A. Baker *et al.*, 1978; Bradshaw, 1981a, 1981b; Aaby, 1983). Such reconstructions of populations and their temporal dynamics are at the scale of vegetational stands as studied by ecologists. They can thus provide insights into questions of vegetational stability and frequency of natural disturbance and regeneration over long time periods. For example, Andersen (1975) demonstrated alternations between mature and senescent stages of forest growth with *Tilia* dominance and pioneer phases of regeneration with *Acer, Corylus,* and *Fraxinus* from transformed pollen spectra of last interglacial age preserved in a very small basin (2.5 m in diameter) in Denmark. He established the frequency of gap-phase regeneration cycles to be about one per 350–400 years, presumably as a result of natural wind-throw and senescence. Such reconstructions can be used to test results of numerical simulations of forest growth at this spatial scale under different boundary conditions of disturbance régime, climate, and soil conditions. Simulation models of forest growth and succession (e.g., Botkin, Janak, and Wallis, 1972) are well developed for this scale. Clearly, local-scale pollen stratigraphical data, suitably transformed into estimates of past population size, can be appropriately and usefully compared with results from simulation experiments, and can assist in the verification of the ecological assumptions implicit in particular simulation models (cf. Botkin, 1981).

At the regional scale, reconstructions of past populations within the presumed pollen-source area of lakes or bogs should be attempted along the lines of, for example, Iversen (1952–1953), Donner (1972), Vuorinen and Tolonen (1975), and Ritchie (1982). Such reconstructions can aid interpretation of pollen stratigraphical data by providing some indication of the likely magnitude of population and vegetational changes through time (e.g., Donner, 1972; Huttunen, 1980). Transformation of pollen-accumulation rates using *R* values or related representation factors based on modern absolute pollen and vegetation data (e.g., Ritchie, 1982) can also provide estimates of absolute plant cover and population sizes. As discussed above, these cover estimates can be particularly valuable in late-glacial studies where bare ground and palynological 'blind spots' may have been important components of the landscape at the spatial scales recorded by pollen assemblages deposited in medium-sized sites. Such transformations will only be useful in population reconstruction if the observed accumulation rates of fossil pollen at the site of interest have not been influenced by site-dependent sedimentary processes, such as sediment focussing, or by changes in pollen-recruitment processes (see Subsection 6.2.4). Late-glacial sediments frequently have a high density because of their high minerogenic content. They are thus not very susceptible to resuspension and focussing, in contrast to organic-rich, less dense post-glacial sediments (H. J. B. Birks, 1981a; M. B. Davis and Ford, 1982; Bennett, 1983b). Transformations of pollen-accumulation rates from late-glacial sediments are likely therefore to provide reliable estimates of past population sizes and absolute plant cover.

At the broad scale of countries (e.g., H. J. B. Birks and Saarnisto, 1975; H. J. B. Birks, Deacon, and Peglar, 1975) or subcontinents (e.g., Bernabo and Webb, 1977), *R* values can be used to transform isopollen maps for particular time intervals into maps of former taxon abundance over large geographical areas (see Huntley and Birks, 1983, for examples of *Tilia* in Europe), so-called dominance (H. R. Delcourt and P. A. Delcourt, 1984) or 'isophyte' maps. These maps portray the changing broad geographical patterns in both abundance and range with time, and can provide a means of testing current biogeographical hypotheses concerning the relationships between geographical range and abundance over long time periods (see Hengeveld and Haeck, 1981, 1982; Hengeveld, 1982).

In all these potential applications, it is essential that modern *R* values be estimated using data from several sites (Wright, 1967; Livingstone, 1968, 1969) whose modern pollen spectra are similar to the fossil spectra of interest (I. C. Prentice, 1982a), that careful consideration be given to the problems of background pollen and pollen-source area (cf. M. B. Davis, 1963), and that, whenever possible, estimates of the variability associated with the estimates of past plant abundances be obtained (Parsons *et al.*, 1983). The techniques developed by Parsons and I. C. Prentice (1981) and discussed in Section 5.4 make it relatively straightforward to estimate *R* values and, if used critically, to avoid serious errors in transforming pollen stratigraphical data into estimates of past plant abundances.

Pollen-representation factors can, we feel, be useful tools in interpreting pollen stratigraphical data by providing a means of reconstructing past plant abundances and population sizes (cf. Faegri, 1966; M. B. Davis, 1969b; Janssen, 1970; D. Walker, 1972; H. J. B. Birks, 1973a, 1973b; A. M. Solomon and Harrington, 1979), if only rather imprecisely at present. Mosimann and Greenstreet (1971, p. 35) comment: 'the jucicious use of corrected pollen diagrams and proportions would seem to offer continued possibilities for the pollen analyst'. We share this optimism.

6.4 Comparing Modern and Fossil Pollen Spectra

We discussed in Chapter 5 various approaches to the numerical analysis of modern pollen data that can be used to characterise present-day vegetation types in terms of their contemporary pollen spectra. If modern pollen–vegetation relationships or 'corresponding patterns' (T. Webb, 1974a) can be established, comparisons can then be made between fossil pollen spectra and modern pollen assemblages from known vegetation on the basis of overall similarities in pollen composition (see Figure 6.1). This is the so-called comparative approach or analogue approach to the reconstruction of past plant communities or vegetation types (see Wright, 1967; Livingstone, 1969; H. J. B. Birks and H. H. Birks, 1980, p. 237; I. C. Prentice, 1982a). Other approaches to vegetational reconstruction are reviewed by H. J. B. Birks and H. H. Birks (1980), T. Webb *et al.* (1981), and Janssen (1981b).

If the fossil and modern assemblages are similar in their pollen composition and proportions, it can be concluded that the spectra were produced by similar vegetation types. A modern vegetational analogue can thus be proposed for the fossil pollen assemblage. If similarities between modern and fossil pollen assemblages can be recognised in a stratigraphical and hence a temporal sequence of fossil pollen spectra, changes in past vegetation over time can be reconstructed in terms of present-day vegetational patterns in space (see Aario, 1940; McAndrews, 1966, 1967; Livingstone and Estes, 1967; Wright *et al.,* 1967; M. B. Davis, 1969b; R. B. Davis *et al.,* 1975). If, on the other hand, no satisfactory match can be made between the modern and fossil spectra and there is no independent evidence for sedimentary disturbance or reworking of pollen and spores from deposits of different ages (e.g., Iversen, 1936; M. B. Davis, 1961), it can be concluded that the past vegetation has no geographically extensive modern analogue, or that the contemporary pollen data are incomplete and that modern analogues should be sought elsewhere. To exclude the latter possibility, comprehensive sets of modern pollen data are required (e.g., Wright *et al.,* 1967; Lichti-Federovich and Ritchie, 1968; R. B. Davis and Webb, 1975; T. Webb and McAndrews, 1976; I. C. Prentice, 1978; Huntley and Birks, 1983).

The model underlying the comparative approach is less restrictive than those underlying the estimation and use of pollen-representation factors. A pollen-source area still has to be specified in establishing modern pollen–vegetation patterns (Section 5.3), but the choice is considerably less critical than it is in deriving pollen-representation factors. The comparative approach utilises the entire pollen assemblage, including minor, rarely occurring tree and shrub pollen types and non-arboreal taxa, all of which are difficult to incorporate satisfactorily in the estimation of modern pollen-representation factors (cf. Ritchie, 1974, 1982; I. C. Prentice, 1982a). The results from the comparative approach are inevitably less quantitative than the results derived from applying modern representation factors to fossil data. D. Walker (1972) has suggested, however, that the comparative approach 'might well prove more useful in the long run'. Results from the comparative approach are frequently adequate for answering particular palaeoecological questions; see McAndrews (1966, 1967), M. B. Davis (1967a), Wright *et al.* (1967), H. J. B. Birks (1973b), Ritchie (1977), and Lamb (1982, 1984) for examples of the use of the comparative approach in solving specific problems in vegetational history. H. J. B. Birks and H. H. Birks (1980) review several examples of its use in the reconstruction of past vegetation types, and Wright (1967), Livingstone (1969), Richard (1976), A. M. Solomon and Harrington (1979), H. J. B. Birks and H. H. Birks (1980), and I. C. Prentice (1982a) discuss the approach in detail and outline its strengths and its weaknesses.

In many applications of the approach, fossil and modern pollen spectra are compared visually (e.g., McAndrews, 1966; Livingstone and Estes, 1967; M. B. Davis, 1967a, 1969b; H. J. B. Birks, 1973b; Sluiter and Kershaw, 1982). Direct comparison of fossil and modern pollen spectra, either as tabulated percentages or as bar diagrams, is difficult when numerous samples with many pollen types are compared. Moreover, bias can easily occur in comparing spectra, particularly in attempting to find a 'match' between modern and fossil spectra. Numerical methods of analysis enable variations in all pollen types to be considered simultaneously, and can prevent subconscious bias in the matching (see Section 3.8).

It is important that all numerical methods of comparison should be able not only to find a match but also to fail to find a match. They must indicate clearly when a fossil pollen spectrum has no convincing modern analogue (I. C. Prentice, 1982a), because the existence of vegetation types in the past with no modern analogues, besides being of considerable ecological interest, is of critical importance in limiting the potential use of pollen-representation factors (see Section 6.3) and modern pollen-climate transfer functions (see Section 6.6). Numerical methods should, moreover, seek not only to find the single modern pollen spectrum that is most similar to a particular fossil spectrum, but also to indicate which *groups* of modern and fossil samples are most similar because, in addition to statistical fluctuations, modern and fossil pollen data are inherently variable both in space and time (see, e.g., R. B. Davis *et al.*, 1969; R. B. Davis, 1974; T. Webb, Laseski, and Bernabo, 1978; T. Webb, 1981, 1982).

There are two broad methods for quantifying the comparative approach, both of which can be used to compare pollen diagrams (see Chapter 4). These approaches are (1) direct comparison of modern and fossil pollen spectra, and (2) combined numerical classifications of modern and fossil pollen spectra using partitioning and scaling methods of analysis. These two approaches are discussed separately, in Subsections 6.4.1 and 6.4.2.

6.4.1 Direct Comparison of Modern and Fossil Spectra

This approach involves initially grouping the available modern pollen spectra according to the plant community, vegetation type, ecological area, or geographical region from which the surface samples are derived (see Subsection 5.3.1), and then calculating an appropriate measure of variability for each group. The fossil spectrum of interest is then added in turn to each group of modern spectra and the variability of each augmented group is calculated. The increase in variability is a measure of the overall difference in composition between the fossil spectrum and the group of modern samples. If some threshold value of variability increase is selected, one can then decide which group or groups of modern spectra resemble the fossil spectrum, or whether the fossil spectrum has no match among these modern samples. The procedure is then repeated for all fossil spectra of interest. Groups of fossil samples (e.g., pollen zones) can also be compared in the same way with either individual modern pollen spectra or with groups of modern spectra. In Section 4.3 we described the application of this approach to the comparison of two groups of fossil spectra from different pollen diagrams.

The use of numerical methods for directly comparing modern and fossil spectra was pioneered by Ogden (1964, 1969, 1977b; see also Whitehead and Tan, 1969; H. J. B. Birks, 1970; A. M. Solomon, Blasing, and J. A. Solomon, 1982). Ogden used Spearman's rank correlation coefficient ρ to measure the similarity among modern spectra, and among a fossil spectrum and groups of modern spectra. In the evaluation of Spearman's ρ, the proportions of pollen in a spectrum are ranked in decreasing order of magnitude; for example, if taxon k is the second most abundant type in spectrum i, the rank a_{ik} takes the value 2; thus, in the absence of tied ranks, $a_{i1}, a_{i2}, \ldots, a_{it}$ take the values $1, 2, \ldots, t$, though not necessarily in that order. Spearman's ρ between spectra i and j is defined by

$$\rho_{ij} = 1 - \frac{6 \sum_{k=1}^{t} (a_{ik} - a_{jk})^2}{t^3 - t}.$$

For the treatment of tied ranks, see M. G. Kendall (1970, Chapter 3). Spearman's ρ takes values between $+1$ (perfect agreement between the ranks of the two spectra) and -1 (perfect disagreement).

Ogden (1969) obtained a measure of the similarity of two sets of pollen spectra by evaluating the average of ρ_{ij} for all pairs of spectra i and j, one from each set. Gordon and Birks (1974, p. 248) indicate how this average value of ρ is related to non-parametric measures of the homogeneities of the two sets of pollen spectra and the combined set of all spectra.

Ogden (1969) showed, for example, that some fossil pollen spectra closely resembled modern assemblages from regions of present-day maple–basswood forests, even though such similarities would probably never have been suggested in the absence of the surface-sample data. Although maple (*Acer*) and basswood (*Tilia*) are both notoriously low pollen producers (see T. Webb *et al.*, 1981), 'the presence of these trees in substantial numbers in a forest distorts the local and regional pollen rain in a statistically characteristic fashion' (Ogden, 1969). This example illustrates very clearly the importance of quantifying the comparative approach (see also A. M. Solomon and Harrington, 1979).

In that it is based only on the ranked values of the taxa in each spectrum, Spearman's ρ is more sensitive to the behaviour of the less common types in the spectrum than more commonly used measures based on the actual amounts of pollen present. On occasion, this can give a useful measure of the resemblance between spectra, but the proportions (and hence also the ranks) of less common taxa will be less precisely estimated, and it is often difficult to justify the loss of information entailed by use of a non-parametric statistic in this situation (see also A. M. Solomon and Harrington, 1979).

Spearman's ρ is, in fact, just Pearson's product-moment correlation coefficient r evaluated on the ranks of the amounts of pollen. Pearson's r has been used by R. B. Davis *et al.* (1975), Bradstreet and Davis (1975), and Ogden (1977b) to compare modern and fossil pollen spectra, by Ritchie (1974) and Lichti-Federovich and Ritchie (1965) to compare modern pollen spectra, and by Adam (1970, 1974), Yarranton and Ritchie (1972), Ritchie (1977), and Ritchie and Yarranton (1978a, 1978b) as a measure of similarity between pairs, or groups, of fossil spectra. Pearson's r gives a measure of the strength of the *linear* relationship between two variables, and is less appropriate for the comparison of two or more sets of proportions in pollen samples (see also I. C. Prentice, 1980).

An alternative and more suitable measure for comparing fossil and modern spectra is Sibson's (1969) information radius (see Section 4.3 and Gordon and Birks, 1974). The information radius H_1 for a group of modern spectra is initially calculated. The information radius H_2 is calculated for the groups of fossil pollen spectra to be compared, and the information radius of the combined set is also calculated ($H(1,2)$). As discussed in Section 4.3, the idea of carrying out a significance test of the hypothesis that the fossil and modern spectra were drawn from the 'same population' is unrealistic and inappropriate. Instead, we proposed (Gordon and Birks, 1974) as a rule of thumb that spectra be regarded as similar if

$$H(1, 2) \leq H_1 + H_2.$$

On occasion, it might be advantageous to relax this inequality, for example, by replacing the right-hand side by a multiple of $H_1 + H_2$.

In the comparison of modern and fossil spectra using information radius, it is advisable that the groups of fossil and modern samples be as homogeneous as possible, and thus show low within-group variability. Numerically defined zones of fossil pollen spectra (Chapter 3), perhaps with transitional levels deleted, and numerically derived partitions of modern surface samples (Subsection 5.3.4) collected from similar vegetation types would be ideally suited for numerical comparisons using information radius.

Other workers (e.g., Whitehead, 1979, 1981; H. R. Delcourt, P. A. Delcourt, Webb, and Overpeck, 1983; Overpeck and Webb, 1983) have used various measures of dissimilarity to compare individual fossil and modern pollen spectra, some of them obtaining empirically critical threshold dissimilarity values beyond which two spectra would not be regarded as similar. I. C. Prentice (1982a) discusses possible measures of dissimilarity for this purpose, and the problem of defining suitable threshold values to enable vegetational reconstruction on the basis of modern pollen analogues.

In contract to Ogden's (1964, 1969) approach, information radius and most dissimilarity measures utilise the quantitative information within the fossil and modern pollen data. However, information radius discards stratigraphical and geographical information present in the *groups* of fossil and modern spectra used for comparison. Such information can clearly be of ecological interest and interpretative value. Nevertheless, information radius can provide a broad over-view of similarities and differences between modern and fossil spectra, and is a useful first step in the reconstruction of past communities and vegetation types by indicating comparisons that merit further study. Dissimilarity measures comparing pairs of *individual* modern and fossil spectra can identify the closest modern pollen analogue for a particular fossil spectrum. However, the 'closest' modern analogue may be different when different dissimilarity measures are used (I. C. Prentice, 1982a). Several measures should be used (e.g., Overpeck and Webb, 1983; H. R. Delcourt, P. A. Delcourt, Webb, and Overpeck, 1983) and the results compared. If similar results are obtained from different dissimilarity measures, one has more confidence in the suggested results.

6.4.2 Combined Numerical Classifications of Modern and Fossil Spectra

In this approach, modern and fossil pollen spectra are classified together using numerical partitioning methods or scaling procedures. This can give a clear indication of whether there are any similarities between the modern and fossil spectra.

Suitable numerical partitioning methods are described in Sections 4.4 and 5.3. Brugam (1980) analysed modern and fossil diatom assemblages together in this manner. He showed that there were some fossil diatom assemblages at Kirchner Marsh, Minnesota, that appeared to have no modern analogue, in the sense that some groups of samples in the partition contained only fossil spectra. The direct comparisons described in Subsection 6.4.1 are similar in spirit to such investigations, attention being restricted to high similarities corresponding to early amalgamations in a dendrogram.

The related approach of obtaining a combined scaling of fossil and modern spectra, although of considerable potential, appears to have been used only once (MacDonald, 1983) to compare modern and fossil assemblages. Instead, geometrical representations of one set of pollen spectra have frequently been obtained, and the other set of spectra has then been inserted at the appropriate positions in the low-dimensional plot. The scaling has usually been carried out on the modern pollen spectra, and the fossil spectra have been added; however, some workers have reversed these rôles for the two sets of spectra.

Geometrical analysis of combined fossil and modern data avoids the chance of spurious matches being indicated for the added spectra, since in a joint analysis such spectra would be assigned their own separate region in the space; however, a study of residual distances of added points from the plane (in the manner described below) reduces this danger.

The first set of data can be analysed by some of the methods described in Section 5.3. If modern spectra are analysed by canonical variates analysis, the required grouping of the data is provided by the vegetation types or geographical regions from which the samples were derived. This approach is commonly used in statistics for the assignment of new objects which are assumed to belong to one of the existing groups. Its use for comparing modern and fossil pollen assemblages is slightly different, in that one wishes to have the option of deciding that the added spectra are not sufficiently similar to any of the groups for a match to be indicated.

Principal components analysis and principal coordinates analysis can also be used to obtain a geometrical representation of one set of spectra. Additional samples can readily be inserted into the plot of a canonical variates analysis or principal components analysis by projecting the extra samples onto the original canonical variates or principal components. The coordinates of the additional samples are simply computed as sample scores or linear combinations of the canonical variate or principal component loadings with the new data values, after any appropriate centring, transformations, or scalings of the new data. Gower

(1968, 1970; see also Wilkinson, 1970) describes how to insert additional points into a principal coordinates analysis plot.

The procedure of inserting unknown fossil samples into geometrical representations of modern samples was pioneered in physical anthropology by Ashton *et al.* (1957) for comparing fossil skulls of ape-like creatures with modern skulls of man and apes (see also Oxnard, 1973). Insertion of fossil pollen spectra into a two-dimensional principal components or principal coordinates analysis representation of modern pollen data has been done by T. Webb (1974c) and Lamb (1982, 1984); conversely, Ritchie (1977) and Ritchie and Yarranton (1978a, 1978b) have inserted modern spectra into a principal components analysis representation derived from fossil pollen data. Adam (1970), H. J. B. Birks (1976b, 1980), and Heide (1981) have inserted fossil spectra into a canonical variates analysis representation of surface-sample data as an aid to the numerical comparison of fossil and modern assemblages. Whiteside (1970) and Mastrogiuseppe, Cridland, and Bogyo (1970) have used related approaches to compare modern and fossil assemblages of chydorids in Danish lake sediments, and modern and fossil *Ginkgo* wood samples, respectively.

Gower (1968) discusses the calculation and interpretation of the residual distances from the true position of the added points to the fitted plane giving the best two-dimensional representation of the objects. The residual distance Δ_a is defined by

$$\Delta_a^2 = d_a^2 - \hat{d}_a^2,$$

where d_a^2 is the squared distance of the added object from the centroid of the set of points, and $\hat{d}_a^2$ is this squared distance as it is represented in the geometrical representation.

It is important to calculate the residual distances when inserting additional fossil samples into any geometrical representation of modern data, or *vice versa,* as new samples may appear to be positioned close to other samples on the first few axes and yet be located some distance from the other samples when further dimensions are considered. Oxnard (1972, 1973) discusses the importance of considering several dimensions before interpreting the positions of inserted fossil samples within a canonical variates analysis or other geometrical representation. Oxnard convincingly shows how erroneous interpretations can result when only the first few dimensions are considered (cf. Day, 1967, 1974; Day and Wood, 1968, 1969).

If the added fossil sample is inserted into the geometrical representation close to modern samples from known vegetation types, *and* the residual distance of the added point is small, interpretation of the added sample in terms of modern vegetation analogues can be made with more confidence (e.g., Adam, 1970; H. J. B. Birks, 1980). In some instances, the added fossil samples may be located some distance away from any present-day samples (e.g., H. J. B. Birks, 1976b) or residual distances may be large; both suggest that modern analogues for the fossil spectra do not exist within the modern surface-pollen data. Ritchie (1977) has, for example, demonstrated the absence of modern analogues for fossil spectra deposited before 6000 B.P. at a site in northern Canada by adding modern pollen samples to a principal components analysis of fossil data and showing the lack of correspondence in the positions of the fossil and modern data points (see also Ritchie and Yarranton, 1978a, 1978b; Lamb, 1982, 1984).

As scalings of large data sets concentrate on portraying broad patterns of resemblance, it can also be valuable to use the methods of analysis described in the last subsection to check potential matches suggested in the geometrical representation.

As an illustration of adding fossil pollen spectra whose vegetational origins are unknown to a scaling of modern pollen samples from known vegetation types, 58 fossil samples of late-Wisconsin age (*ca.* 20,500–8500 B.P.) from Wolf Creek, Minnesota (H. J. B. Birks, 1976b), are added (Figure 6.5) to the canonical variates analysis representation of the 131 modern samples collected from the western interior of Canada by Lichti-Federovich and Ritchie (1968). Canonical variates analysis of these modern samples was discussed in Subsection 5.3.3 (see Figures 5.5 and 5.6). As the first two canonical variate axes account for 72% of the total variation, the fossil samples are inserted into an efficient low-dimensional representation. Their residual distances from the plane of the first and second canonical variate axes are all small, suggesting that the positions of the fossil samples shown in Figure 6.5 are unlikely to be misleading. Fossil samples from the Compositae-Cyperaceae and *Pinus-Pteridium* assemblage zones do not match any of the modern spectra, whereas a few, but by no means all, of the samples from the *Picea-Larix* assemblage zone match modern spectra from the closed coniferous forests of central Manitoba. These results suggest that there are no modern analogues within the Manitoba surface samples for many of the fossil spectra at Wolf Creek. H. J. B. Birks (1976b) discusses the problems of reconstructing past vegetation from such fossil spectra that have no modern analogues.

A related method for comparing modern and fossil pollen spectra is presented by R. B. Davis *et al.* (1975) (see also Bradstreet and Davis, 1975). They constructed contour maps of Pearson's product-moment correlation coefficients between fossil spectra averaged for each pollen zone or a selected time interval at the site of interest and every individual modern spectrum (about 400 in total) then available from eastern North America (see R. B. Davis and Webb, 1975). Some spatial smoothing of the modern spectra and temporal smoothing of the fossil spectra are implicit in this mapping procedure. Given a more appropriate measure of similarity or dissimilarity between modern and fossil spectra (see I. C. Prentice, 1980, 1982a), this approach can effectively summarise large sets of fossil and modern data. It allows one to see at a glance any similarities between modern and fossil assemblages, and to appreciate the probable extent of vegetation change in the past by the distance the vegetation type or region with the highest similarity has moved since it surrounded the site where the fossil spectra were deposited (see M. B. Davis, 1969b). It is important, however, to define a reliable threshold similarity between modern and fossil spectra (e.g. Overpeck and Webb, 1983), so

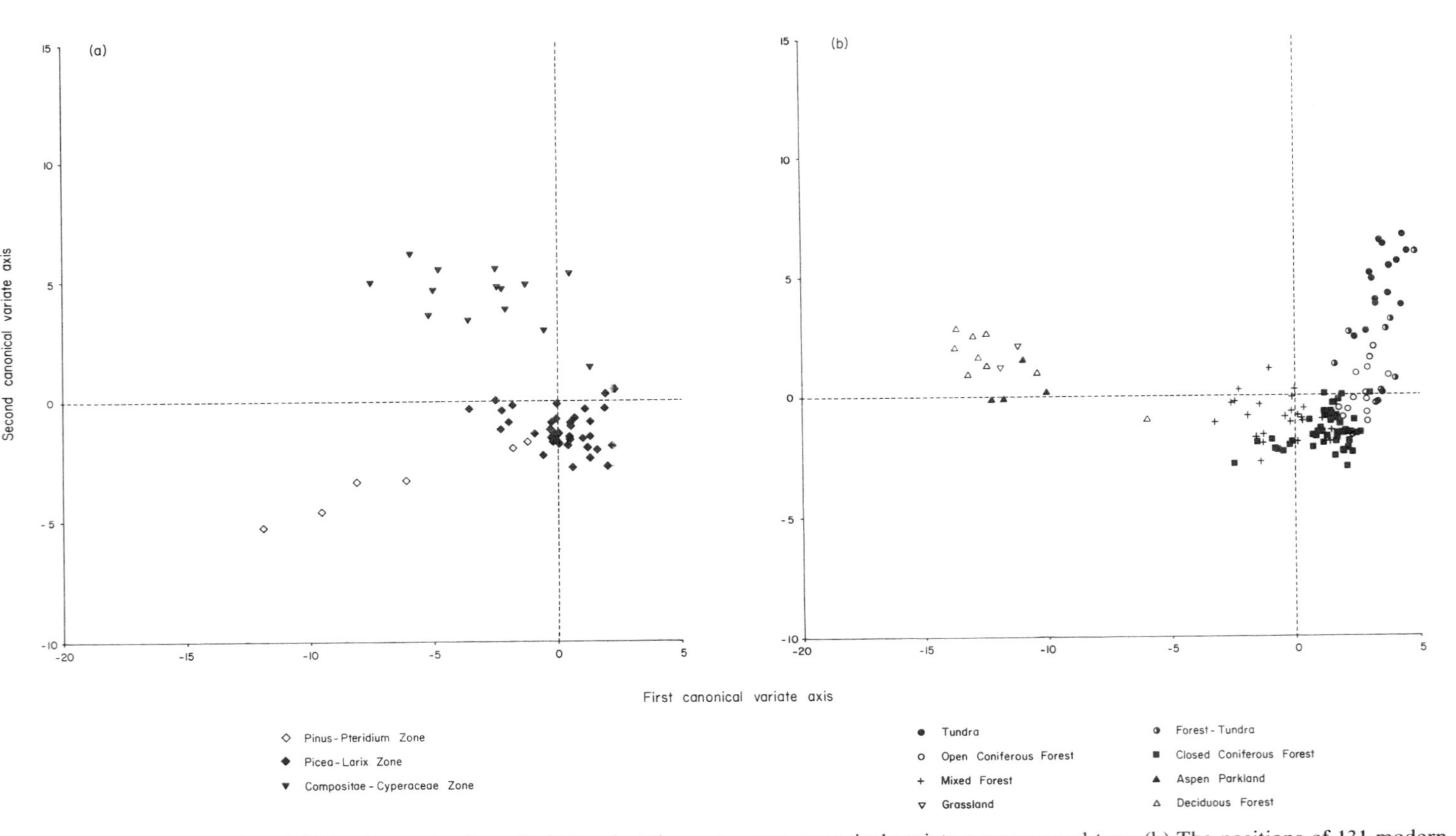

Figure 6.5 (a) Interpolation of 58 fossil samples from Wolf Creek, Minnesota, onto canonical variate axes one and two. (b) The positions of 131 modern pollen samples from eight major vegetation–landform units in the western interior of Canada Manitoba surface samples on the same two axes (see also Figure 5.4). (After H. J. B. Birks, 1976b.)

that any fossil assemblages with no modern analogues can also be clearly displayed. It would also be of ecological interest to reverse the procedure and to see what modern spectra have no fossil counterparts, at least at the scale of mapping adopted by R. B. Davis *et al.* (1975).

6.4.3 Discussion

There are several advantages and disadvantages of the comparative approach, irrespective of whether the comparisons are made visually or numerically (see H. J. B. Birks and H. H. Birks, 1980, pp. 251–255). Problems include the importance of local within-site and between-site factors and the pollen-depositional character of each site (R. B. Davis *et al.,* 1969; T. Webb, 1974c; Jacobson and Bradshaw, 1981), the rôle of man in influencing the composition and structure of modern vegetation (T. Webb, 1973; M. B. Davis *et al.,* 1973; Brubaker, 1975; I. C. Prentice, 1982a), the importance of comparing only fossil and modern spectra collected from the same sedimentary environment (Oldfield, 1970; Ritchie, 1974; H. J. B. Birks, 1977b), the problem that two or more modern vegetation types may produce similar contemporary pollen assemblages due to palynological 'blind spots' (*sensu* M. B. Davis, 1963, 1967a) in the vegetation, the difficulties in evaluating the significance of variations in pollen proportions between samples (M. B. Davis, 1967a), the problems of selecting appropriate measures of similarity between modern and fossil spectra and of assessing the observed similarities (Prentice, 1982a), and the possibility that some vegetation types of the past have no modern analogues (Janssen, 1970, 1981b).

The comparative approach is most useful in the reconstruction of past communities and vegetation when the fossil assemblages can be matched with modern pollen spectra which

1. are distinctive from other modern spectra,
2. refer to characteristic, homogeneous, and widespread vegetation types (e.g., vegetation formations), and
3. are derived from comparable sites of pollen deposition in topographically similar areas.

Modern pollen spectra and the comparative approach are thus most appropriate for vegetational reconstruction at the scale of vegetation formations or vegetation–landform units (e.g., McAndrews, 1966, 1967; Wright *et al.,* 1967; Wright, Bent, Hansen, and Maher, 1973; Rampton, 1971; R. G. Baker, 1976; Ritchie, 1976, 1977; Heusser, 1977; Lamb, 1980, 1982, 1984; Barnosky, 1981). If no satisfactory match between fossil and modern spectra can be found after extensive comparisons with comprehensive surface-sample data, it must be concluded that no extensive modern vegetational analogue exists. The absence of such modern analogues provides support for Gleason's (1939) individualistic concept of the nature of vegetation, which proposes that vegetation varies continuously in space and time. There are now several demonstrations that 'our present plant communities have no long history in the Quaternary, but are merely temporary aggregations under given conditions of climate, other environmental factors, and historical factors' (West, 1964, p. 55). See, for example, Iversen (1954), Andersen (1961), Cushing (1963, 1965), Janssen (1970), Watts (1973), H. J. B. Birks (1973b, 1976b, 1981a), Ritchie (1977, 1981, 1982), D. Walker (1978), Cwynar (1982), Brubaker , Garfinkel, and Edwards, (1983), and Lamb (1984) for discussions of no-analogue vegetation types in the past. In such instances, vegetational reconstructions from fossil pollen assemblages are extremely difficult. Reconstructions from pollen stratigraphical data of no-analogue vegetation types and of vegetation types and mosaics at finer scales than the formation or vegetation–landform unit must rely on the detection and interpretation of stratigraphical trends of individual taxa (see Section 6.2), on estimates of the past abundances of taxa (see Section 6.3), on the use of 'indicator species' (*sensu* Janssen, 1967b, 1970, 1981b; H. J. B. Birks, 1973b; D. Walker, 1978), and on the derivation of 'recurrent groups' of fossil pollen and spores. We consider the numerical detection of recurrent groups in the next section as an additional approach to the reconstruction of past plant communities.

6.5 Recurrent Groups

The model underlying this approach to the reconstruction of past communities proposes that a fossil assemblage consistently occurring together or numerically correlated in a series of samples within and between stratigraphical sequences represents a past life assemblage or community, here defined as a group of organisms that lived together in the same place and at the same time. Such groups of associated or correlated fossil taxa are called *recurrent groups*. The approach assumes that the unit of study, namely, the fossil assemblage, is closely related in space and time to the life assemblage. It is potentially most useful when fossil assemblages cannot be recognised as resembling any known modern communities (H. J. B. Birks and H. H. Birks, 1980, pp. 27–28).

The recurrent-group approach (see Figure 6.1) has not been widely used in Quaternary palynology because the comparative and indicator-species approaches are available (H. J. B. Birks and H. H. Birks, 1980), and because the model underlying the approach is not particularly appropriate for most pollen stratigraphical data (see Subsection 6.5.2). The use of recurrent groups has, however, assisted the reconstruction of past plant communities from Quaternary pollen assemblages when the assemblages appear to have no satisfactory modern analogue, such as late-glacial assemblages (e.g., R. H. Johnson, Tallis, and Pearson, 1972; H. J. B. Birks and H. H. Birks, 1980, p. 233) or last-glacial spectra (e.g., Sears and Clisby, 1955; P. S. Martin and Mosimann, 1965), or when the pollen assemblages are very diverse (e.g., A. R. H. Martin, 1968; Harris and Norris, 1972).

Recurrent groups can be derived visually by arranging together fossils with similar stratigraphical occurrences (e.g., Janssen, 1967b, 1972; Rybniček and Rybničkova, 1968). Alternatively, one can define a measure of the similarity between taxa based on their occurrences or relative abundances, for example, the association between taxa recorded as present or absent in a sample, or the correlation between taxa recorded quantitatively. Cheetham and Hazel (1969) and Reyment (1971) discuss some such measures, and Buzas (1969) reviews the application of numerical methods in the delimitation of recurrent groups. If many taxa are considered together, it can be difficult to assess by eye the information contained in matrices of similarities. This assessment can be helped by numerical classifications of the taxa, using the methods described and illustrated in Section 5.3 for the analysis of modern pollen spectra. Thus, taxa that are associated or correlated with one another will be grouped together by a numerical partitioning or cluster analysis, and positioned near to each other by a scaling procedure such as principal coordinates analysis.

A variety of clustering procedures has been used to delimit recurrent groups within fossil assemblages of marine invertebrates (e.g., Valentine and Peddicord, 1967; Fox, 1968; Mello and Buzas, 1968; Tipper, 1975), microplankton (e.g., D. K. Goodman, 1979), plant macrofossils (e.g., P. D. Moore, 1973; Robichaux and Taylor, 1977; H. J. B. Birks and H. H. Birks, 1980), and pollen (e.g., Oltz, 1969, 1971; Clapham, 1972; Boulter and Hubbard, 1982; Hubbard and Boulter, 1983). Several scaling methods have also been used as an aid to the recognition of recurrent groups in fossil assemblages of marine invertebrates (e.g., Reyment, 1963a, 1976a; Shaffer and Wilke, 1965; Park, 1968; Symons and De Meuter, 1974), plant macrofossils (e.g., Robichaux and Taylor, 1977; Spicer and Hill, 1979), and pollen (e.g., Clapham, 1972; H. J. B. Birks and H. H. Birks, 1980; Gordon, 1982c; Boulter and Hubbard, 1982).

As these methods have been illustrated earlier in the book we concentrate in this section on describing an alternative method of representation: association or correlation nets or plexuses (McIntosh, 1973), which can incorporate more formal significance tests of the existence of correlation between pollen types.

6.5.1 Correlation Nets

In a correlation net, each taxon is represented by a point in a two-dimensional plot. If two taxa are positively associated with each other, the corresponding pair of points is joined by an unbroken line. If two taxa are negatively associated with each other (in the sense that they rarely occur together, or their abundances are largely complementary), the corresponding pair of points is joined by a broken line (see Figure 6.6).

This approach has similarities to numerical classification procedures: a recurrent group comprises taxa linked by unbroken lines, and these taxa will tend to be placed together in the same group in a cluster analysis, or be represented by points which are close together in the results of a scaling method. Indeed, if many taxa are being analysed, a scaling procedure could be used to provide the positions of the points representing the taxa, in order to simplify the appearance of the correla-

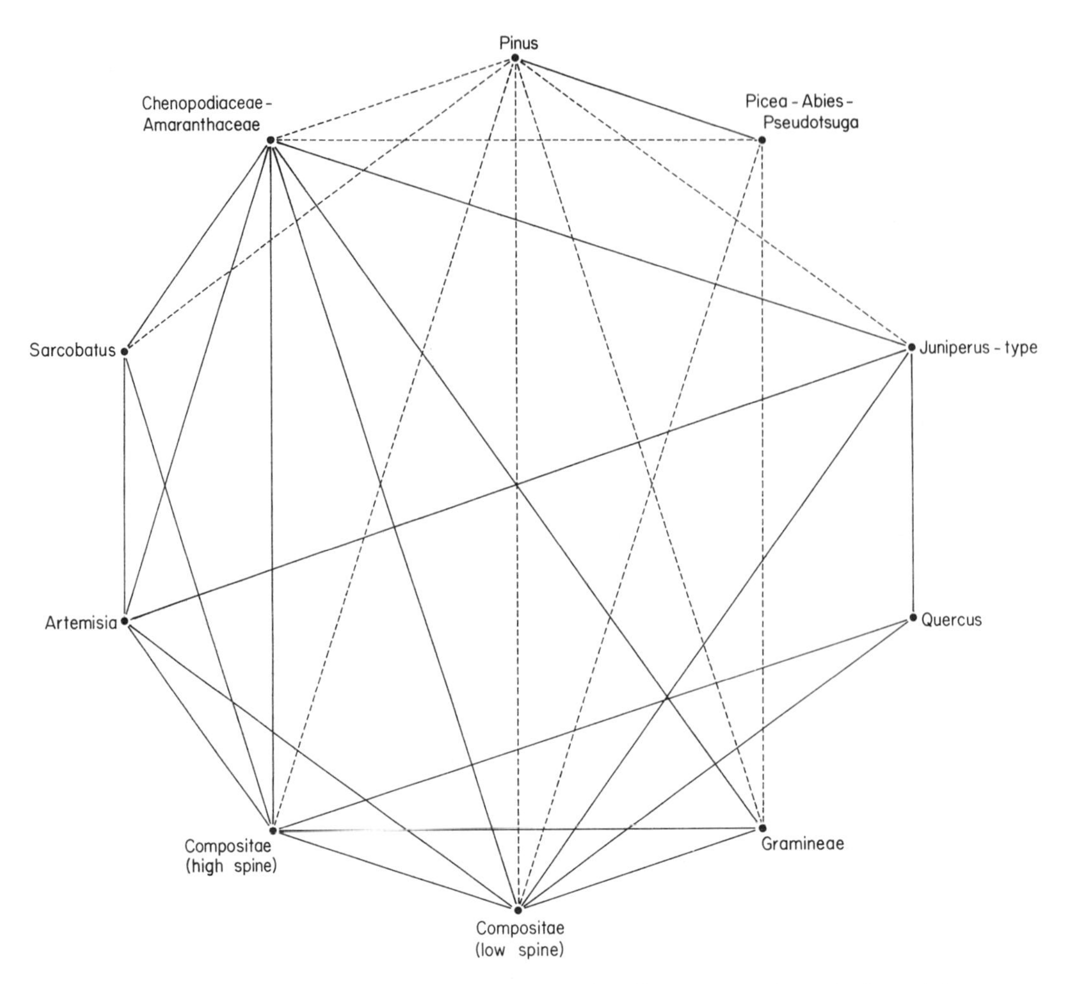

Figure 6.6 Correlation net for 10 major pollen types in 133 samples at Lake Cochise, Arizona. Significant (0.05 level) constraint-positive correlations are shown as solid lines, significant constraint-negative correlations are shown as broken lines. (After P. S. Martin and Mosimann, 1965.)

tion net. One additional feature of a correlation net is that details are given about groups that are negatively associated with each other; this information is available only indirectly from clustering or scaling procedures.

This approach has been widely used to delimit recurrent groups within fossil assemblages of marine invertebrates (e.g., R. G. Johnson, 1962; Valentine and Mallory, 1965), conodonts (e.g., Kohut, 1969), microplankton (e.g., Brideaux, 1971), Permian pollen and spores (e.g., Clapham, 1969, 1970, 1972), late Tertiary pollen and spores (e.g., Sluiter and Kershaw, 1982), and Quaternary pollen and spores (e.g., Sears and Clisby, 1955; P. S. Martin and Mosimann, 1965; A. R. H. Martin, 1968; Harris and Norris, 1972). A simple illustrative example is given in this section.

Some thought has to be given to deciding when two taxa should be regarded as being positively or negatively associated. P. S. Martin and Mosimann (1965; see also Mosimann, 1962, 1963, 1970; Taillie, Ord, Mosimann, and Patil, 1979) address this problem in the comparison of proportions of two pollen types by the product-moment correlation coefficient. The difficulty arises because the pollen proportions must sum to 1, and hence one would expect some inbuilt negative correlation between two pollen types: if the proportion of one type increases, the proportion of the other type would be expected to decrease. Mosimann (1963) and P. S. Martin and Mosimann (1965) discuss two 'null' models for the behaviour of pollen types, the aim being to detect significant departures from such null models of no relationship between the taxa.

In the first null model, the underlying proportions of the t taxa are assumed to be constant throughout the time period being considered: if u_{ik} denotes the underlying proportion in the sediment corresponding to the ith spectrum which belongs to pollen type k, it is assumed that, for each value of k,

$$u_{1k} = u_{2k} = \cdots = u_{nk}, \qquad (6.5.1)$$

where n denotes the number of spectra in the period of time under consideration. The pollen counts will thus be samples from a multinomial distribution (see Section 2.2).

In the second null model, the u_{ik}'s are allowed to vary through time, but only in the manner specified below. In this model, $y_{ik} (k = 1, 2, \ldots, t)$ is defined to be the total number of grains of taxon k deposited in the ith period of time; thus

$$u_{ik} = y_{ik}/y_{i\cdot}, \qquad \text{where} \quad y_{i\cdot} \equiv \sum_{l=1}^{t} y_{il}. \qquad (6.5.2)$$

It is assumed that $y_{1k}, \ldots, y_{nk}$ are observations from the same distribution, that the y variables are independent of one another, and that each u_{ik} is independent of the total number of grains $y_{i\cdot}$. Thus, for example, it would be assumed that the number of *Pinus* pollen grains deposited at the site during a given time period has no effect on the number of *Picea* grains deposited, and that there are no density-dependent effects, the underlying proportions being unrelated to the total number of grains deposited.

Under both null models, the correlation ρ_{kl}^0 between the proportions of the kth and lth pollen types is given by

$$\rho_{kl}^0 = -\sqrt{\frac{u_k u_l}{(1 - u_k)(1 - u_l)}}, \qquad (6.5.3)$$

where u_k is the common value of the u_{ik} in expression (6.5.1) under the first model, or the expected value of the corresponding random variable under the second model (Mosimann, 1962).

If p_k is an estimate of $u_k (k = 1, \ldots, t)$, ρ_{kl}^0 can be estimated by r_{kl}^0, defined by

$$r_{kl}^0 = -\sqrt{\frac{p_k p_l}{(1 - p_k)(1 - p_l)}}. \qquad (6.5.4)$$

This value, rather than 0, should thus be regarded as the 'point of departure' from which to assess any correlation.

For example, Mosimann (1962) re-analysed some data from Bellas Artes, Mexico (Clisby and Sears, 1955), restricting attention to the amounts of four taxa (*Pinus, Abies, Quercus,* and *Alnus*) in $n = 73$ stratigraphical levels. When the proportions of these taxa in each level were evaluated with respect to a pollen sum comprising just these four taxa, it was found that the correlation between *Pinus* and *Alnus* was −0.69. The mean proportions of *Pinus* and *Alnus* in the data set were 0.863 and 0.032, respectively; by substituting these values for p_k and p_l into Equation (6.5.4), the correlation to be expected under either null model works out at approximately −0.46. Mosimann (1962) described an approximate procedure for testing whether an observed correlation r differs significantly from the value r^0 to be expected under the null models; the correlations are subjected to Fisher's z transformation (Morrison, 1976, Chapter 3):

$$z = \frac{1}{2} \log\left(\frac{1 + r}{1 - r}\right), \tag{6.5.5}$$

and $\sqrt{n - 3}\,(z - z^0)$ is compared with a standard normal deviate. In the above example,

$$\sqrt{n - 3}\,(z - z^0) = \sqrt{70}\,(-0.85 + 0.50) = -2.93,$$

which is highly significant; hence one would reject the null hypothesis of no correlation in favour of the hypothesis that *Pinus* and *Alnus* pollen were negatively correlated.

The comparisons between all pairs of taxa analysed by Mosimann (1962) are summarised in Table 6.4. Differences between individual observed and expected null correlation coefficients which are significant at the 1% level are indicated by asterisks in the final subtable. However, it would be misleading to regard the methodology described above as providing rigorous statistical tests of the significance of each of a set of observed correlation coefficients; apart from the approximation involved, there is the problem that many related pairwise comparisons are being made (Aitchison, 1981). Nevertheless, this approach provides a useful way of specifying the links that should be included in a correlation net.

In Table 6.4, the observed correlations of *Abies* pollen with the other pollen types are not significantly different from the expected null coefficients, suggesting that the observed correlations may result solely from the proportion constraints. *Quercus* and *Alnus* proportions show a significant positive correlation, whereas *Pinus* is negatively correlated with both *Quercus* and *Alnus* to a greater extent than would be expected due to the proportion constraints alone. Three groups of taxa can thus be delimited on the basis of their correlations—*Alnus* + *Quercus, Pinus, Abies*—and the corresponding correlation net would have an unbroken line linking *Alnus* and *Quercus,* and broken lines linking both of these taxa to *Pinus*.

P. S. Martin and Mosimann (1965) used this approach to determine significant correlations between 12 major pollen types in 133 samples from a 42-m long core

Table 6.4

Observed and Expected Null Correlation Coefficients between Four Pollen Taxa in 73 Samples in the Bellas Artes Core, Mexico[a]

	Observed correlation coefficients				Expected null correlation coefficients			
	Pinus	*Abies*	*Quercus*	*Alnus*	*Pinus*	*Abies*	*Quercus*	*Alnus*
Pinus	1.0				1.0			
Abies	−0.30	1.0			−0.30	1.0		
Quercus	−0.90	+0.09	1.0		−0.79	−0.04	1.0	
Alnus	−0.69	+0.08	+0.36	1.0	−0.46	−0.06	−0.02	1.0
	Corresponding z values				Corresponding z values			
	Pinus	*Abies*	*Quercus*	*Alnus*	*Pinus*	*Abies*	*Quercus*	*Alnus*
Pinus	—				—			
Abies	−0.31	—			−0.31	—		
Quercus	−1.47	+0.09	—		−1.07	−0.04	—	
Alnus	−0.85	+0.08	+0.38	—	−0.50	−0.06	−0.02	—
	Differences of z values				Significance results[b]			
	Pinus	*Abies*	*Quercus*	*Alnus*	*Pinus*	*Abies*	*Quercus*	*Alnus*
Pinus	—				—			
Abies	0.00	—			ns	—		
Quercus	−0.40	+0.13	—		−*	ns	—	
Alnus	−0.35	+0.14	+0.40	—	−*	ns	+*	—

[a] Data from Mosimann (1962) and Clisby and Sears (1955). $n = 73$; $\sqrt{n-3} = 8.37$

[b] ns = not significant at 0.05 level; * = significant at 0.01 level.

from pluvial Lake Cochise in Arizona. *Pinus* pollen (see Figure 6.6) is positively correlated, when allowance is made for proportion constraints, only with *Picea* + *Abies* + *Pseudotsuga* pollen. They are negatively correlated with most of the other taxa, which in turn are positively associated among themselves. Martin and Mosimann interpret the patterns of correlation as reflecting shifts between woodland of various types and yellow-pine parkland. *Picea, Abies,* and *Pseudotsuga* proportions increase with pine during the time when yellow-pine parkland expanded. Martin and Mosimann also derived similar patterns of correlation between pollen types within closely sampled (contiguous 6-mm samples) intervals and within pollen zones.

A. R. H. Martin (1968) used P. S. Martin and Mosimann's (1965) approach to delimit recurrent groups in a species-rich pollen diagram from Groenvlei, a site situated near the ecotone between forest and semi-arid heath vegetation in South Africa. The correlation net that A. R. H. Martin (1968) derived suggested that there were two large recurrent groups, representing *Podocarpus* forest and *Rhus* scrub, respectively. An *Artemisia–Stoebe* Compositae group representing dry sand-dune communities was strongly negatively correlated with the *Podocarpus* and *Rhus* groups. Three small, negatively correlated groups characterised by Leguminosae, Ericaceae, and *Anthospermum* were all negatively correlated with the other recurrent groups. Today, Leguminosae, Ericaceae, and *Anthospermum* grow together in Cape-heath communities. The correlation patterns from the Groenvlei pollen profile suggest the existence in the past of vegetation types that have no modern analogues today.

Figure 6.6 suggests some limitations to the use of correlation nets: a net with many taxa and many links can be difficult to interpret. It is thus important to position the points so that similar or correlated taxa are close together, and to have a clear idea about the degree of linking considered necessary for the definition of a recurrent group (presumably not all the taxa in a recurrent group need to be linked to one another; see Figure 6.6). Correlation nets would seem to be most useful for the investigation of a fairly small number of taxa; larger data sets could profitably be analysed by classification methods.

6.5.2 Discussion

In applying the recurrent-group approach to the reconstruction of past plant communities from pollen stratigraphical data, it is assumed throughout that the unit of study, the fossil pollen assemblage, is closely related in space and time to the life assemblage or past plant community that we are trying to reconstruct. Many of the pollen grains and spores that enter a site of deposition and subsequently form the fossil assemblage are derived from an undefinable pollen-source area and hence life assemblage (H. J. B. Birks and H. H. Birks, 1980). Many changes in the pollen assemblage may occur, due, for example, to processes of dispersal, sedimentation, and preservation, before it is finally preserved as a fossil pollen assemblage. There is thus no simple spatial relationship between a fossil pollen assemblage and the plant communities that produce it (see Subsection 6.3.3). Recurrent groups of fossil pollen and spores can only, therefore, reflect 'associations' in time, whereas a Quaternary palaeoecologist is concerned with associations of taxa both in time and space. More locally derived fossils—such as pollen from aquatic and mire taxa, seeds, and fruits, and many animal fossils—will tend to form fossil assemblages that are more closely linked in space to the life assemblages from which they were derived (H. J. B. Birks and H. H. Birks, 1980). The recurrent-group approach can thus be a valuable aid to reconstructing past communities from such fossil assemblages (e.g., Janssen, 1967b, 1972, 1973).

Despite this limitation, the recurrent-group approach can be of use in some Quaternary pollen analytical studies, particularly in instances in which the vegetation of the past appears to have no modern vegetational analogue, for example in interglacials, the late-glacial, and the early post-glacial (see Subsection 6.4.3).

6.6 Environmental Reconstructions

The final question in a palaeoecological study involving pollen analysis asks what was the environment in which the plant communities existed in the past (see

Figure 6.1). The reconstruction of past environments is the ultimate aim of many, if not all, Quaternary palaeoecological studies. It can be answered effectively only after all the available biological and lithological data have been assembled, evaluated, and correlated. Climatic, edaphic, and biotic (including human) changes are the most important determinants of late-Quaternary vegetational history. The reason why pollen analysis provides a potential means for reconstructing past environments is that a pollen assemblage at a particular time and place is a function of the regional flora and vegetation, and because the flora and vegetation of a region are largely controlled by the regional environment, there is therefore a relationship, admittedly a complex and indirect one, between a pollen assemblage and regional environment.

All environmental reconstructions from fossil assemblages, such as pollen spectra, require some knowledge of the present-day ecological tolerances and requirements of the organisms that are found as fossils in the stratigraphical record. The basic assumption of all environmental reconstructions is thus methodological uniformitarianism (see Section 1.1; Rymer, 1978; and H. J. B. Birks and H. H. Birks, 1980, for discussions of uniformitarianism), namely, that modern-day observations and relationships can be used as a model for past conditions and, more specifically, that the relationships between plants and their environment have not changed with time, at least for the time period of the late-Quaternary (see Faegri, 1950; D. Walker, 1978).

As in the reconstruction of past plant communities, reconstruction of past environments can be attempted using either a single 'indicator-species' approach or a multivariate assemblage approach. The environmental reconstructions that can be inferred from adopting an indicator-species approach are often quantitative (e.g., Iversen, 1944; Churchill, 1968), but we will not discuss this approach further here. Faegri (1950), M. B. Davis (1978), D. Walker (1978), T. Webb (1980), H. J. B. Birks and H. H. Birks (1980), and H. J. B. Birks (1981c) discuss the approach in detail and consider its assumptions, advantages, and limitations.

The assemblage approach attempts to establish empirical, often quantitative, relationships between modern assemblages of pollen and spores and present-day environmental variables. Such relationships are based on correlations between modern pollen and environmental data. Because of the large amount and type of data, numerical methods of data analysis are essential in deriving these quantitative relationships. The modern relationships are then applied to fossil assemblages to derive reconstructions of the past environment (Figure 6.1). We now discuss the assemblage approach, its application to environmental reconstruction from Quaternary pollen analytical data, and its assumptions and limitations.

6.6.1 Multivariate Assemblage Approach

The approach considers the pollen assemblage as a whole and the numerical proportions of the different pollen types within the assemblage. It assumes that a pollen assemblage can be related to the environment around the site of deposition by some function called a transfer or calibration function (Sachs, Webb, and Clark, 1977; Imbrie and Webb, 1981).

Transfer functions are a means of relating spatial arrays of assemblages of organisms to their environments. The models which will be described are (in the terminology presented in Section 2.1) empirical rather than explanatory. They can be summarised in the matrix equation

$$E_m = X_m T_m,$$

where X_m is a matrix of specified biological responses, such as the proportions of the taxa of interest within a defined domain of space and time; E_m is a matrix of environmental variables measured over the same time–space domain and assumed to be functionally related to X_m; and T_m is a matrix of modern transfer functions. This rather concise matrix equation will be explained more fully below; for the present, we need only regard the matrix T_m as a means of transforming the values taken by a set of biological variables X_m into values of environmental variables E_m. We stress that by expressing the relationship in this form, we are *not* implying that biological variables cause the environmental variables to take particular values, but rather that the environmental variables can be predicted from the biological variables by means of the matrix T_m. If the transfer functions are assumed to be invariant in time and space, $\hat{E}_f$, the estimates of the past environment, can be derived from a fossil assemblage X_f by

$$\hat{E}_f = X_f T_m.$$

The assemblage approach to environmental reconstruction has been used in an intuitive, non-quantitative way by pollen analysts since von Post (1918). For example, pollen assemblages may be interpreted as reflecting tundra, shrub–tundra, forest–tundra, birch forest, or pine forest. Inferences about past environment, particularly past climate, are based on the present-day environment in which these broad vegetation types are found (e.g., Cushing, 1967b; H. J. B. Birks, 1976b, 1981a). Traditionally, pollen analysts have attempted environmental reconstructions in two stages, as discussed by T. Webb (1971, 1983) and T. Webb and Bryson (1972). Firstly, the past vegetation V_f is reconstructed from fossil pollen assemblages P_f using either qualitative reasoning (as above) or quantitative modern pollen-representation factors R_m that compensate for the differential pollen representation of the taxa within the fossil assemblage (see Section 6.3), that is,

$$\hat{V}_f = P_f R_m.$$

(Note that R_m is used here in a slightly different sense from the R values of Sections 5.4 and 6.3). In the second stage, the vegetational reconstruction $\hat{V}_f$ is used as a basis for reconstructing the past climate C_f by

$$\hat{C}_f = \hat{V}_f D_m,$$

where D_m is a set of ecological functions that expresses the relationships between modern vegetation and its environment. The equations for these two stages can, however, be combined (T. Webb and Bryson, 1972) into

$$\hat{C}_{\mathrm{f}} = P_{\mathrm{f}}(R_{\mathrm{m}} D_{\mathrm{m}}).$$

Although the relationship R_{m} between modern pollen and vegetation has often been quantified (see Section 5.4), the relationship D_{m} between modern vegetation and its environment has rarely been quantified (cf. Bryson, 1966; Larsen, 1971). Because of the difficulty of estimating D_{m}, H. S. Cole and Bryson (1968), H. S. Cole (1969), T. Webb (1971), T. Webb and Bryson (1972), and later workers have adopted an alternative approach and attempted to correlate modern pollen assemblages directly with contemporary climate by means of a modern transfer function T_{m}, namely,

$$C_{\mathrm{m}} = P_{\mathrm{m}} T_{\mathrm{m}}.$$

The transfer function T_{m} is thus equivalent to the product $(R_{\mathrm{m}} D_{\mathrm{m}})$, but there is no longer the necessity to estimate R_{m} and D_{m} separately. Past climate is then reconstructed from fossil pollen assemblages by assuming that T_{m} is invariant in time and space; thus, estimates of past climate $\hat{C}_{\mathrm{f}}$ are obtained directly from the equation

$$\hat{C}_{\mathrm{f}} = P_{\mathrm{f}} T_{\mathrm{m}}. \tag{6.6.1}$$

Several mathematical methods have been used to estimate modern quantitative pollen–climate transfer functions, and these are reviewed by T. Webb and Clark (1977) and Sachs *et al.* (1977). Most of them are variants of the basic multiple linear regression model

$$c_{i1} = \beta_{01} + \sum_{k=1}^{t} \beta_{k1} p_{ik} + \varepsilon_{i1} \qquad (i = 1, \ldots, n), \tag{6.6.2}$$

where c_{i1} is some measure of the modern climate (such as mean July temperature) at site i, p_{ik} is a measure of the modern abundance of pollen taxon k at site i, ε_{11}, ..., ε_{n1} are error terms, and β_{01}, β_{11}, ..., β_{t1} are regression coefficients. For the specified values of the climatic and pollen variables, the estimates of the regression coefficients have been obtained by minimising

$$\sum_{i=1}^{n} \left(c_{i1} - \beta_{01} - \sum_{k=1}^{t} \beta_{k1} p_{ik} \right)^2. \tag{6.6.3}$$

If the coefficients which minimise expression (6.6.3) are denoted by $\hat{\beta}_{01}$, $\hat{\beta}_{11}$, ..., $\hat{\beta}_{t1}$, one can estimate the climatic variable c_{i1} by $\hat{c}_{i1}$, defined by

$$\hat{c}_{i1} = \hat{\beta}_{01} + \sum_{k=1}^{t} \hat{\beta}_{k1} p_{ik} \qquad (i = 1, \ldots, n). \tag{6.6.4}$$

If p_{ik} refers now to the abundance of the kth taxon in the ith stratigraphical sample, $\hat{c}_{i1}$ is an estimate of the climatic variable at the corresponding period of time.

This set of n equations can be expressed concisely in the matrix formula

(6.6.1), where T_m is a column vector holding the regression coefficients. If l different climatic variables are measured at each site, each of these would have a separate set of regression coefficients; the relationship could again be represented by the formula (6.6.1), where T_m is now a matrix with $t + 1$ rows and l columns.

The mathematics of multiple linear regression are discussed by Draper and Smith (1966), Cooley and Lohnes (1971), and Chatterjee and Price (1977). The mathematical assumptions of the method, as applied to pollen–climate transfer functions, are reviewed in detail by Howe and Webb (1977, 1983) and Bartlein and Webb (1985). Use of the method to derive estimates of the regression coefficients requires the following assumptions:

1. the correct pollen variables, the relative abundances of which are assumed to be a function of climate, are included on the right-hand side of Equation (6.6.2);
2. the functional relationship between modern pollen and present-day climate is correctly specified;
3. the modern pollen and climatic data are representative and adequate—further observations should not significantly alter the estimates—and the regression coefficients are constant throughout the area of study;
4. the values of the independent palynological variables are known without error;
5. for each climatic variable, the error component ε_{il} in one observation is statistically independent of the error terms in all other observations;
6. for each climatic variable, the error component has the same variance for all observations (for hypothesis-testing purposes, it is often further assumed that the error component in the climatic variables is normally distributed).

These assumptions cannot be completely justified in the analysis of palynological data. Problems include the following.

1. There may be considerable difficulty in satisfying assumption 2. For example, should the (p_{ik}) in Equation (6.6.2) denote the pollen proportions or some transformation of these proportions, in which the aim is to convert non-linear but monotonic relationships into linear ones, and should higher-degree polynomial terms be included in the model? Relevant investigative procedures are described by Box and Tidswell (1962) and Cook and Weisberg (1982), and applications to palynological data are presented by Howe and Webb (1977, 1983) and Bartlein and Webb (1985).

2. Concerning assumption 4, it is clear that the quantities and types of pollen released into the atmosphere can be influenced by the values taken by relevant climatic variables in the locality. However, the pollen variables p_{ik} used in expressions (6.6.3) and (6.6.4) are not these quantities, but are relative proportions based on sample counts from a site in the locality. Thus, in addition to sampling variability, extra sources of error introduced relate to the efficiency of dispersal of differ-

ent types of pollen grains and the manner in which they are incorporated into the sediment matrix (see T. Webb and McAndrews, 1976). The errors in the pollen variables could thus be of considerable magnitude (cf. Howe and Webb, 1983), and obtaining the estimates of the set of β_{kj}'s by least-squares linear regression as in expression (6.6.3) could lead to inaccurate predictions. This problem of calibration is one which has received some attention, and a discussion of some methods of analysis is presented by Brown (1982).

3. The values taken by a climatic variable at different locations in the same area will be related, making it questionable whether condition 5 will be satisfied. Howe and Webb (1983) describe a procedure for testing for the existence of any spatial autocorrelation of residual terms, noting that there was not strong evidence of such correlation in data from Lower Michigan which they analysed.

Despite these problems, one hopes that any strongly marked regional climatic variations can be modelled reasonably accurately for the broad spatial and temporal scales of interest, as long as the assumptions are only mildly violated (see I. C. Prentice, 1983b; Bartlein, Webb, and Fleri, 1984; Bartlein and Webb, 1985).

The question of which pollen variables should be included in Equation (6.6.2) (assumption 1) is one which has attracted much attention. Some aspects of variation within pollen data may have little or no relevance to regional climate. On occasion, one can argue on *a priori* ecological grounds that some pollen types should be excluded from the model, as they are likely to result in misleading predictions (see H. S. Cole, 1969; T. Webb and Bryson, 1972). Examples include pollen types that have high modern values as a result of recent human disturbance of vegetation (e.g., *Ambrosia*-type), that are produced locally at the site (e.g., *Sphagnum,* cf. Andrews and Nichols, 1981; Kay and Andrews, 1983; Cyperaceae, Gramineae, *Alnus*), or that are extra-regional in their origin (e.g., *Picea* and *Alnus* in the Canadian Arctic; cf. Andrews and Nichols, 1981).

We describe below three different numerical methods that attempt to detect and isolate any irrelevant, non-climatic variation within the pollen data, all of which—sometimes in combination with each other—have been used to derive pollen–climate transfer functions (see T. Webb and Clark, 1977).

Firstly, investigators have considered the step-wise addition or deletion of pollen variables, in an attempt to obtain a set of r (where r is less than t) variables which effectively model the variations in the climatic variables. Arigo, Howe, and Webb (1982), Howe and Webb (1983), and Bartlein and Webb (1985) describe the implementation of this approach in detail and discuss the computer programs required.

Secondly, a reduction in the number of pollen variables has been achieved by carrying out a principal components analysis (or other scaling) of the pollen data, and then using the first r components as the new variables to be substituted for p_{ik} in the right-hand side of Equation (6.6.2) (e.g., H. S. Cole, 1969; T. Webb and Clark, 1977). While such components may represent informative summaries of modern pollen data, it is not obvious that the major variations in the pollen data will always necessarily represent the most important factors for prediction of climatic variables.

These two methods model separately each climatic variable, using a variant of Equation (6.6.4). If several different climatic variables are to be modelled simultaneously using the same set of pollen data, one can employ a third numerical method, namely, a combined process of canonical correlation and regression analysis (T. Webb, 1971; T. Webb and Bryson, 1972). Canonical correlation analysis was outlined at the end of Section 5.3 (see Morrison, 1976, Chapter 7). Given a set of t pollen variables $(p_1, p_2, \ldots, p_t)$ and a set of l climatic variables $(c_1, c_2, \ldots, c_l)$, one seeks the linear combinations

$$g_j \equiv \sum_{k=1}^{t} a_{jk} p_k \quad \text{and} \quad h_j \equiv \sum_{k=1}^{l} b_{jk} c_k,$$

where $j = 1, 2, \ldots$, for which the correlation between g_j and h_j is maximised, subject to g_j and h_j being uncorrelated with all g_m and h_m (where m is smaller than j). The canonical variable pairs $\{(g_j, h_j)\ (j = 1, 2, \ldots)\}$ indicate which linear combinations of pollen and climatic variables are most closely related to one another. The jth climatic canonical variable, h_j, has then been linearly regressed on the jth pollen canonical variable, g_j (although other methods of relating the pairs of canonical variables merit study). Since each climatic variable can be expressed in terms of the climatic canonical variables (h_j), one has thus modelled the climatic variables in terms of the pollen canonical variables. By restricting attention to the first few pairs of canonical variables, one might hope to obtain a model for the climatic variables which is based on the relevant variation in the palynological data set. The mathematics of this procedure are described by Glahn (1968) and in the appendix of T. Webb and Bryson (1972). In addition, one can use the associated procedure of redundancy analysis (Stewart and Love, 1968; Cooley and Lohnes, 1971) to investigate the importance and reliability of sets of variables for predictive purposes (for examples, see Reyment, 1972, 1976a; Gittins, 1979, 1981; Huntley and Birks, 1979). Bryson and Kutzbach (1974, Table 1) showed, for example, that although four canonical variables accounted for 89% of the variance in their five-variable climatic data set, only 48% of the variance in the 10-variable modern pollen data was accounted for by these canonical variables. However, a higher proportion of the variance in the climate set is expected *a priori* to be accounted for, because there are twice as many variables in the pollen data set as in the climate set; such factors should be borne in mind when evaluating the results of any canonical correlation analysis. The remaining 52% of the variance in the pollen data may result from non-climatic sources such as soil factors, natural disturbances, anthropogenic effects, and local site factors, or from non-linear relationships between modern pollen and regional climate (T. Webb and Clark, 1977).

The above methods, and variants of them, have been used to derive pollen–climate transfer functions and to reconstruct past climate from pollen stratigraphical data by, for example, H. S. Cole and Bryson (1968), H. S. Cole (1969), T. Webb (1971, 1980, 1983), T. Webb and Bryson (1972), Bryson and Kutzbach

(1974), Howe and Webb (1977, 1983), T. Webb and Clark (1977), Kay (1979), Heusser and Streeter (1980), Heusser, Heusser, and Streeter (1980), Andrews, Mode, and Davis (1980), Andrews, Davis, Mode, Nichols, and Short (1981), Andrews and Diaz (1981), Andrews and Nichols (1981), Bernabo (1981), Bryson and Swain (1981), Heusser and Heusser (1981), Heusser, Streeter, and Stuiver (1981), R. W. Mathewes and Heusser (1981), Swain, Kutzbach, and Hastenrath (1983), Kay and Andrews (1983), Bartlein *et al.* (1984), and Bartlein and Webb (1985). Similar investigations have been conducted in marine palaeoecology by, for example, Imbrie and Kipp (1971), Imbrie, Van Donk, and Kipp (1973), Kipp (1976), Hutson (1977, 1978), Malmgren and Kennett (1978a, 1978b), and Molfino, Kipp, and Morley (1982); and in correlating climate and tree-ring thickness by, for example, Fritts *et al.* (1971), Fritts (1976), and Hughes, Kelly, Pilcher, and LaMarche (1982).

When we described attempts to use palynological data to predict vegetational composition (Sections 5.4 and 6.3), we stressed the importance of testing any models on a range of *independent* modern data before applying them to the interpretation of pollen stratigraphical data. A similar exercise, in which pollen–climate transfer functions are used with independent modern pollen data to predict contemporary climate at sites where the independent data were collected, is clearly just as important in the modelling and testing of modern pollen–climate relationships. T. Webb and Bryson (1972) and Bartlein *et al.* (1984) present the results of such tests. Further tests are required, at a variety of spatial scales, to evaluate modern pollen–climate transfer functions prior to their use in reconstructing past climate from pollen stratigraphical data. Such tests should include a study of the spatial distribution of residuals (differences between observed and predicted climatic variables) in order to highlight any areas where a particular transfer function performs poorly (see T. Webb and Bryson, 1972; Arigo, Howe, and Webb, 1982; Bartlein *et al.*, 1984; Bartlein and Webb, 1985).

6.6.2 Discussion

Although much has been written about the detailed mathematics of estimating transfer functions and about the mathematical assumptions of the different methods, there has been comparatively little discussion of the biological assumptions of the transfer-function approach, particularly when applied to pollen stratigraphical data. T. Webb and Bryson (1972), T. Webb and Clark (1977), Sachs *et al.* (1977), T. Webb (1980), and Imbrie and Webb (1981) outline the major biological assumptions (see also Imbrie and Kipp, 1971). They are as follows:

1. The ecological system of interest has not changed significantly since the time the fossil assemblage P_f was deposited (Ogden, 1977a). The taxa in P_m are assumed to be biologically similar to those in P_f, and their responses to individual environmental factors are assumed to be invariant with time (see Faegri, 1950; D. Walker, 1978). Interactions among taxa are also assumed not to have changed with time. Constancy in environmental response and

biological interactions for genera such as *Quercus, Pinus,* and *Betula* that contain many species with different ecological tolerances but morphologically similar pollen is only possible if the species complement remains constant through time (A. M. Solomon and Harrington, 1979). This seems unlikely for time spans of more than a few thousand years.

2. The assemblage of pollen types in P_j is systematically related to the environmental variables represented by C_j (where j indexes the modern set m or the fossil set f).
3. The environmental variables represented by C_j either are, or are linearly related to, the environmental factors that actually control the past and present distribution and abundance of the taxa represented in P_j.
4. Modern surface-sample data P_{m} provide relevant and adequate information for interpreting fossil data P_{f}, and thus the observed modern geographical patterns of covariance between pollen assemblages P_{m} and climate C_{m} are adequate for interpreting pollen stratigraphical changes through time. It is assumed that any correspondence between the spatial distribution of modern biological assemblages and geographical patterns of modern climate is equivalent to the effects of climatic change at a single geographical locality over hundreds or thousands of years (T. Webb and Clark, 1977). In other words, a direct correspondence between spatial patterns today and temporal changes in the past is assumed to exist.
5. The mathematical equations incorporating linear combinations of taxa model the responses of these taxa to environmental change accurately enough to yield reliable transfer functions.
6. Climate is the ultimate cause of change in the fossil pollen stratigraphical record, at least at the spatial and temporal scales of interest in a given study. At these scales, vegetation and species distributions and abundances, both past and present, are assumed to be in equilibrium with climate (H. J. B. Birks, 1981c).

This final assumption is currently a topic of controversy within Quaternary palaeoecology; see, for example, M. B. Davis (1978, 1981), T. Webb (1980), H. J. B. Birks (1981c), and I. C. Prentice (1983b). One hypothesis, favoured by T. Webb (1980, 1983) and implicit in the use of transfer functions for reconstructing climate from pollen stratigraphical data (e.g., T. Webb and Bryson, 1972; Howe and Webb, 1983; Bartlein *et al.,* 1984; Bartlein and Webb, 1985), proposes that climatic change is the primary determinant of many, if not all, of the major observed pollen stratigraphical changes during the late-Quaternary, and thus that vegetation and species distributions are in equilibrium with climate over long periods of time.

To provide evidence to support this hypothesis, T. Webb (1980, 1983) and Bartlein and Webb (1985) argue that the spatial patterns of pollen stratigraphical changes should be mapped. As broad-scale geographical patterns of modern pollen are similar to contemporary vegetation patterns, for example, at the scale of vegetation formations (see Section 5.3), and as broad-scale vegetation patterns often resemble broad-scale patterns of modern climate (see Bryson, 1966; Hare and Ritchie, 1972; cf. Rowe, 1966), geographically coherent and consistent patterns of pollen stratigraphical changes over large areas in the past are, according to T. Webb (1980, 1983), Howe and Webb (1983), and I. C. Prentice (1983b), likely to be the result of regional climatic change. M. B. Davis (1976, 1981) has mapped the ages of the first arrival and expansion of several tree genera in eastern North America since the last glaciation. These maps demonstrate broad-scale, coherent geographical patterns that fulfill the criteria of T. Webb's (1980) test. Their interpretation in terms of direct climatic control or non-climatic, migrational lags remains, however, an open question despite their apparent geographical coherency.

An alternative hypothesis concerning the major determinants of pollen stratigraphical change was proposed by Faegri (1950, 1963) and Iversen (1960) and adopted and developed by Watts (1973, 1982), Hare (1976), Wright (1976), M. B. Davis (1976, 1978, 1981), Ritchie (1977, 1981), A. M. Solomon and Harrington (1979), H. J. B. Birks (1981c, 1985), and Brubaker *et al.* (1983). It suggests that climatic change may not be the major determinant of many of the major pollen stratigraphical changes during the last 10,000–12,000 years in northwestern Europe or much of eastern North America, and that many of these changes may simply reflect the progressive migration and subsequent expansion of trees into new areas from their glacial refugia (see Faegri, 1963, 1974; M. B. Davis, 1976, 1981; Huntley and Birks, 1983; H. J. B. Birks, 1985). Such migrations were almost certainly stimulated initially by large and possibly rapid changes in climate at the end of the last glaciation (Iversen, 1960; Hare, 1976; Watts, 1982). Each species may then have migrated in different dirctions over many thousands of kilometres and at different rates of 100 to 1000 m per year (M. B. Davis, 1981; Huntley and Birks, 1983; H. J. B. Birks, 1985) until it reached its natural climatic limits and attained an equilibrium with climate. Some species may have reached these limits soon after the onset of the post-glacial, whereas others may not yet have reached their climatic limits even after the 10,000 years since deglaciation. Such differences between species may result from differences in, for example, inherent rates of spread, location of glacial refugia, and ecological and competitive tolerances (Rowe, 1966; M. B. Davis, 1981; Watts, 1982). It is thus possible that the present-day geographical distribution of some taxa are not in equilibrium with regional climate of today (cf. assumptions 3, 4, and 6 above). Similarly, in the past some species distributions may not have been in equilibrium with climate, even though they may be in equilibrium at the present day (see Tsukada, 1981, 1982a). Thus pollen stratigraphical changes that record the unidirectional arrival and expansion into an area of species that were limited in their spread by their rates of migration cannot be assumed to reflect regional climatic changes that were synchronous, or nearly so, with the observed palynological changes (Faegri, 1963, 1974). However, once a species has reached its climatic limits, changes in its distribution (advance or retreat) may then be controlled by climatic change (see, e.g., Faegri, 1950, 1963; Hyvärinen, 1975; M. B. Davis, 1978; Jacobson, 1979; H. J. B. Birks, 1981c; Bartlein *et al.*, 1984).

At present the falsification of one or both of these hypotheses represents a critically important problem in terrestrial Quaternary palaeoecology. Faegri (1950) warned against mistakes that could be made 'by confusing migration limits with climatic ones.' I. C. Prentice (1983b) discusses possible approaches and suggests the use of simulation models in an attempt to disentangle the effects of migrational lags from the effects of climatic change on the observed changes in pollen stratigraphy.

Further problems can arise when there have been delays in migration due not only to inherently slow rates of spread but also to barriers to migration, such as mountain ranges, unfavourable climate, and competition from existing vegetation. As a result of delayed migration, combinations and abundances of taxa and the interactions among them may have been different in the past, resulting in vegetation types that have no modern analogues. Such no-analogue vegetation types are not uncommon in the late-Quaternary (see Section 6.4; cf. H. R. Delcourt, P. A. Delcourt, Webb, and Overpeck, 1983; Overpeck and Webb, 1983). No-analogue vegetation types indicate that some taxa may once have been more abundant than they are today (H. J. B. Birks, 1981c), that some taxa grew together even though they do not occur together today (Watts, 1973), and that some taxa may possibly have had wider ecological amplitudes at the end of the last glaciation than they have today, possibly because of the absence of species that migrated later (D. Walker, 1978). The fossil pollen assemblages may, on the other hand, reflect vegetation types that developed in response to a different climatic seasonality or to a unique combination of climatic and other environmental factors at the time of rapid ice retreat at the end of the last glaciation (see Iversen, 1954; Cushing, 1963; T. Webb and Bryson, 1972; H. J. B. Birks, 1976b, 1981a; D. Walker, 1978; Amundson and Wright, 1979). We should clearly be cautious (see also T. Webb and Clark, 1977; Barnosky, 1981; Bartlein *et al.*, 1984) in accepting any reconstruction of past climate based on quantitative transfer functions applied to fossil pollen assemblages that have no satisfactory modern analogues (cf. T. Webb and Bryson, 1972; Bryson and Kutzbach, 1974; Heusser, Heusser, and Streeter, 1980; Heusser and Heusser, 1981; Heusser *et al.*, 1981).

Some workers have attempted quantitative climatic reconstructions only for the last 3000–6000 years of the post-glacial, a period when all the major trees concerned had migrated into the area of interest (e.g., Bernabo, 1981; Andrews *et al.*, 1981; Howe and Webb, 1983), thereby avoiding the no-analogue problem that is so prevalent in late-glacial and early post-glacial vegetational history. In such studies, the reconstructed changes in climate are often small (0.5–2° C) relative to the standard deviations associated with the estimates, which can be as high as 1.3° C. Such standard deviations should be regarded as providing conservative estimates of the likely errors, as they are dependent on the model of pollen–climate relationships having been accurately specified (cf. Howe and Webb, 1977, 1983; Bartlein and Webb, 1985). It can be a useful exercise to obtain estimates of the past climatic variables from several different models (e.g., Hutson, 1977; T. Webb and Clark, 1977; Kay and Andrews, 1983), different groups of organisms (e.g., Molfino *et al.*, 1982), and independent palaeoclimatic records (T. Webb, 1980, 1983; H. J. B. Birks, 1981c; Howe and Webb, 1983; Bartlein *et al.*, 1984). If the predictions do not differ markedly, one may have more confidence in the reliability of the estimates.

Pollen–climate transfer functions, carefully estimated so that violations of the assumptions of a linear calibration model are minimised and critically applied to well-dated contemporaneous pollen data from many sites within a climatically sensitive area, provide an important means of reconstructing quantitatively the spatial and temporal patterns of climatic change. Bartlein *et al.*'s (1984) study in the American Midwest exemplifies this approach, and provides climatic reconstructions in a form that permits direct comparison with results of palaeoclimatic simulations from general circulation models (e.g., Kutzbach, 1981). The estimates of past climate based on pollen stratigraphical data presented by Bartlein *et al.* (1984) are corroborated by independent evidence for climatic change from fluctuations in fluvial activity and lake levels. The importance of using independent evidence, wherever possible, to test climatic reconstructions from pollen data is also illustrated by Swain *et al.* (1983), where pollen and lake-level data provide independent but broadly comparable estimates of precipitation changes between 3500 and 10,500 B.P. in northwest India.

The main limitation of the assemblage approach (and the indicator-species approach) is that, although it is possible to derive statistically significant correlations between modern pollen assemblages and a variety of climatic variables, such correlations *on their own* do not prove that there is a cause-and-effect relationship between modern pollen and contemporary climate (Faegri, 1950; H. J. B. Birks, 1981c). L. C. Cole (1957) and Scott (1979) provide valuable (and amusing) discussions of the problem of so-called spurious or nonsense correlations (*sensu* Yule, 1926) in biology. To establish cause-and-effect relationships we need to learn considerably more than we currently know about the ecology, physiology, and population biology of the taxa concerned and about the response of plants to climatic change. The detailed work by Black and Bliss (1980) on the ecology of *Picea mariana* and by Pigott and Huntley (1981) on *Tilia cordata* and its reproductive biology in relation to climate are particularly important contributions to Quaternary palaeoecology. Both studies demonstrate that there is no simple relationship between tree distribution and climate, and that a complex of factors, both climatic and non-climatic, appears to influence the abundance, performance, and distribution of these taxa at a variety of spatial and temporal scales. Similarly, Loubère (1982) discusses the importance of biological factors in influencing the distribution and abundance of foraminifers in the oceans today, and outlines some of the problems in interpreting marine fossil assemblages in terms of past climates. Faegri (1950), Carter and Prince (1981), Hengeveld and Haeck (1981, 1982), and Hengeveld (1982) discuss models of the distribution and abundance of species in equilibrium with regional climate. These models illustrate the difficulties in predicting and understanding the limits of distribution even when the distribution can be assumed to be in equilibrium with climate. Clearly much has to be discovered about the ecological control of the present-day geographical distribution and abundance of taxa before we can use modern distributions and abundances, either of individual taxa or assemblages, as a basis for the reconstruction of past climate.

Besides reconstructing regional climate from palaeoecological data, numerical methods can also be used to reconstruct other environmental factors such as local conditions within the site of deposition, for example, water depth, water chemistry, and substrate type (see Figure 6.1). Although we are unaware of detailed quantitative reconstructions in Quaternary palaeoecology, work by Burnaby (1961) and Reyment (1976a, 1979, 1980a) on Cretaceous foraminifers and ostracods in relation to water depth and chemistry are elegant examples of quantitative palaeoenvironmental reconstructions (see Fox, 1968 and Wilson, 1980, for related examples). Binford (1982) has used the results of numerical analyses of late-Quaternary ostracod and cladoceran assemblages to reconstruct, in a semi-quantitative way, changes in water level, salinity, and substrate over the last 12,000 years at Lake Valencia, Venezuela (see Section 3.7). These studies are guides to how quantitative local environmental reconstructions can be made using cladoceran, molluscan, diatom, or plant macrofossils preserved in late-Quaternary sequences.

In some instances, numerical methods can clearly aid the palaeoecologist in the task of reconstructing past environments, but as in all quantitative Quaternary palaeoecology, mathematical methods must not be viewed as a substitute for sound and critical ecological knowledge. It is important not only that the numerical methods of analysis employed be appropriate for the type of data and the aims of the study, but also that any quantitative environmental reconstructions from Quaternary pollen stratigraphical data be based on a critical interaction between quantitative methods, knowledge of the ecology of the taxa concerned and of environmental patterns and processes, and common sense. Pollen analysts should always remember Faegri's (1966, p. 138) comments about the interpretation of pollen diagrams: 'It presumes a very intimate knowledge of the ecology and sociology of the vegetation types concerned. ... pollen analysis of any vegetation type without such knowledge is bound to become at its best a lifeless stratigraphical tool, at its worst useless altogether'.

Appendix: The Program ZONATION

The classification procedures described in Sections 3.3 to 3.5 have been implemented in the FORTRAN IV program ZONATION. This program has five main subroutines:

1. PREP: Reads in the pollen counts, and carries out some preliminary manipulations on the data.
2. CSLINK: Carries out a constrained single link analysis of the data (Section 3.3).
3. SPLINF: Carries out a binary divisive analysis of the data, using the information content criterion (Section 3.4).
4. SPLSQ: Carries out a binary divisive analysis of the data, using the sum-of-squares criterion (Section 3.4).
5. DNAMIC: Obtains the overall optimal sum-of-squares partition, using a dynamic programming algorithm (Section 3.5).

A listing of the program is given in Figure A1. The required format for input to the program is as follows:

Title line
NPOL, NSAM, NITERS, DET, WEIGHT in format (3I3, 2A1)
[NPOL = number of different pollen types
NSAM = number of samples taken
NITERS = number of divisions to be undertaken in the subroutines SPLINF, SPLSQ, DNAMIC; if this is left blank, a default value of 10 is chosen
DET = parameter controlling the amount of information which is provided about the constrained single link analysis: DET = blank: sufficient information is provided to allow construction of the constrained single link dendrogram; DET = any non-blank character: complete information is provided on each amalgamation.

```
C PROGRAM ZONATION - VERSION OF 29.5.79
C
C MAXIMUM OF 89 SAMPLES AND 18 POLLEN TYPES
C
        INTEGER T
        COMMON NPOL,NSAM,PERCT(18,99),B,HOLD(99,2),
     1  W(20),SPLIT(99,4),NMARK,DC(5000),DET,WEIGHT,NITERS,
     2  NBIT(100),DIAG(100),SPLURG(100),LEAST(100),
     3  NCOUNT,NLEV,NCLUST,TINY,HUGE,T(100),DEND(100),PREV
        CALL PREP
        CALL CSLINK
        CALL SPLINF
        CALL SPLSQ
        CALL DNAMIC
        STOP
        END


        SUBROUTINE PREP
        INTEGER T
        COMMON NPOL,NSAM,PERCT(18,99),B,HOLD(99,2),
     1  W(20),SPLIT(99,4),NMARK,DC(5000),DET,WEIGHT,NITERS,
     2  NBIT(100),DIAG(100),SPLURG(100),LEAST(100),
     3  NCOUNT,NLEV,NCLUST,TINY,HUGE,T(100),DEND(100),PREV
        DIMENSION POL(2,18),NDATA(18,99),NSUM(99),ID(20)
        DATA BL/1H /
C
C READS IN POLLEN COUNTS
C
        READ(5,503)  (ID(I), I = 1,20)
        WRITE(6,602)  (ID(I), I = 1,20)
        READ(5,501)  NPOL,NSAM,NITERS,DET,WEIGHT
        WRITE(6,603) NPOL,NSAM
        IF (WEIGHT.NE.BL) GO TO 8
        DO 9 I = 1,NPOL
    9   W(I) = 1.0
        GO TO 10
C
C READS IN TAXA WEIGHTS FOR SPLITLSQ
C
    8   READ(5,504) (W(I), I = 1,NPOL)
C
C NUMBER OF DIVISIONS FOR SPLITINF, SPLITLSQ, AND DYNAMIC :
C DEFAULT VALUE IS 10; MAXIMUM ALLOWED IS 20
C
   10   IF (NITERS) 11,11,12
   11   NITERS = 10
   12   CONTINUE
        DO 1 I = 1,NPOL
    1   READ(5,502) POL(1,I),POL(2,I),(NDATA(I,J),J=1,NSAM)
C
C CONVERTS DATA TO PROPORTIONS
C
        DO 2 J = 1,NSAM
        NSUM(J) = 0
        DO 2 I = 1,NPOL
    2   NSUM(J) = NSUM(J) + NDATA(I,J)
        WRITE(6,601)
        DO 3 J = 1,NSAM
        K = NSUM(J)
        X = 1.0/FLOAT(K)
        DO 4 I = 1,NPOL
        L = NDATA(I,J)
    4   PERCT(I,J) = X*FLOAT(L)
    3   WRITE(6,604) J,NSUM(J),(PERCT(I,J),BL,I = 1,NPOL)
        WRITE(6,605) (POL(1,I),POL(2,I),BL, I = 1,NPOL)
C
C CONSTRUCTS DISSIMILARITY MATRIX FOR CONSLINK
C
        KK = 0
        DO 5 I = 2,NSAM
        I1 = I - 1
        DO 6 J = 1,I1
        J1 = J + KK
        TOT = 0.0
        DO 7 K = 1,NPOL
        X = PERCT(K,I) - PERCT(K,J)
    7   TOT = TOT + ABS(X)
    6   DC(J1) = TOT
    5   KK = KK + I1
        RETURN
  501   FORMAT(3I3,2A1)
  502   FORMAT(A2,A4,2X,18I4/(20I4))
  503   FORMAT(20A4)
  504   FORMAT(16F5.1)
  601   FORMAT('1SAMPLE PROPORTIONS',//)
  602   FORMAT(1H1,20X,20A4,//)
  603   FORMAT(' NUMBER OF TAXA =',I4,4X,'NUMBER OF SAMPLES =',I4,//)
  604   FORMAT(I3,I5,16(F6.3,A1)/(12X,15(F6.3,A1)))
  605   FORMAT(' NO  SUM ',16(A2,A4,A1)/(12X,15(A2,A4,A1)))
        END


        SUBROUTINE CSLINK
        INTEGER T
        COMMON NPOL,NSAM,PERCT(18,99),B,HOLD(99,2),
     1  W(20),SPLIT(99,4),NMARK,DC(5000),DET,WEIGHT,NITERS,
     2  NBIT(100),DIAG(100),SPLURG(100),LEAST(100),
     3  NCOUNT,NLEV,NCLUST,TINY,HUGE,T(100),DEND(100),PREV
C
C CONSTRAINED SINGLE LINK ANALYSIS
C
        WRITE(6,601)
        CALL SETUP
        NSAM1 = NSAM - 1
        DO 1 NUMBER      =      1,NSAM1
        CALL MINIM
        CALL GROUP
        IF (NCLUST - 1) 2,2,1
    1   CONTINUE
    2   CONTINUE
        RETURN
  601   FORMAT('1RESULTS OF CONSLINK',//)
        END


        SUBROUTINE SETUP
        INTEGER T
        COMMON NPOL,NSAM,PERCT(18,99),B,HOLD(99,2),
     1  W(20),SPLIT(99,4),NMARK,DC(5000),DET,WEIGHT,NITERS,
     2  NBIT(100),DIAG(100),SPLURG(100),LEAST(100),
     3  NCOUNT,NLEV,NCLUST,TINY,HUGE,T(100),DEND(100),PREV
        DATA STAR,SLASH,BLANK/1H*,1H/,1H /
C
C PRELIMINARIES TO CONSTRAINED SINGLE LINK ANALYSIS :
C GROUP BOUNDARIES HELD IN THE ARRAY NBIT
C
        HUGE     =       0.0
        K        =       0
        DO 1 I   =       2,NSAM
        I1       =       I - 1
        T(I)     =       K
        K1       =       K + 1
        KI1      =       K + I1
        DIAG(I1)         =       DC(KI1)
        SPLURG(I1)       =       SLASH
        NBIT(I1)         =       0
        HUGE     =       HUGE + DIAG(I1)
```

```
1         K          =          KI1
          NBIT(NSAM) = 0
          IF (DET.NE.BLANK) GO TO 2
          WRITE(6,601)
2         CONTINUE
          PREV       =          0.0
          NCLUST     =          NSAM
          NLEV       =          0
          RETURN
601       FORMAT(//,' DETAILS OF EACH AMALGAMATION ARE SUPPRESSED')
          END


          SUBROUTINE MINIM
          INTEGER T
          COMMON NPOL,NSAM,PERCT(18,99),B,HOLD(99,2),
     1     W(20),SPLIT(99,4),NMARK,DC(5000),DET,WEIGHT,NITERS,
     2     NBIT(100),DIAG(100),SPLURG(100),LEAST(100),
     3     NCOUNT,NLEV,NCLUST,TINY,HUGE,T(100),DEND(100),PREV
C
C IDENTIFIES NEXT AMALGAMATION(S)
C
          NSAM1      =          NSAM - 1
          NCOUNT     =          1
          LEAST(1)              =          1
          TINY       =          DIAG(1)
          DO 1 I     =          2,NSAM1
          IF (TINY - DIAG(I)) 1,2,3
3         NCOUNT     =          1
          TINY       =          DIAG(I)
          LEAST(1)              =          I
          GO TO 1
2         NCOUNT     =          NCOUNT + 1
          LEAST(NCOUNT)         =          I
1         CONTINUE
          RETURN
          END


          SUBROUTINE GROUP
          INTEGER T
          COMMON NPOL,NSAM,PERCT(18,99),B,HOLD(99,2),
     1     W(20),SPLIT(99,4),NMARK,DC(5000),DET,WEIGHT,NITERS,
     2     NBIT(100),DIAG(100),SPLURG(100),LEAST(100),
     3     NCOUNT,NLEV,NCLUST,TINY,HUGE,T(100),DEND(100),PREV
          DATA STAR,SLASH,BLANK/1H*,1H/,1H /
C
C CARRIES OUT NEXT AMALGAMATION(S), UPDATING ARRAYS
C
          DO 8 I     =          1,NCOUNT
          J          =          LEAST(I)
          SPLURG(J)             =          STAR
          DIAG(J)               =          HUGE
          IF (NBIT(J)) 999,2,3
2         NBIT(J)               =          1
          J1         =          J
          GO TO 5
3         NBIT(J)               =          0
          DO 4 K     =          1,J
          J1         =          J - K
          IF (NBIT(J1)) 999,4,5
4         CONTINUE
          GO TO 999
5         J11        =          J1 - 1
          JJ         =          J + 1
          IF (NBIT(JJ)) 999,12,13
12        NBIT(JJ)              =          1
          J2         =          JJ
          GO TO 15
13        NBIT(JJ)              =          0
          NSAMJ      =          NSAM - JJ
          DO 14 K               =          1,NSAMJ
          J2         =          JJ + K
          IF (NBIT(J2)) 999,14,15
14        CONTINUE
          GO TO 999
15        J21        =          J2 + 1
          IF (J11) 999,21,22
22        DO 23 NUM             =          JJ,J2
          IWT        =          T(NUM) + J11
          D          =          DC(IWT)
          IF (D - DIAG(J11)) 24,23,23
24        DIAG(J11)             =          D
23        CONTINUE
21        IF (J2 - NSAM) 25,1,999
25        DO 27 NUM             =          J1,J
          JWT        =          T(J21) + NUM
          D          =          DC(JWT)
          IF (D - DIAG(J2)) 28,27,27
28        DIAG(J2)              =          D
27        CONTINUE
1         CONTINUE
          IF (TINY - PREV) 6,7,7
6         DEND(J)               =          PREV
          GO TO 8
7         DEND(J)               =          TINY
8         CONTINUE
          NSAM1      =          NSAM - 1
          NLEV       =          NLEV + 1
          NCLUST     =          NCLUST - NCOUNT
          IF (DET.EQ.BLANK) GO TO 16
C
C DETAILED INFORMATION ON AMALGAMATION(S)
C
          WRITE(6,602) NLEV,TINY,(I,SPLURG(I), I = 1,NSAM1),NSAM
          WRITE(6,603) NCLUST
          DO 26 I               =          1,NCOUNT
          J          =          LEAST(I)
26        SPLURG(J)             =          BLANK
16        IF (NCLUST - 1) 999,31,32
31        WRITE(6,604) NLEV
C
C PRINTS BRANCHES OF SINGLE LINK DENDROGRAM
C
          WRITE(6,601)
          DO 9 I     =          1,NSAM1
9         WRITE(6,605) I,DEND(I)
          WRITE(6,605) NSAM
          RETURN
32        CONTINUE
          IF (PREV - TINY) 10,10,11
10        PREV       =          TINY
11        CONTINUE
999       RETURN
601       FORMAT('1CONSTRAINED SINGLE LINK DENDROGRAM',//)
602       FORMAT(/,' AMALGAMATION NO',I4,' AT ',F8.4/' GROUPS ARE'//
     1     (15(I4,1X,A1,1X)))
603       FORMAT(' NUMBER OF GROUPS AT THIS STAGE IS ',I5,//)
604       FORMAT(' TOTAL NUMBER OF AMALGAMATIONS IS ',I5)
605       FORMAT(I4,1X,F10.3)
          END


          SUBROUTINE SPLINF
          COMMON NPOL,NSAM,PERCT(18,99),B,HOLD(99,2),
     1     W(20),SPLIT(99,4),NMARK,DC(5000),DET,WEIGHT,NITERS,
     2     NBIT(100),DIAG(100),SPLURG(100),LEAST(100),
     3     NCOUNT,NLEV,NCLUST,TINY,HUGE,T(100),DEND(100),PREV
```

Figure A1 Listing of the program ZONATION.

```
      DIMENSION NVEC(20),SUMT1(18),SUMT2(18)
BINARY DIVISION USING INFORMATION CONTENT CRITERION
      B = 1.0/0.69314718
      WRITE(6,600)
      CALL INF(1,NSAM,TOTINF,SUMT1,SUMT2)
      WRITE(6,601) TOTINF
      GTOT = 0.01*TOTINF
      ITER = 1
      CALL INF2(1,NSAM,TOTINF,1)
      TOTINF = HOLD(1,1)
      H = HOLD(1,2)
      NVEC(1) = IFIX(H)
      NMARK = 1
      PERC = TOTINF/GTOT
      WRITE(6,602) TOTINF,PERC,NVEC(1)
      L1 = 1
      L2 = NVEC(NMARK)
      L3 = L2 + 1
      L4 = NSAM
      NGP = 1
4     NGP = NGP + 1
      IF (L2 - L1 - 1) 1,2,7
2     CONTINUE
      CALL INF(L1,L2,TOTF1,SUMT1,SUMT2)
      SPLIT(1,1) = FLOAT(L1)
      SPLIT(1,2) = FLOAT(L2)
      SPLIT(1,3) = FLOAT(L1)
      SPLIT(1,4) = TOTF1
      GO TO 8
1     NGP = NGP - 1
      NGP1 = NGP - 1
      DO 9 NUM = 1,NGP1
      NUM1 = NUM + 1
      DO 9 L = 1,4
9     SPLIT(NUM,L) = SPLIT(NUM1,L)
      GO TO 8
7     CONTINUE
      CALL INF(L1,L2,TOTF1,SUMT1,SUMT2)
      CALL INF2(L1,L2,TOTF1,1)
8     IF (L4 - L3 - 1) 11,12,10
12    CONTINUE
      CALL INF(L3,L4,TOTF2,SUMT1,SUMT2)
      SPLIT(NGP,1) = FLOAT(L3)
      SPLIT(NGP,2) = FLOAT(L4)
      SPLIT(NGP,3) = FLOAT(L3)
      SPLIT(NGP,4) = TOTF2
      GO TO 13
11    NGP = NGP - 1
      GO TO 13
10    CONTINUE
      CALL INF(L3,L4,TOTF2,SUMT1,SUMT2)
      CALL INF2(L3,L4,TOTF2,NGP)
13    CONTINUE
      CALL RSORT(SPLIT,20,4,4,NGP,1)
      IF (ITER - NITERS) 3,5,999
3     TOTINF = TOTINF - SPLIT(1,4)
      NMARK = NMARK + 1
      S1 = SPLIT(1,1)
      S2 = SPLIT(1,2)
      S3 = SPLIT(1,3)
      L1 = IFIX(S1)
      L2 = IFIX(S3)
      L3 = L2 + 1
      L4 = IFIX(S2)
      NVEC(NMARK) = L2
      CALL RADSRT(NVEC,NMARK)
      PERC = TOTINF/GTOT
      WRITE(6,602) TOTINF,PERC,(NVEC(NUM),NUM = 1,NMARK)
      ITER = ITER + 1
      GO TO 4
5     WRITE(6,603)
      DO 6 NUM = 1,NGP
      S1 = SPLIT(NUM,1)
      S2 = SPLIT(NUM,2)
      S3 = SPLIT(NUM,3)
      L1 = IFIX(S1)
      L2 = IFIX(S2)
      L3 = IFIX(S3)
      PERC = SPLIT(NUM,4)/GTOT
6     WRITE(6,602)  SPLIT(NUM,4),PERC,L1,L2,L3
999   RETURN
600   FORMAT('1RESULTS OF SPLITINF',//)
601   FORMAT(' INF CONTENT    PERCENT OF TOTAL         MARKERS',/,2X,
     1  F10.6)
602   FORMAT(2X,F10.6,8X,F6.2,8X,20I4/(38X,18I4))
603   FORMAT('0FURTHER SPLITS ARE')
      END

      SUBROUTINE INF(L1,L2,TOT,TOT1,TOT2)
      COMMON NPOL,NSAM,PERCT(18,99),B,HOLD(99,2),
     1  W(20),SPLIT(99,4),NMARK,DC(5000),DET,WEIGHT,NITERS,
     2  NBIT(100),DIAG(100),SPLURG(100),LEAST(100),
     3  NCOUNT,NLEV,NCLUST,TINY,HUGE,T(100),DEND(100),PREV
      DIMENSION TOT1(18),TOT2(18)
C
C FINDS THE INFORMATION CONTENT OF THE GROUP CONSISTING
C OF LEVELS L1 TO L2 INCLUSIVE
C
      TOT = 0.0
      DO 1 I = 1,NPOL
      TOT1(I) = 0.0
      TOT2(I) = 0.0
      DO 2 J = L1,L2
      P = PERCT(I,J)
      IF (P) 999,2,3
3     TOT1(I) = TOT1(I) + P
      TOT2(I) = TOT2(I) + P*B*ALOG(P)
2     CONTINUE
      Z = TOT1(I)
      IF (Z) 999,1,4
4     TOT = TOT + TOT2(I)-Z*B*ALOG(Z/FLOAT(L2-L1+1))
1     CONTINUE
999   RETURN
      END

      SUBROUTINE INF2(L1,L2,TOTF1,KK)
      COMMON NPOL,NSAM,PERCT(18,99),B,HOLD(99,2),
     1  W(20),SPLIT(99,4),NMARK,DC(5000),DET,WEIGHT,NITERS,
     2  NBIT(100),DIAG(100),SPLURG(100),LEAST(100),
     3  NCOUNT,NLEV,NCLUST,TINY,HUGE,T(100),DEND(100),PREV
      DIMENSION SUMT1(18),SUMT2(18),SUMB1(18),SUMB2(18)
      DATA C/0.999/
C
C FINDS OPTIMAL (INFORMATION CONTENT) BINARY DIVISION OF THE GROUP
C L1 TO L2, GIVING INFORMATION ABOUT 'NEAR-OPTIMAL' DIVISIONS
C
      CALL INF(L1,L1,TOT,SUMT1,SUMT2)
      L7 = L1 + 1
      L8 = L2 - 1
      CALL INF(L7,L2,TOTAL,SUMB1,SUMB2)
      HOLD(1,1) = TOT + TOTAL
      HOLD(1,2) = FLOAT(L1)
      DO 1 NUM = L7,L8
      TOT = 0.0
```

```
      TOTAL = 0.0
      DO 2 I = 1,NPOL
      P = PERCT(I,NUM)
      IF (P) 999,8,6
    6 SUMT1(I) = SUMT1(I) + P
      SUMT2(I) = SUMT2(I) + P*B*ALOG(P)
      SUMB1(I) = SUMB1(I) - P
      SUMB2(I) = SUMB2(I) - P*B*ALOG(P)
    8 IF (SUMT1(I)) 9,9,10
   10 X = SUMT1(I)/FLOAT(NUM - L1 + 1)
      TOT = TOT + SUMT2(I) - SUMT1(I)*B*ALOG(X)
    9 IF (SUMB1(I)) 2,2,7
    7 X = SUMB1(I)/FLOAT(L2 - NUM)
      TOTAL = TOTAL + SUMB2(I) - SUMB1(I)*B*ALOG(X)
    2 CONTINUE
      NUM1 = NUM - L1 + 1
      HOLD(NUM1,1) = TOT + TOTAL
    1 HOLD(NUM1,2) = FLOAT(NUM)
      LL = L2 - L1
      CALL RSORT(HOLD,99,2,1,LL,0)
      SPLIT(KK,1) = FLOAT(L1)
      SPLIT(KK,2) = FLOAT(L2)
      SPLIT(KK,3) = HOLD(1,2)
      SPLIT(KK,4) = TOTF1 - HOLD(1,1)
      K2 = 2
      H = HOLD(1,2)
      K3 = IFIX(H)
    4 IF (HOLD(1,1) - HOLD(K2,1)) 11,5,999
    5 H = HOLD(K2,2)
      K1 = IFIX(H)
      WRITE(6,601) K1,L1,L2
      K2 = K2 + 1
      GO TO 4
   11 IF (HOLD(1,1) - C*HOLD(K2,1)) 3,12,12
   12 H = HOLD(K2,2)
      K1 = IFIX(H)
      WRITE(6,602) K3,K1,L1,L2,HOLD(1,1),HOLD(K2,1)
      K2 = K2 + 1
      GO TO 4
    3 CONTINUE
  999 RETURN
  601 FORMAT(' ALTERNATIVELY SI MARKER =',I4,' IN GROUP',I4,' -',I4)
  602 FORMAT('  *** INVESTIGATE',2I4,' IN GROUP',I4,' -',I4,/,
     1 ' CONTENTS ARE '2F10.4)
      END


      SUBROUTINE SPLSQ
      COMMON NPOL,NSAM,PERCT(18,99),B,HOLD(99,2),
     1 W(20),SPLIT(99,4),NMARK,DC(5000),DET,WEIGHT,NITERS,
     2 NBIT(100),DIAG(100),SPLURG(100),LEAST(100),
     3 NCOUNT,NLEV,NCLUST,TINY,HUGE,T(100),DEND(100),PREV
      DIMENSION NVEC(20),SUMT1(18),SUMT2(18)
      DATA BL/1H /
C
C BINARY DIVISION USING SUM OF SQUARES CRITERION
C
      WRITE(6,600)
      IF (WEIGHT.EQ.BL) GO TO 14
      WRITE(6,604) (W(I), I = 1,NPOL)
      GO TO 15
   14 WRITE(6,605)
   15 CONTINUE
      CALL SSQ(1,NSAM,TOTSSQ,SUMT1,SUMT2)
      WRITE(6,601) TOTSSQ
      GTOT = 0.01*TOTSSQ
      ITER = 1
      CALL SSQ2(1,NSAM,TOTSSQ,1)
      TOTSSQ = HOLD(1,1)
      H = HOLD(1,2)
      NVEC(1) = IFIX(H)
      NMARK = 1
      PERC = TOTSSQ/GTOT
      WRITE(6,602) TOTSSQ,PERC,NVEC(1)
      L1 = 1
      L2 = NVEC(NMARK)
      L3 = L2 + 1
      L4 = NSAM
      NGP = 1
    4 NGP = NGP + 1
      IF (L2 - L1 - 1) 1,2,7
    2 CONTINUE
      CALL SSQ(L1,L2,TOTQ1,SUMT1,SUMT2)
      SPLIT(1,1) = FLOAT(L1)
      SPLIT(1,2) = FLOAT(L2)
      SPLIT(1,3) = FLOAT(L1)
      SPLIT(1,4) = TOTQ1
      GO TO 8
    1 NGP = NGP - 1
      NGP1 = NGP - 1
      DO 9 NUM = 1,NGP1
      NUM1 = NUM + 1
      DO 9 L = 1,4
    9 SPLIT(NUM,L) = SPLIT(NUM1,L)
      GO TO 8
    7 CONTINUE
      CALL SSQ(L1,L2,TOTQ1,SUMT1,SUMT2)
      CALL SSQ2(L1,L2,TOTQ1,1)
    8 IF (L4 - L3 - 1) 11,12,10
   12 CONTINUE
      CALL SSQ(L3,L4,TOTQ2,SUMT1,SUMT2)
      SPLIT(NGP,1) = FLOAT(L3)
      SPLIT(NGP,2) = FLOAT(L4)
      SPLIT(NGP,3) = FLOAT(L3)
      SPLIT(NGP,4) = TOTQ2
      GO TO 13
   11 NGP = NGP - 1
      GO TO 13
   10 CONTINUE
      CALL SSQ(L3,L4,TOTQ2,SUMT1,SUMT2)
      CALL SSQ2(L3,L4,TOTQ2,NGP)
   13 CONTINUE
      CALL RSORT(SPLIT,20,4,4,NGP,1)
      IF (ITER - NITERS) 3,5,999
    3 TOTSSQ = TOTSSQ - SPLIT(1,4)
      NMARK = NMARK + 1
      S1 = SPLIT(1,1)
      S2 = SPLIT(1,2)
      S3 = SPLIT(1,3)
      L1 = IFIX(S1)
      L2 = IFIX(S3)
      L3 = L2 + 1
      L4 = IFIX(S2)
      NVEC(NMARK) = L2
      CALL BADSRT(NVEC,NMARK)
      PERC = TOTSSQ/GTOT
      WRITE(6,602) TOTSSQ,PERC,(NVEC(NUM), NUM = 1,NMARK)
      ITER = ITER + 1
      GO TO 4
    5 WRITE(6,603)
      DO 6 NUM = 1,NGP
      S1 = SPLIT(NUM,1)
      S2 = SPLIT(NUM,2)
      S3 = SPLIT(NUM,3)
      L1 = IFIX(S1)
      L2 = IFIX(S2)
      L3 = IFIX(S3)
      PERC = SPLIT(NUM,4)/GTOT
```

Figure A1 *(Continued)*

```
    6     WRITE(6,602) SPLIT(NUM,4),PERC,L1,L2,L3
  999     RETURN
  600     FORMAT('1RESULTS OF SPLITLSQ',//)
  601     FORMAT(' SUM OF SQUARES    PERCENT OF TOTAL     MARKERS',/,2X,
     1     F10.6)
  602     FORMAT(2X,F10.6,8X,F6.2,6X,20I4/(36X,18I4))
  603     FORMAT(' FURTHER SPLITS ARE ')
  604     FORMAT(46H WEIGHTED LEAST SQUARES ANALYSIS USING WEIGHTS,/,
     1     16F5.1/(16F5.1))
  605     FORMAT(34H UNWEIGHTED LEAST SQUARES ANALYSIS)
          END


          SUBROUTINE SSQ(L1,L2,TOT,TOT1,TOT2)
          COMMON NPOL,NSAM,PERCT(18,99),B,HOLD(99,2),
     1     W(20),SPLIT(99,4),NMARK,DC(5000),DET,WEIGHT,NITERS,
     2     NBIT(100),DIAG(100),SPLURG(100),LEAST(100),
     3     NCOUNT,NLEV,NCLUST,TINY,HUGE,T(100),DEND(100),PREV
          DIMENSION TOT1(18),TOT2(18)
C
C FINDS THE SUM OF SQUARES OF THE GROUP CONSISTING
C OF LEVELS L1 TO L2 INCLUSIVE
C
          TOT = 0.0
          DO 1 I = 1,NPOL
          TOT1(I) = 0.0
          TOT2(I) = 0.0
          DO 2 J = L1,L2
          P = PERCT(I,J)
          TOT1(I) = TOT1(I) + P
    2     TOT2(I) = TOT2(I) + P*P
    1     TOT = TOT + W(I)*(TOT2(I)-TOT1(I)*TOT1(I)/FLOAT(L2-L1+1))
          RETURN
          END


          SUBROUTINE SSQ2(L1,L2,TOTQ1,KK)
          COMMON NPOL,NSAM,PERCT(18,99),B,HOLD(99,2),
     1     W(20),SPLIT(99,4),NMARK,DC(5000),DET,WEIGHT,NITERS,
     2     NBIT(100),DIAG(100),SPLURG(100),LEAST(100),
     3     NCOUNT,NLEV,NCLUST,TINY,HUGE,T(100),DEND(100),PREV
          DIMENSION SUMT1(18),SUMT2(18),SUMB1(18),SUMB2(18)
          DATA C/0.999/
C
C FINDS OPTIMAL (SUM OF SQUARES) BINARY DIVISION OF THE GROUP
C L1 TO L2, GIVING INFORMATION ABOUT 'NEAR-OPTIMAL' DIVISIONS
C
          CALL SSQ(L1,L1,TOT,SUMT1,SUMT2)
          L7 = L1 + 1
          L8 = L2 - 1
          CALL SSQ(L7,L2,TOTAL,SUMB1,SUMB2)
          HOLD(1,1) = TOT + TOTAL
          HOLD(1,2) = FLOAT(L1)
          DO 1 NUM = L7,L8
          TOT = 0.0
          TOTAL = 0.0
          DO 2 I = 1,NPOL
          P = PERCT(I,NUM)
          SUMT1(I) = SUMT1(I) + P
          SUMT2(I) = SUMT2(I) + P*P
          SUMB1(I) = SUMB1(I) - P
          SUMB2(I) = SUMB2(I) - P*P
          TOT = TOT + W(I)*(SUMT2(I)-SUMT1(I)*SUMT1(I)/FLOAT(NUM-L1+1))
    2     TOTAL=TOTAL+W(I)*(SUMB2(I)-SUMB1(I)*SUMB1(I)/FLOAT(L2-NUM))
          NUM1 = NUM - L1 + 1
          HOLD(NUM1,1) = TOT + TOTAL
    1     HOLD(NUM1,2) = FLOAT(NUM)
          LL = L2 - L1
          CALL RSORT(HOLD,99,2,1,LL,0)
          SPLIT(KK,1) = FLOAT(L1)
          SPLIT(KK,2) = FLOAT(L2)
          SPLIT(KK,3) = HOLD(1,2)
          SPLIT(KK,4) = TOTQ1 - HOLD(1,1)
          K2 = 2
          H = HOLD(1,2)
          K3 = IFIX(H)
    4     IF (HOLD(1,1) - HOLD(K2,1)) 11,5,999
    5     H = HOLD(K2,2)
          K1 = IFIX(H)
          WRITE(6,601) K1,L1,L2
          K2 = K2 + 1
          GO TO 4
   11     IF (HOLD(1,1) - C*HOLD(K2,1)) 3,12,12
   12     H = HOLD(K2,2)
          K1 = IFIX(H)
          WRITE(6,602) K3,K1,L1,L2,HOLD(1,1),HOLD(K2,1)
          K2 = K2 + 1
          GO TO 4
    3     CONTINUE
  999     RETURN
  601     FORMAT(' ALTERNATIVELY LSQ MARKER =',I4,' IN GROUP',I4,' -',I4)
  602     FORMAT('  *** INVESTIGATE',2I4,' IN GROUP',I4,' -',I4,/,
     1     ' CONTENTS ARE ',2F10.4)
          END


          SUBROUTINE DNAMIC
          INTEGER T
          COMMON NPOL,NSAM,PERCT(18,99),B,HOLD(99,2),
     1     W(20),SPLIT(99,4),NMARK,DC(5000),DET,WEIGHT,NITERS,
     2     NBIT(100),DIAG(100),SPLURG(100),LEAST(100),
     3     NCOUNT,NLEV,NCLUST,TINY,HUGE,T(100),DEND(100),PREV
          DIMENSION S(99,21),S2(99),TOT(20),TOTL(20)
          NGROUP = NITERS + 1
C
C FILLS S(I,1) = SUM OF SQUARES OF GROUP COMPRISING
C OBJECTS 1 TO I INCLUSIVE
C
          TOT2 = 0.0
          DO 1 J = 1,NPOL
    1     TOT(J) = 0.0
          DO 2 I = 1,NSAM
          HOLF = 0.0
          F = FLOAT(I)
          DO 3 J = 1,NPOL
          TEMP = PERCT(J,I)
          TOT(J) = TOT(J) + TEMP
          TOT2 = TOT2 + TEMP*TEMP
    3     HOLF = HOLF - TOT(J)*TOT(J)/F
          S2(I) = TOT2
    2     S(I,1) = TOT2 + HOLF
          N1 = NSAM + 1
          DO 9 J = 1,NPOL
    9     PERCT(J,N1) = 0.0
          K = 0
          DO 8 I = 1,NSAM
          T(I) = K
```

```
      K1 = K + 1
      DC(K1) = 0.0
    8 K = K + NSAM - I
      WRITE(6,601)
      GTOT = S(NSAM,1)
      WRITE(6,602) GTOT
      GTOT = 100.0/GTOT
C
C FILLS 'DC(I,J)' = SUM OF SQUARES OF GROUP COMPRISING
C OBJECTS I TO J INCLUSIVE
C
      DO 5 I1 = 3,NSAM
      I2 = NSAM + 3 - I1
      I3 = I2 - 1
      I4 = I2 + 1
      DO 4 J = 1,NPOL
      TOT(J) = TOT(J) - PERCT(J,I4)
    4 TOTL(J) = TOT(J)
      HOLF = S2(I2)
      DO 6 K = 2,I3
      WAIT = 0.0
      K1 = K - 1
      K2 = I2 - K1
      F = FLOAT(K2)
      DO 7 J = 1,NPOL
      TEMP = PERCT(J,K1)
      TOTL(J) = TOTL(J) - TEMP
      HOLF = HOLF - TEMP*TEMP
    7 WAIT = WAIT - TOTL(J)*TOTL(J)/F
      KI2 = T(K) + I2
    6 DC(KI2) = WAIT + HOLF
    5 CONTINUE
C
C BUILDS UP AND PRINTS OUT S(I,K) = SUM OF SQUARES WHEN
C OBJECTS 1 TO I ARE OPTIMALLY DIVIDED INTO K GROUPS
C
      DO 11 K = 2,NGROUP
      WRITE(6,603) K
      S(K,K) = 0.0
      K1 = K + 1
      K2 = K - 1
      DO 12 I = K1,NSAM
      KI = T(K) + I
      SMIN = DC(KI)
      NMARK = K2
      DO 13 J = K1,I
      J1 = J - 1
      JI = T(J) + I
      STRY = S(J1,K2) + DC(JI)
      IF (STRY - SMIN) 14,15,13
   15 WRITE(6,605) J1,NMARK
      GO TO 13
   14 SMIN = STRY
      NMARK = J1
   13 CONTINUE
      S(I,K) = SMIN
      WRITE(6,604) I,NMARK
   12 CONTINUE
      SPR = GTOT*SMIN
   11 WRITE(6,606) K,SMIN,SPR
      RETURN
  601 FORMAT('1RESULTS OF DYNAMIC',//)
  602 FORMAT('0TOTAL SSQ = ',E14.6)
  603 FORMAT('1OPTIMAL DIVISION OF OBJECTS 1 TO L INTO',I3,
     1  ' GROUPS',//,'  L   MARKER')
  604 FORMAT(I3,3X,I3)
  605 FORMAT(' *** EQUALITY - ',I3,' IS ALTERNATIVE TO ',I3,' ***')
  606 FORMAT(' OPTIMAL DIVISION OF DATA INTO',I4,' GROUPS HAS SSQ =',
     1  E14.6,' WHICH IS ',F5.2,' PERCENT')
      END

      SUBROUTINE BADSRT(KE,N)
      DIMENSION KE(20)
C
C SORTS ELEMENTS OF ARRAY KE IN INCREASING ORDER OF MAGNITUDE
C
      N1 = N - 1
      DO 1 I = 1,N1
      NSMALL = KE(I)
      I1 = I + 1
      DO 1 J = I1,N
      IF (NSMALL - KE(J)) 1,1,2
    2 NHOLD = NSMALL
      KE(I) = KE(J)
      KE(J) = NHOLD
      NSMALL = KE(I)
    1 CONTINUE
      RETURN
      END

      SUBROUTINE RSORT(X,NR,NC,IC,MR,INDEX)
      DIMENSION Y(99,2),X(99,4)
C
C SORTS THE FIRST MR ROWS OF THE (NR X NC) ARRAY X IN
C INCREASING ORDER OF THE VALUES IN COLUMN IC
C
      DO 3 I = 1,MR
      Y(I,1) = X(I,IC)
    3 Y(I,2) = FLOAT(I)
      MR1 = MR - 1
      DO 1 I = 1,MR1
      SMALL = Y(I,1)
      I1 = I + 1
      DO 1 J = I1,MR
      IF (SMALL - Y(J,1)) 1,1,2
    2 HOLD1 = SMALL
      Y(I,1) = Y(J,1)
      Y(J,1) = HOLD1
      SMALL = Y(I,1)
      HOLD2 = Y(I,2)
      Y(I,2) = Y(J,2)
      Y(J,2) = HOLD2
    1 CONTINUE
      IF (INDEX) 11,11,21
   11 DO 14 J = 1,NC
      DO 13 I = 1,MR
   13 Y(I,1) = X(I,J)
      DO 14 I = 1,MR
      Z = Y(I,2)
      K = IFIX(Z)
   14 X(I,J) = Y(K,1)
      RETURN
   21 DO 24 J = 1,NC
      DO 23 I = 1,MR
   23 Y(I,1) = X(I,J)
      DO 24 I = 1,MR
      I1 = MR + 1 - I
      Z = Y(I1,2)
      K = IFIX(Z)
   24 X(I,J) = Y(K,1)
      RETURN
      END
```

Figure A1 *(Continued)*

WEIGHT = parameter controlling the weights assigned to the different pollen types:
WEIGHT = blank: pollen types are weighted equally;
WEIGHT = any non-blank character: weights will be provided by the user on the next line].

(W(K), K=1, NPOL) in format (16 F5.1)

[Weights for the different pollen types. Note that this line should only be present if the parameter WEIGHT is specified on the previous line to be some non-blank character].

Then NPOL sets of lines, each set being of the following form:

POL(1,K), POL(2,K), (NDATA(K,I), I=1,NSAM)
in format (A2, A4, 2X, 18I4/(20I4))

[This gives the pollen counts for each taxon. The first six characters on the first line give the label for the taxon; the last 72 spaces on the first line, and up to all 80 spaces on subsequent lines, give the counts for this taxon, recorded in order down the core. Each count is recorded in I4, so the first line will contain counts from levels 1 to 18, the second line (only present if NSAM is greater than 18) will contain counts from levels 19 to 38, etc.].

In order to illustrate this, the input required for the analysis of the 1974 Abernethy Forest test data set is shown in Figure A2. The output obtained from analysing these data is shown in Figure A3. Small differences will be noted between these results and the detailed analyses described in Chapter 3. These differences are due to rounding errors; the analyses described in Chapter 3 were based on the proportions given in Table 1.3, which were rounded to three decimal places. Although these proportions agree with those shown in the output in Figure A3, the computer analysis will have carried the extra decimal places into the subsequent computations.

The constrained single link dendrogram, which has been sketched in Figure A3, is not produced in the computer output, but can be obtained as follows: opposite each level draw a horizontal line with length proportional to the value shown (this value gives the height at which the level amalgamates with the group containing the level below it). The line opposite the last level should be longer

```
 ABERNETHY FOREST 1974 TEST DATA SET
  9 10   5
BETULA    59 175 365 390 394 311  18  23 183 100
PINUS    425 317   9  12  13   4   7  11   6   1
CORYLU    12  37 150  98  12   1   1   1   4   1
JUNIPE     0   2   0   8  64 132   1   0  13   0
EMPETR     0   0   4   3  11  31  10   8 200 190
GRAMIN     2  11  19  14  34  78  13  19  69  81
CYPERA     3   3   5   5  13  12  18  14  47  36
ARTEMI     0   0   0   0   0   1 214 130   8   6
RUMEX      0   0   0   1   5   1  11  12  26  97
```

Figure A2 Required input for the analysis of the Abernethy Forest 1974 test data set.

ABERNETHY FOREST 1974 TEST DATA SET

NUMBER OF TAXA = 9 NUMBER OF SAMPLES = 10

SAMPLE PROPORTIONS

NO	SUM	BETULA	PINUS	CORYLU	JUNIPE	EMPETR	GRAMIN	CYPERA	ARTEMI	RUMEX
1	501	0.118	0.848	0.024	0.000	0.000	0.004	0.006	0.000	0.000
2	545	0.321	0.582	0.068	0.004	0.000	0.020	0.006	0.000	0.000
3	552	0.661	0.016	0.272	0.000	0.007	0.034	0.009	0.000	0.000
4	531	0.734	0.023	0.185	0.015	0.006	0.026	0.009	0.000	0.002
5	546	0.722	0.024	0.022	0.117	0.020	0.062	0.024	0.000	0.009
6	571	0.545	0.007	0.002	0.231	0.054	0.137	0.021	0.002	0.002
7	293	0.061	0.024	0.003	0.003	0.034	0.044	0.061	0.730	0.038
8	218	0.106	0.050	0.005	0.000	0.037	0.087	0.064	0.596	0.055
9	556	0.329	0.011	0.007	0.023	0.360	0.124	0.085	0.014	0.047
10	512	0.195	0.002	0.002	0.000	0.371	0.158	0.070	0.012	0.189

RESULTS OF CONSLINK

DETAILS OF EACH AMALGAMATION ARE SUPPRESSED
TOTAL NUMBER OF AMALGAMATIONS IS 9

CONSTRAINED SINGLE LINK DENDROGRAM

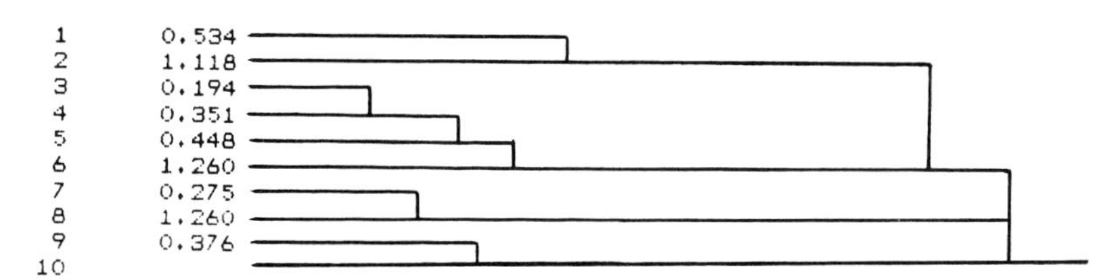

Figure A3 Output obtained from analysing the Abernethy Forest 1974 test data set using the program ZONATION.

```
RESULTS OF SPLITINF


INF CONTENT       PERCENT OF TOTAL          MARKERS
  10.705766
   5.788173           54.07            6
   3.068021           28.66            2   6
   1.168523           10.91            2   6   8
   0.441228            4.12            2   4   6   8
   0.306488            2.86            1   2   4   6   8

FURTHER SPLITS ARE
   0.126359            1.18            9  10   9
   0.105102            0.98            5   6   5
   0.041558            0.39            7   8   7
   0.033470            0.31            3   4   3




RESULTS OF SPLITLSQ


UNWEIGHTED LEAST SQUARES ANALYSIS
SUM OF SQUARES      PERCENT OF TOTAL       MARKERS
   2.542179
   1.621644          63.79           2
   0.781960          30.76           2   6
   0.206122           8.11           2   6   8
   0.121454           4.78           2   4   6   8
   0.064126          .2.52           1   2   4   6   8
FURTHER SPLITS ARE
   0.025873           1.02           5   6   5
   0.020212           0.80           9  10   9
   0.011390           0.45           7   8   7
   0.006650           0.26           3   4   3




RESULTS OF DYNAMIC


TOTAL SSQ =     0.254218E+01


OPTIMAL DIVISION OF OBJECTS 1 TO L INTO  2 GROUPS

 L  MARKER
 3     1
 4     2
 5     2
 6     2
 7     2
 8     6
 9     2
10     2
OPTIMAL DIVISION OF DATA INTO   2 GROUPS HAS SSQ =  0.162165E+01 WHICH IS 63.79
PERCENT
```

Figure A3 (*Continued*)

```
OPTIMAL DIVISION OF OBJECTS 1 TO L INTO  3 GROUPS

 L  MARKER
 4     2
 5     2
 6     4
 7     6
 8     6
 9     6
10     6
OPTIMAL DIVISION OF DATA INTO   3 GROUPS HAS SSQ =  0.781961E+00 WHICH IS 30.76
PERCENT


OPTIMAL DIVISION OF OBJECTS 1 TO L INTO  4 GROUPS

 L  MARKER
 5     3
 6     4
 7     6
 8     6
 9     8
10     8
OPTIMAL DIVISION OF DATA INTO   4 GROUPS HAS SSQ =  0.206124E+00 WHICH IS  8.11
PERCENT


OPTIMAL DIVISION OF OBJECTS 1 TO L INTO  5 GROUPS

 L  MARKER
 6     4
 7     6
 8     6
 9     8
10     8
OPTIMAL DIVISION OF DATA INTO   5 GROUPS HAS SSQ =  0.121457E+00 WHICH IS  4.78
PERCENT


OPTIMAL DIVISION OF OBJECTS 1 TO L INTO  6 GROUPS

 L  MARKER
 7     6
 8     7
 9     8
10     8
OPTIMAL DIVISION OF DATA INTO   6 GROUPS HAS SSQ =  0.641289E-01 WHICH IS  2.52
PERCENT
```

Figure A3 (*Continued*)

than any of the other lines. Now, from the end of each of the first (NSAM-1) lines, drop a perpendicular until it hits some line below it. This yields the constrained single link dendrogram; by centring branches after each amalgamation, a more conventional dendrogram, as shown in Figure 3.4, is obtained.

The SPLINF and SPLSQ output is self-explanatory. The DNAMIC output requires some additional work on the part of the investigator, in tracing back all the markers; the program only prints out the highest marker, but the rest can be obtained in the manner described in Section 3.5.

As listed, the program is restricted to analysing data sets containing no more than 89 sampling levels and 18 pollen types, but these limits can readily be extended by modification of the dimensions of the appropriate arrays.

References

Aaby, B. (1983). Forest development, soil genesis and human activity illustrated by pollen and hypha analysis of two neighbouring podzols in Draved Forest, Denmark. *Danmarks geolgiske Undersøgelse* Series II 114, 116 pp.

Aaby, B., Jacobsen, J., and Jacobsen, O. S. (1979). Pb-210 dating and lead deposition in the ombrotrophic peat bog, Draved Mose, Denmark. *Danmarks geologiske Undersøgelse Årbog* 1978, 45–68.

Aaby, B., and Tauber, H. (1975). Rates of peat formation in relation to degree of humification and local environment as shown by studies of a raised bog in Denmark. *Boreas* 4, 1–17.

Aalto, M., Taavitsainen, J-P., and Vuorela, I. (1980). Palaeobotanical investigations at the site of a sledge runner find, dated to about 4900 B.P. in Noormarkku, S.W. Finland. *Suomen Museo* 1980, 41–65.

Aario, L. (1940). Waldgrenzen und subrezenten pollen spektren in Petsamo, Lappland. *Annales Academie Scientiarium Fennicae Series A* 54(8), 120 pp.

Abele, L. G., and Walters, K. (1979a). Marine benthic diversity: a critique and alternative explanation. *Journal of Biogeography* 6, 115–126.

Abele, L. G., and Walters, K. (1979b). The stability-time hypothesis: reevaluation of the data. *American Naturalist* 114, 559–568.

Adam, D. P. (1970). Some palynological applications of multivariate statistics. Ph.D. thesis, University of Arizona.

Adam, D. P. (1974). Palynological applications of principal component and cluster analyses. *Journal of Research of the United States Geological Survey* 2, 727–741.

Adam, D. P. (1975). A late Holocene pollen record from Pearson's Pond, Weeks Creek landslide, San Francisco Peninsula, California. *Journal of Research of the United States Geological Survey* 3, 721–731.

Aitchison, J. (1981). A new approach to null correlations of proportions. *Journal of Mathematical Geology* 13, 175–189.

Aitchison, J. (1982). The statistical analysis of compositional data. *Journal of the Royal Statistical Society Series B* 44, 139–177.

American Commission on Stratigraphic Nomenclature (1961). Code of stratigraphic nomenclature. *Bulletin of the American Association of Petroleum Geologists* 45, 645–665.

Ammann, B. (1979). Palynology in some lakes of the northern Alpine piedmont (Switzerland). *Acta Universitatis Ouluensis Series A*, 82, *Geologica* 3, 89–96.

Ammann-Moser, B. (1975). Vegetationskundliche und Pollenanalytische Untersuchungen auf dem Heidenweg in Bielersee. *Beiträge zur Geobotanischen Landesaufnahme der Schweiz* 56, 76 pp.

Amundson, D. C., and Wright, H. E., Jr. (1979). Forest changes in Minnesota at the end of the Pleistocene. *Ecological Monographs* 49, 1–16.
Anderberg, M. R. (1973). *Cluster Analysis for Applications*. Academic Press, London and New York.
Andersen, S. Th. (1960). Silicone oil as a mounting medium for pollen grains. *Danmarks geologiske Undersøgelse* Series IV 4(1), 24 pp.
Andersen, S. Th. (1961). Vegetation and its environment in Denmark in the Early Weichselian Glacial. *Danmarks geologiske Undersøgelse* Series II 75, 175 pp.
Andersen, S. Th. (1964). Interglacial plant successions in the light of environmental changes. *Report VIth International Congress on the Quaternary, Warsaw 1961* Vol. II, 359–367.
Andersen, S. Th. (1966). Interglacial vegetational succession and lake development in Denmark. *Palaeobotanist* 15, 117–127.
Andersen, S. Th. (1967). Tree pollen rain in a mixed deciduous forest in south Jutland (Denmark). *Review of Palaeobotany and Palynology* 3, 267–275.
Andersen, S. Th. (1969). Interglacial vegetation and soil development. *Meddelelser Dansk geologisk Förening* 19, 90–102.
Andersen, S. Th. (1970). The relative pollen productivity and representation of North European trees, and correction factors for tree pollen spectra. *Danmarks geologiske Undersøgelse* Series II 96, 99 pp.
Andersen, S. Th. (1973). The differential pollen productivity of trees and its significance for the interpretation of a pollen diagram from a forested region. In *Quaternary Plant Ecology* (eds. H. J. B. Birks and R. G. West), Blackwell Scientific Publications, Oxford.
Andersen, S. Th. (1974). Wind conditions and pollen deposition in a mixed deciduous forest II. Seasonal and annual pollen deposition 1967–1972. *Grana* 14, 64–77.
Andersen, S. Th. (1975). The Eemian freshwater deposit at Egernsund, South Jylland, and the Eemian landscape development in Denmark. *Danmarks geologiske Undersøgelse Årbog* 1974, 49–70.
Andersen, S. Th. (1978a). Local and regional vegetational development in eastern Denmark in the Holocene. *Danmarks geologiske Undersøgelse Årbog* 1976, 5–27.
Andersen, S. Th. (1978b). On the size of *Corylus avellana* L. pollen mounted in silicone oil. *Grana* 17, 5–13.
Andersen, S. Th. (1980a). Early and Late Weichselian chronology and birch pollen assemblages in Denmark. *Boreas* 9, 53–69.
Andersen, S. Th. (1980b). Influence of climatic variation on pollen season severity in wind-pollinated trees and shrubs. *Grana* 19, 47–52.
Andersen, S. Th. (1980c). The relative pollen productivity of the common forest trees in the early Holocene in Denmark. *Danmarks geologiske Undersøgelse Årbog* 1979, 5–19.
Anderson, A. J. B. (1971a). Numeric examination of multivariate soil samples. *Journal of Mathematical Geology* 3, 1–14.
Anderson, A. J. B. (1971b). Ordination methods in ecology. *Journal of Ecology* 59, 713–726.
Anderson, R. Y., and Kirkland, D. W. (1966). Intrabasin varve calibration. *Geological Society of America Bulletin* 77, 241–256.
Andrews, J. T., Davis, P. T., Mode, W. N., Nichols, H., and Short, S. K. (1981). Relative departures in July temperatures in northern Canada for the past 6000 years. *Nature, London* 289, 164–167.
Andrews, J. T., and Diaz, H. F. (1981). Eigenvector analysis of reconstructed Holocene July temperature departures over northern Canada. *Quaternary Research* 16, 373–389.
Andrews, J. T., Mode, W. N., and Davis, P. T. (1980). Holocene climate based on pollen transfer functions, eastern Canadian Arctic. *Arctic and Alpine Research* 12, 41–64.
Andrews, J. T., and Nichols, H. (1981). Modern pollen deposition and Holocene paleotemperature reconstructions, central Northern Canada. *Arctic and Alpine Research* 13, 387–408.
Andrews, J. T., Webber, P. J., and Nichols, H. (1979). A late Holocene pollen diagram from Pangnirtung Pass, Baffin Island, Canada. *Review of Palaeobotany and Palynology* 27, 1–28.
Anonymous (1972). Making pollen diagrams. *Nature, London* 240, 324.
Arigo, R., Howe, S. E., and Webb, T., III. (1982). Computer programs for climatic calibration of

pollen data. In *Palaeohydrological Changes in the Temperate Zone in the last 15000 years. Subproject B. Lake and Mire Environments. Volume III. Specific Methods* (ed. B. E. Berglund), University of Lund, Lund.

Ashton, E. H., Healy, M. J. R., and Lipton, S. (1957). The descriptive use of discriminant functions in physical anthropology. *Proceedings of the Royal Society of London Series B* 146, 552–572.

Austin, M. P., and Greig-Smith, P. (1968). The application of quantitative methods to vegetation survey. II. Some methodological problems of data from rain forest. *Journal of Ecology* 56, 827–844.

Baker, C. A., Moxey, P. A., and Oxford, P. M. (1978). Woodland continuity and change in Epping Forest. *Field Studies* 4, 645–669.

Baker, R. G. (1976). Late Quaternary vegetation history of the Yellowstone Lake Basin, Wyoming. *United States Geological Survey Professional Paper* 729–E, 48 pp.

Barkley, F. A. (1934). The statistical theory of pollen analysis. *Ecology* 15, 283–289.

Barnosky, C. W. (1981). A record of late Quaternary vegetation from Davis Lake, southern Puget Lowland, Washington. *Quaternary Research* 16, 221–239.

Barry, R. G., Elliott, D. L., and Crane, R. G. (1981). The palaeoclimatic interpretation of exotic pollen peaks in Holocene records from the Eastern Canadian Arctic: a discussion. *Review of Palaeobotany and Palynology* 33, 153–167.

Bartlein, P. J., and Webb, T., III. (1985). Paleoclimatic interpretation of Holocene pollen data: statistical considerations. In *Paleoecological Uses of Pollen Data* (ed. J. Lentin), American Association of Stratigraphic Palynologists, Contribution Series (in press).

Bartlein, P. J., Webb, T., III, and Fleri, E. (1984). Holocene climatic changes in the northern Midwest: pollen-derived estimates. *Quaternary Research* 22, 361–374.

Beales, P. W. (1976). Palaeolimnological studies of a Shropshire mere. Ph.D. thesis, University of Cambridge.

Beckett, S. C. (1979). Pollen influx in peat deposits: values from raised bogs in the Somerset levels, south-western England. *New Phytologist* 83, 839–847.

Bellman, R. E., and Dreyfus, S. E. (1962). *Applied Dynamic Programming*. Princeton University Press, Princeton, New Jersey.

Bement, T. R., and Waterman, M. S. (1977). Locating maximum variance segments in sequential data. *Journal of Mathematical Geology* 9, 55–61.

Bennett, K. D. (1982). Tree population history in the Flandrian of East Anglia. Ph.D. thesis, University of Cambridge.

Bennett, K. D. (1983a). Devensian late-glacial and Flandrian vegetational history at Hockham Mere, Norfolk, England I. Pollen percentages and concentrations. *New Phytologist* 95, 457–487.

Bennett, K. D. (1983b). Devensian late-glacial and Flandrian vegetational history at Hockham Mere, Norfolk, England II. Pollen accumulation rates. *New Phytologist* 95, 489–504.

Bennett, K. D. (1983c). Postglacial population expansion of forest trees in Norfolk, UK. *Nature, London* 303, 164–167.

Benninghoff, W. S. (1962). Calculation of pollen and spore density in sediments by addition of exotic pollen in known quantities. *Pollen et Spores* 4, 332–333.

Bent, A. M., and Wright, H. E., Jr. (1963). Pollen analyses of surface materials and lake sediments from the Chuska Mountains, New Mexico. *Geological Society of America Bulletin* 74, 491–500.

Benzécri, J.-P. (1969). Statistical analysis as a tool to make patterns emerge from data. In *Methodologies of Pattern Recognition* (ed. S. Watanabe), Academic Press, London and New York.

Benzécri, J.-P. (Ed.) (1973). *L'Analyse des Données. 2. L'Analyse des Correspondances*. Dunod, Paris.

Berglund, B. E. (1966). Late Quaternary vegetation in eastern Blekinge, south-eastern Sweden. A pollen-analytical study. II. Post-glacial time. *Opera Botanica* 12(2), 190 pp.

Berglund, B. E. (1973). Pollen dispersal and deposition in an area of southeastern Sweden—some preliminary results. In *Quaternary Plant Ecology* (eds. H. J. B. Birks and R. G. West), Blackwell Scientific Publications, Oxford.

Berglund, B. E. (1979a). Definition of investigation areas. In *Palaeohydrological Changes in the Temperate Zone in the last 15000 years. Subproject B. Lake and Mire Environments. Volume I. General Project Description* (ed. B. E. Berglund), University of Lund, Lund.

Berglund, B. E. (1979b). Pollen analysis. In *Palaeohydrological Changes in the Temperate Zone in the last 15000 years. Subproject B. Lake and Mire Environments. Volume II. Specific Methods* (ed. B. E. Berglund), University of Lund, Lund.

Berglund, B. E. (1983). Palaeohydrological studies in lakes and mires—a palaeoecological research strategy. In *Background to Palaeohydrology* (ed. K. J. Gregory), J. Wiley and Sons, Chichester.

Berglund, B. E., and Digerfeldt, G. (1976). Environmental changes during the Holocene—a geological correlation project on a Nordic basis. *Newsletters Stratigraphy* 5, 80–85.

Bernabo, J. C. (1981). Quantitative estimates of temperature changes over the last 2700 years in Michigan based on pollen data. *Quaternary Research* 15, 143–159.

Bernabo, J. C., and Webb, T., III. (1977). Changing patterns in the Holocene pollen record of northeastern North America: a mapped summary. *Quaternary Research* 8, 64–96.

Binford, M. W. (1982). Ecological history of Lake Valencia, Venezuela: interpretation of animal microfossils and some chemical, physical, and geological factors. *Ecological Monographs* 52, 307–333.

Birks, H. H. (1970). Studies in the vegetational history of Scotland. I. A pollen diagram from Abernethy Forest, Inverness-shire. *Journal of Ecology* 58, 827–846.

Birks, H. H. (1972). Studies in the vegetational history of Scotland. II. Two pollen diagrams from the Galloway Hills, Kirkcudbrightshire. *Journal of Ecology* 60, 183–217.

Birks, H. H. (1973). Modern macrofossil assemblages in lake sediments in Minnesota. In *Quaternary Plant Ecology* (eds. H. J. B. Birks and R. G. West), Blackwell Scientific Publications, Oxford.

Birks, H. H. (1975). Studies in the vegetational history of Scotland. IV. Pine stumps in Scottish blanket peats. *Philosophical Transactions of the Royal Society of London Series B* 270, 181–226.

Birks, H. H., and Mathewes, R. W. (1978). Studies in the vegetational history of Scotland. V. Late Devensian and early Flandrian pollen and macrofossil stratigraphy at Abernethy Forest, Inverness-shire. *New Phytologist* 80, 455–484.

Birks, H. J. B. (1970). Inwashed pollen spectra at Loch Fada, Isle of Skye. *New Phytologist* 69, 807–820.

Birks, H. J. B. (1973a). Modern pollen rain studies in some arctic and alpine environments. In *Quaternary Plant Ecology* (eds. H. J. B. Birks and R. G. West), Blackwell Scientific Publications, Oxford.

Birks, H. J. B. (1973b). *Past and Present Vegetation of the Isle of Skye—a Palaeoecological Study*. Cambridge University Press, London.

Birks, H. J. B. (1974). Numerical zonations of Flandrian pollen data. *New Phytologist* 73, 351–358.

Birks, H. J. B. (1976a). The distribution of European pteridophytes: a numerical analysis. *New Phytologist* 77, 257–287.

Birks, H. J. B. (1976b). Late-Wisconsinan vegetational history at Wolf Creek, central Minnesota. *Ecological Monographs* 46, 395–429.

Birks, H. J. B. (1977a). The Flandrian forest history of Scotland: a preliminary synthesis. In *British Quaternary Studies—Recent Advances* (ed. F. W. Shotton), Clarendon Press, Oxford.

Birks, H. J. B. (1977b). Modern pollen rain and vegetation of the St. Elias Mountains, Yukon Territory. *Canadian Journal of Botany* 55, 2367–2382.

Birks, H. J. B. (1978). Geographic variation of *Picea abies* (L.) Karsten pollen in Europe. *Grana* 17, 149–160.

Birks, H. J. B. (1980). Modern pollen assemblages and vegetational history of the moraines of the Klutlan Glacier and its surroundings, Yukon Territory, Canada. *Quaternary Research* 14, 101–129.

Birks, H. J. B. (1981a). Late Wisconsin vegetational and climatic history at Kylen Lake, northeastern Minnesota. *Quaternary Research* 16, 322–355.

Birks, H. J. B. (1981b). Long-distance pollen in Late Wisconsin sediments of Minnesota, U.S.A.: a quantitative analysis. *New Phytologist* 87, 630–661.

Birks, H. J. B. (1981c). The use of pollen analysis in the reconstruction of past climates: a review. In *Climate and History* (eds. T. M. L. Wigley, M. J. Ingram, and G. Farmer), Cambridge University Press, Cambridge.

Birks, H. J. B. (1982a). Holocene (Flandrian) chronostratigraphy of the British Isles: a review. *Striae* 16, 99–105.

Birks, H. J. B. (1982b). Mid-Flandrian forest history of Roudsea Wood National Nature Reserve, Cumbria. *New Phytologist* 90, 339–354.

Birks, H. J. B. (1985). Flandrian (post-glacial) isochrone maps and tree migration patterns in the British Isles. Submitted for publication.

Birks, H. J. B., and Berglund, B. E. (1979). Holocene pollen stratigraphy of southern Sweden: a reappraisal using numerical methods. *Boreas* 8, 257–279.

Birks, H. J. B., and Birks, H. H. (1980). *Quaternary Palaeoecology*. Edward Arnold, London.

Birks, H. J. B., Deacon, J., and Peglar, S. M. (1975). Pollen maps for the British Isles 5000 years ago. *Proceedings of the Royal Society of London Series B* 189, 87–105.

Birks, H. J. B., and Huntley, B. (1978). Program POLLDATA.MK5 documentation relating to FORTRAN IV program of 26 June 1978. Sub-Department of Quaternary Research, University of Cambridge.

Birks, H. J. B., and Madsen, B. J. (1979). Flandrian vegetational history of Little Loch Roag, Isle of Lewis, Scotland. *Journal of Ecology* 67, 825–842.

Birks, H. J. B., and Peglar, S. M. (1979). Interglacial pollen spectra from Sel Ayre, Shetland. *New Phytologist* 83, 559–575.

Birks, H. J. B., and Peglar, S. M. (1980). Identification of *Picea* pollen of late Quaternary age in eastern North America: a numerical approach. *Canadian Journal of Botany* 58, 2043–2058.

Birks, H. J. B., and Ransom, M. E. (1969). An interglacial peat at Fugla Ness, Shetland. *New Phytologist* 68, 777–796.

Birks, H. J. B., and Saarnisto, M. (1975). Isopollen maps and principal components analysis of Finnish pollen data for 4000, 6000, and 8000 years ago. *Boreas* 4, 77–96.

Birks, H. J. B., Webb, T., III, and Berti, A. A. (1975). Numerical analysis of pollen samples from central Canada: a comparison of methods. *Review of Palaeobotany and Palynology* 20, 133–169.

Björck, S. (1979). Late Weichselian stratigraphy of Blekinge, S.E. Sweden, and water level changes in the Baltic Ice Lake. Ph.D. thesis, University of Lund.

Björck, S. (1981). A stratigraphic study of Late Weichselian deglaciation, shore displacement and vegetation history in south-eastern Sweden. *Fossils and Strata* 14, 1–93.

Björck, S., and Persson, T. (1981). Late Weichselian and Flandrian biostratigraphy and chronology from Hochstetter Forland, Northeast Greenland. *Meddelelser om Grønland Geoscience* 5, 19 pp.

Black, R. A. and Bliss, L. C. (1980). Reproductive ecology of *Picea mariana* (Mill) B.S.P., at treeline near Inuvik, Northwest Territories, Canada. *Ecological Monographs* 50, 331–354.

Blackith, R. E., and Reyment, R. A. (1971). *Multivariate Morphometrics*. Academic Press, London and New York.

Blashfield, R. K. (1976). Mixture model tests of cluster analysis: accuracy of four agglomerative hierarchical methods. *Psychological Bulletin* 83, 377–388.

Bonny, A. P. (1972). A method for determining absolute pollen frequencies in lake sediments. *New Phytologist* 71, 393–405.

Bonny, A. P. (1976). Recruitment of pollen to the seston and sediment of some Lake District lakes. *Journal of Ecology* 64, 859–887.

Bonny, A. P. (1978). The effect of pollen recruitment processes on pollen distribution over the sediment surface of a small lake in Cumbria. *Journal of Ecology* 66, 385–416.

Bonny, A. P. (1980). Seasonal and annual variation over five years in contemporary airborne pollen trapped at a Cumbrian lake. *Journal of Ecology* 68, 421–441.

Bostwick, L. G. (1978). An environmental framework for cultural change in Maine: pollen influx and percentage diagrams from Monhegan Island. M.S. thesis, University of Maine.

Botkin, D. B. (1981). Causality and succession. In *Forest Succession—Concepts and Applications* (eds. D. C. West, H. H. Shugart, and D. B. Botkin), Springer Verlag, New York.

Botkin, D. B., Janak, J. F., and Wallis, J. R. (1972). Some ecological consequences of a computer model of forest growth. *Journal of Ecology* 60, 849–872.

Boulter, M. C., and Hubbard, R. N. L. B. (1982). Objective paleoecological and biostratigraphic interpretation of Tertiary palynological data by multivariate statistical analysis. *Palynology* 6, 55–68.

Bowman, P. W. (1931). Study of a peat bog near the Matamek River, Quebec, Canada, by the method of pollen analysis. *Ecology* 12, 694–708.

Box, G. E. P. and Tidwell, P. W. (1962). Transformation of the independent variables. *Technometrics* 4, 531–550.

Bradshaw, R. H. W. (1978). Modern pollen representation factors and recent woodland history in S.E. England. Ph.D. thesis, University of Cambridge.

Bradshaw, R. H. W. (1981a). Modern pollen-representation factors for woods in south-east England. *Journal of Ecology* 69, 45–70.

Bradshaw, R. H. W. (1981b). Quantitative reconstruction of local woodland vegetation using pollen analysis from a small basin in Norfolk, England. *Journal of Ecology* 69, 941–955.

Bradshaw, R. H. W., and Webb, T., III. (1985). Relationships between contemporary pollen and vegetation data in Wisconsin and Michigan, U.S.A. *Ecology* 66, 721–737.

Bradstreet, T. E., and Davis, R. B. (1975). Mid-postglacial environments in New England with emphasis on Maine. *Arctic Anthropology* 12, 7–22.

Brenner, G. J. (1974). Palynostratigraphy of the lower Cretaceous Gevar'am and Talme Yafe formations in the Gevar'am 2 Well (Southern Coastal Plain, Israel). *Bulletin of the Israel Geological Survey* 59, 27 pp.

Brideaux, W. W. (1971). Recurrent species groupings in fossil microplankton assemblages. *Palaeogeography, Palaeoclimatology, Palaeoecology* 9, 101–122.

Broecker, W. S. (1965). Isotope geochemistry and the Pleistocene climatic record. In *The Quaternary of the United States* (eds. H. E. Wright, Jr. and D. G. Frey), Princeton University Press, Princeton, New Jersey.

Broecker, W. S., and Walton, A. (1959). The geochemistry of C^{14} in freshwater systems. *Geochimia et Cosmochimica Acta* 16, 15–38.

Brookes, D., and Thomas, K. W. (1967). The distribution of pollen grains on microscope slides. Part I. The non-randomness of the distribution. *Pollen et Spores* 9, 621–629.

Brothwell, D., and Krzanowski, W. (1974). Evidence of biological differences between early British populations from Neolithic to Medieval times, as revealed by eleven commonly available cranial vault measurements. *Journal of Archaeological Science* 1, 249–260.

Brown, P. J. (1982). Multivariate calibration. *Journal of the Royal Statistical Society Series B* 44, 287–321.

Brubaker, L. B. (1975). Postglacial forest patterns associated with till and outwash in northcentral Upper Michigan. *Quaternary Research* 5, 499–527.

Brubaker, L. B., Garfinkel, H. L., and Edwards, M. E. (1983). A Late Wisconsin and Holocene vegetation history from the Central Brooks Range: implications for Alaskan paleoecology. *Quaternary Research* 20, 194–214.

Brugam, R. B. (1980). Postglacial diatom stratigraphy of Kirchner Marsh, Minnesota. *Quaternary Research* 13, 133–146.

Bruno, M. G., and Lowe, R. L. (1980). Differences in the distribution of some bog diatoms: a cluster analysis. *American Midland Naturalist* 104, 70–79.

Brush, G. S., and DeFries, R. S. (1981). Spatial distributions of pollen in surface sediments of the Potomac estuary. *Limnology and Oceanography* 26, 295–309.

Bryson, R. A. (1966). Air masses, streamlines, and the Boreal forest. *Geographical Bulletin* 8, 228–269.

Bryson, R. A., and Kutzbach, J. E. (1974). On the analysis of pollen-climate canonical transfer functions. *Quaternary Research* 4, 162–174.

Bryson, R. A., and Swain, A. M. (1981). Holocene variations of monsoon rainfall in Rajasthan. *Quaternary Research* 16, 135–145.

Burnaby, T. P. (1961). The palaeoecology of the foraminifera of the Chalk Marl. *Palaeontology* 4, 599–608.

Buzas, M. A. (1967). An application of canonical analysis as a method for comparing faunal areas. *Journal of Animal Ecology* 36, 563–577.

Buzas, M. A. (1969). On the quantification of biofacies. *Proceedings of the North American Paleontological Convention B* 101–116.

Buzas, M. A. (1972). Biofacies analysis of presence or absence data through canonical variate analysis. *Journal of Paleontology* 46, 55–57.

Cain, S. A. (1940). The identification of species in fossil pollen of *Pinus* by size-frequency determinations. *American Journal of Botany* 27, 301–308.

Cain, S. A. (1948). Palynological studies at Sodon Lake. I. Size-frequency studies of fossil spruce pollen. *Science* 108, 115–117.

Cain, S. A., and Cain, L. G. (1948). Palynological studies at Sodon Lake. II. Size-frequency studies of pine pollen, fossil and modern. *American Journal of Botany* 35, 583–591.

Campbell, M. M. (1978). Classification of some pollen data subject to contiguity constraints. Honours dissertation, University of St. Andrews.

Campbell, N. A. (1982). Robust procedures in multivariate analysis II. Robust canonical variate analysis. *Applied Statistics* 31, 1–8.

Campbell, N. A., and Atchley, W. R. (1981). The geometry of canonical variates analysis. *Systematic Zoology* 30, 268–280.

Carleton, T. J., and Maycock, P. F. (1978). Dynamics of the boreal forest south of James Bay. *Canadian Journal of Botany* 56, 1157–1173.

Carleton, T. J., and Maycock, P. F. (1980). Vegetation of the boreal forests south of James Bay: non-centered component analysis of the vascular flora. *Vegetatio* 61, 1199–1212.

Carney, H. J. (1982). Algal dynamics and trophic interactions in the recent history of Frains Lake, Michigan. *Ecology* 63, 1814–1826.

Carter, R. N., and Prince, S. D. (1981). Epidemic models used to explain biogeographical distribution limits. *Nature, London* 293, 644–645.

Caseldine, C. J. (1981). Surface pollen studies across Bankhead Moss, Fife, Scotland. *Journal of Biogeography* 8, 7–25.

Caseldine, C. J., and Gordon, A. D. (1978). Numerical analysis of surface pollen spectra from Bankhead Moss, Fife. *New Phytologist* 80, 435–453.

Caswell, H. (1982). Life history tactics and the equilibrium status of populations. *American Naturalist* 120, 317–339.

Chatfield, C. (1980). *The Analysis of Time Series: an Introduction.* (Second edition). Chapman and Hall, London and New York.

Chatterjee, S., and Price, B. (1977). *Regression Analysis by Example.* J. Wiley and Sons, London and New York.

Chayes, F., and Kruskal, W. H. (1966). An approximate statistical test for correlations between proportions. *Journal of Geology* 74, 692–702.

Cheetham, A. H., and Hazel, J. E. (1969). Binary (presence–absence) similarity coefficients. *Journal of Paleontology* 43, 1130–1136.

Christensen, B. B. (1946). Measurement as a means of identifying fossil pollen. *Danmarks geologiske Undersøgelse* Series IV 3(2), 20 pp.

Christopher, R. A. (1978). Quantitative palynologic correlation of three Campanian and Maestrichtian sections (Upper Cretaceous) from the Atlantic coastal plain. *Palynology* 2, 1–27.

Churchill, D. M. (1968). The distribution and prehistory of *Eucalyptus diversicolor* F. Muell., *E. marginata* Donn ex Sm. and *E. calophylla* R. Br. in relation to rainfall. *Australian Journal of Botany* 16, 125–151.

Cisne, J. L., and Rabe, B. D. (1978). Coenocorrelation: gradient analysis of fossil communities and its application in stratigraphy. *Lethaia* 11, 341–364.

Clapham, W. B. (1969). Evolution of Upper Permian terrestrial floras in Oklahoma as determined from pollen and spores. *Proceedings of the North American Paleontological Convention E* 411–427.

Clapham, W. B. (1970). Nature and paleogeography of Middle Permian floras of Oklahoma as inferred from their pollen record. *Journal of Geology* 78, 153–171.
Clapham, W. B. (1972). Numerical analysis and group formation in palynology. *Geoscience and Man* 4, 73–85.
Clark, R. L. (1982). Point count estimation of charcoal in pollen preparations and thin sections of sediments. *Pollen et Spores* 24, 523–535.
Clisby, K. H., and Sears, P. B. (1955). Palynology in southern North America. III. Microfossil profiles under Mexico City correlated with the sedimentary profiles. *Geological Society of America Bulletin* 66, 511–520.
Cole, H. S. 1969. Objective reconstruction of the paleoclimatic record through the application of eigenvectors of present-day pollen spectra and climate to the late-Quaternary pollen stratigraphy. Ph.D. thesis, University of Wisconsin.
Cole, H. S., and Bryson, R. A. (1968). The application of eigenvector techniques to the climatic interpretation of pollen diagrams: an initial study. I. Bog D Pond, Itasca. Center for Climatic Research, University of Wisconsin.
Cole, L. C. (1957). Biological clock in the Unicorn. *Science* 125, 874–876.
Colinvaux, P. (1978). On the use of the word "absolute" in pollen statistics. *Quaternary Research* 9, 132–133.
Colinvaux, P. A. (1983). The meaning of palaeoecology. *Quarterly Review of Archaeology* 8, 8–9.
Comanor, P. L. (1968). Forest vegetation and the pollen spectrum: an examination of the usefulness of the R-value. *Bulletin of the New Jersey Academy of Sciences* 13, 7–19.
Connell, J. H., and Sousa, W. P. (1983). On the evidence needed to judge ecological stability or persistence. *American Naturalist* 121, 789–824.
Cook, R. D., and Weisberg, S. (1982). *Residuals and Influence in Regression*. Chapman and Hall, New York and London.
Cooley, W. W., and Lohnes, P. R. (1971). *Multivariate Data Analysis*. J. Wiley and Sons, London and New York.
Cormack, R. M. (1971). A review of classification. *Journal of the Royal Statistical Society Series A* 134, 321–367.
Cox, D. R., and Lewis, P. A. W. (1966). *The Statistical Analysis of Series of Events*. Methuen, London.
Crabtree, K. (1968). Pollen analysis. *Science Progress* 56, 83–101.
Craig, A. J. (1972). Pollen influx to laminated sediments: a pollen diagram from northeastern Minnesota. *Ecology* 53, 46–57.
Craig, A. J. (1978). Pollen percentage and influx analyses in south-east Ireland: a contribution to the ecological history of the late-glacial period. *Journal of Ecology* 66, 297–324.
Crain, I. K. (1970). Computer interpolation and contouring of two-dimensional data: a review. *Geoexploration* 8, 71–86.
Crowder, A., and Starling, R. N. (1980). Contemporary pollen in the Salmon River basin, Ontario. *Review of Palaeobotany and Palynology* 30, 11–26.
Curtis, J. T. (1959). *The Vegetation of Wisconsin*. University of Wisconsin Press, Madison.
Cushing, E. J. (1963). Late-Wisconsin pollen stratigraphy in east-central Minnesota. Ph.D. thesis, University of Minnesota.
Cushing, E. J. (1964). Application of the Code of Stratigraphic Nomenclature to pollen stratigraphy. (Unpublished manuscript).
Cushing, E. J. (1965). Problems in the Quaternary phytogeography of the Great Lakes Region. In *The Quaternary of the United States* (eds. H. E. Wright, Jr. and D. G. Frey), Princeton University Press, Princeton, New Jersey.
Cushing, E. J. (1967a). Evidence for differential pollen preservation in late Quaternary sediments in Minnesota. *Review of Palaeobotany and Palynology* 4, 87–101.
Cushing, E. J. (1967b). Late-Wisconsin pollen stratigraphy and the glacial sequence in Minnesota. In

Quaternary Paleoecology (eds. E. J. Cushing and H. E. Wright, Jr.), Yale University Press, New Haven and London.
Cushing, E. J. (1979). Program POLDATA, version 5. (Unpublished manuscript).
Cushing, E. J., and Wright, H. E., Jr. (1967). Introduction. In *Quaternary Paleoecology* (eds. E. J. Cushing and H. E. Wright, Jr.), Yale University Press, New Haven and London.
Cwynar, L. C. (1978). Recent history of fire and vegetation from laminated sediment of Greenleaf Lake, Algonquin Park, Ontario. *Canadian Journal of Botany* 56, 10–21.
Cwynar, L. C. (1982). A late-Quaternary vegetation history from Hanging Lake, northern Yukon. *Ecological Monographs* 52, 1–24.
Dabrowski, M. F. (1975). Tree pollen rain and the vegetation of the Białowieza National Park. *Biułetyn Geologiczny* 19, 157–172.
Dale, M. B., and Walker, D. (1970). Information analysis of pollen diagrams I. *Pollen et Spores* 12, 21–37.
Damblon, F., and Schumacker, R. (1971). New prospects for study of palynological data: the use of computers. *Pollen et Spores* 13, 609–614.
Davenport, M., and Studdert-Kennedy, G. (1972). The statistical analysis of aesthetic judgement: an exploration. *Applied Statistics* 21, 324–333.
David, M., Campiglio, C., and Darling, R. (1974). Progress in R- and Q-mode analysis: correspondence analysis and its application to the study of geological processes. *Canadian Journal of Earth Science* 11, 131–146.
David, M., and Dagbert, M. (1975). Correspondence analysis in geology: the method and a review of several applications. *Rapport EP-75-R-39, Département de Génie Minéral, Ecole Polytechnie de Montréal,* 73 pp.
Davis, A. M. (1980). Modern pollen spectra from the tundra-boreal forest transition in northern Newfoundland, Canada. *Boreas* 9, 89–100.
Davis, J. C. (1973). *Statistics and Data Analysis in Geology*. J. Wiley and Sons, London and New York.
Davis, J. C., and McCullagh, M. J. (1975). *Display and Analysis of Spatial Data*. J. Wiley and Sons, London and New York.
Davis, M. B. (1961). The problem of rebedded pollen in late-glacial sediments at Taunton, Massachusetts. *American Journal of Science* 259, 211–222.
Davis, M. B. (1963). On the theory of pollen analysis. *American Journal of Science* 261, 897–912.
Davis, M. B. (1965a). A method for determination of absolute pollen frequency. In *Handbook of Paleontological Techniques* (eds. B. Kummel and D. Raup), W. H. Freeman, San Francisco.
Davis, M. B. (1965b). Phytogeography and palynology of northeastern United States. In *The Quaternary of the United States* (eds. H. E. Wright, Jr. and D. G. Frey), Princeton University Press, Princeton, New Jersey.
Davis, M. B. (1967a). Late-glacial climate in northern United States: a comparison of New England and the Great Lakes Region. In *Quaternary Paleoecology* (eds. E. J. Cushing and H. E. Wright, Jr.), Yale University Press, New Haven and London.
Davis, M. B. (1967b). Pollen accumulation rates at Rogers Lake, Connecticut, during late- and post-glacial time. *Review of Palaeobotany and Palynology* 2, 219–230.
Davis, M. B. (1967c). Pollen deposition in lakes as measured by sediment traps. *Bulletin of the Geological Society of America* 78, 849–858.
Davis, M. B. (1968). Pollen grains in lake sediments: redeposition caused by seasonal water circulation. *Science* 162, 796–799.
Davis, M. B. (1969a). Climatic changes in southern Connecticut recorded by pollen deposition at Rogers Lake. *Ecology* 50, 409–422.
Davis, M. B. (1969b). Palynology and environmental history during the Quaternary Period. *American Scientist* 57, 317–332.
Davis, M. B. (1973). Redeposition of pollen grains in lake sediment. *Limnology and Oceanography* 18, 44–52.

Davis, M. B. (1976). Pleistocene biogeography of temperate deciduous forests. *Geoscience and Man* 13, 13–26.

Davis, M. B. (1978). Climatic interpretation of pollen in Quaternary sediments. In *Biology and Quaternary Environments* (eds. D. Walker and J. C. Guppy), Australian Academy of Science. Canberra.

Davis, M. B. (1981). Quaternary history and the stability of forest communities. In *Forest Succession—Concepts and Applications* (eds. D. C. West, H. H. Shugart, and D. B. Botkin), Springer Verlag, New York.

Davis, M. B., and Brubaker, L. B. (1973). Differential sedimentation of pollen grains in lakes. *Limnology and Oceanography* 18, 635–646.

Davis, M. B., Brubaker, L. B., and Beiswenger, J. M. (1971). Pollen grains in lake sediments: pollen percentages in surface sediments from southern Michigan. *Quaternary Research* 1, 450–467.

Davis, M. B., Brubaker, L. B., and Webb, T., III. (1973). Calibration of absolute pollen influx. In *Quaternary Plant Ecology* (eds. H. J. B. Birks and R. G. West), Blackwell Scientific Publications, Oxford.

Davis, M. B., and Deevey, E. S. (1964). Pollen accumulation rates: estimates from late-glacial sediment of Rogers Lake. *Science* 145, 1293–1295.

Davis, M. B., and Ford, M. S. (1982). Sediment focusing in Mirror Lake, New Hampshire. *Limnology and Oceanography* 27, 137–150.

Davis, M. B., and Goodlett, J. C. (1960). Comparison of the present vegetation with pollen-spectra in surface samples from Brownington Pond, Vermont. *Ecology,* 41, 346–357.

Davis, R. B. (1974). Stratigraphic effects of tubificids in profundal lake sediments. *Limnology and Oceanography* 19, 466–488.

Davis, R. B., Bradstreet, T. E., Stuckenrath, R., and Borns, H. W. (1975). Vegetation and associated environments during the past 14,000 years near Moulton Pond, Maine. *Quaternary Research* 5, 435–465.

Davis, R. B., Brewster, L. A., and Sutherland, J. (1969). Variation in pollen spectra within lakes. *Pollen et Spores* 11, 557–571.

Davis, R. B., and Norton, S. A. (1978). Paleolimnologic studies of human impact on lakes in the United States, with emphasis on recent research in New England. *Polskie Archiwum Hydrobiologii* 25, 99–115.

Davis, R. B., and Webb, T., III. (1975). The contemporary distribution of pollen in eastern North America: a comparison with vegetation. *Quaternary Research* 5, 395–434.

Day, M. H. (1967). Olduvai Hominid 10: a multivariate analysis. *Nature, London* 215, 323–324.

Day, M. H. (1974). The interpolation of isolated fossil foot bones into a discriminant function analysis—a reply. *American Journal of Physical Anthropology* 41, 222–236.

Day, M. H., and Wood, B. A. (1968). Functional affinities of the Olduvai Hominid 8 talus. *Man* 3, 440–455.

Day, M. H., and Wood, B. A. (1969). Hominoid tali from East Africa. *Nature, London* 222, 591–592.

Dayton, P. K. (1979). Ecology: a science and a religion. In *Ecological Processes in a Coastal and Marine System* (ed. R. J. Livingston), Plenum Press, New York.

Dean, W. E., and Anderson, R. Y. (1974). Application of some correlation coefficient techniques to time-series analysis. *Journal of Mathematical Geology* 6, 363–372.

Dearing, J. (1982). Core correlations and total sediment influx. In *Palaeohydrological Changes in the Temperate Zone in the last 15000 years. Subproject B. Lake and Mire Environments. Volume III. Specific Methods* (ed. B. E. Berglund), University of Lund, Lund.

Deevey, E. S. (1967). Introduction. In *Pleistocene Extinctions* (eds. P. S. Martin and H. E. Wright, Jr.), Yale University Press, New Haven and London.

Deevey, E. S., Gross, M. S., Hutchinson, G. E., and Kraybill, H. L. (1954). The natural C^{14} contents of material from hard-water lakes. *Proceedings of the National Academy of Science* 40, 285–288.

Deevey, E. S., and Potzger, J. E. (1951). Peat samples for radiocarbon analysis: problems in pollen statistics. *American Journal of Science* 249, 473–511.

Delcoigne, A., and Hansen, P. (1975). Sequence comparison by dynamic programming. *Biometrika* 62, 661–664.

Delcourt, H. R., and Delcourt, P. A. (1984). Late-Quaternary history of the spruce–fir ecosystem in the Southern Appalachian mountain region. In *Proceedings of the 1983 Conference on the Spruce–Fir Ecosystem in the Southern Appalachian Mountains* (eds. P. S. White and J. Wood), United States Department of Interior, National Park Service Report SER-71, 22–35.

Delcourt, H. R., Delcourt, P. A., and Spiker, E. C. (1983). A 12000-year record of forest history from Cahaba Pond, St. Clair County, Alabama. *Ecology* 64, 874–887.

Delcourt, H. R., Delcourt, P. A., Webb, T., III, and Overpeck, J. T. (1983). Modern analogs for full-glacial vegetation of Tennessee. *Bulletin of the Ecological Society of America* 64, 156.

Delcourt, P. A., Delcourt, H. R., and Davidson, J. L. (1983). Mapping and calibration of modern pollen-vegetation relationships in the southeastern United States. *Review of Palaeobotany and Palynology* 39, 1–45.

Delfiner, P., and Delhomme, J. P. (1975). Optimum interpolation by kriging. In *Display and Analysis of Spatial Data* (eds. J. C. Davis and M. J. McCullagh), J. Wiley and Sons, London and New York.

Digerfeldt, G. (1977). The Flandrian development of Lake Flarken. Regional vegetation history and palaeolimnology. Report 13, University of Lund, Department of Quaternary Geology.

Dodson, J. R. (1972). Computer programs for the pollen analyst. *Pollen et Spores* 14, 455–465.

Dodson, J. R. (1983). Modern pollen rain in southeastern New South Wales, Australia. *Review of Palaeobotany and Palynology* 38, 249–268.

Donner, J. J. (1971). Towards a stratigraphical division of the Finnish Quaternary. *Commentationes Physico-Mathematicae, Societas Scientiarum Fennica* 41, 281–305.

Donner, J. J. (1972). Pollen frequencies in the Flandrian sediments of Lake Vakojärvi, south Finland. *Commentationes Biologicae, Societas Scientiarum Fennica* 53, 19 pp.

Donner, J. J., Alhonen, P., Eronen, M., Jungner, H., and Vuorela, I. (1978). Biostratigraphy and radiocarbon dating of the Holocene lake sediments of Tytötjärvi and the peats in the adjoining bog Varrassuo west of Lahti in southern Finland. *Annales Botanici Fennici* 15, 258–280.

Donner, J. J., Jungner, H., and Vasari, Y. (1971). The hard-water effect on radiocarbon measurements of samples from Säynäjalampi, north-east Finland. *Commentationes Physico-Mathematicae, Societas Scientiarum Fennica* 41, 307–310.

Draper, N. R., and Smith, H. (1966). *Applied Regression Analysis*. J. Wiley and Sons, London and New York.

Drury, W. H., and Nisbet, I. C. (1973). Succession. *Journal of the Arnold Arboretum* 54, 331–368.

Dudnik, E. E. (1971). *SYMAP User's Reference Manual for Synagraphic Computer Mapping*. University at Chicago Circle, Illinois.

Eccles, M., Hickey, M., and Nichols, H. (1979). Computer techniques for the presentation of palynological and paleoenvironmental data. *Institute of Arctic and Alpine Research Occasional Paper* 16, 139 pp.

Edwards, A. W. F. (1972). *Likelihood. An Account of the Statistical Concept of Likelihood and its Application to Scientific Inference*. Cambridge University Press, London.

Edwards, M. E. (1980). Ecology and historical ecology of oakwoods in North Wales. Ph.D. thesis, University of Cambridge.

Elliot-Fisk, D. L., Andrews, J. T., Short, S. K., and Mode, W. N. (1982). Isopoll maps and an analysis of the distribution of the modern pollen rain, eastern and central Northern Canada. *Géographie physique et Quaternaire* 36, 91–108.

Elner, J. K., and Happey-Wood, C. M. (1980). The history of two linked but contrasting lakes in North Wales from a study of pollen, diatoms and chemistry in sediment cores. *Journal of Ecology* 68, 95–121.

Eneroth, O. (1951). Undersöknig rörande möjlighterna alt i fossilt material urskilja de olika *Betula*-arternas pollen. *Geologiska Föreningens i Stockholm Förhandlingar* 73, 343–405.

Erdtman, G. (1943). *An Introduction to Pollen Analysis*. Chronica Botanica, Waltham, Massachusetts.

Everitt, B. S. (1978). *Graphical Techniques for Multivariate Data*. Heinemann, London.

Everitt, B. (1980). *Cluster Analysis* (Second edition). Heinemann, London.

Eyster-Smith, N. M. (1977). Holocene pollen stratigraphy of Lake St. Croix, Minnesota–Wisconsin, and some aspects of the depositional history. M.S. thesis, University of Minnesota.

Faegri, K. (1950). On the value of palaeoclimatological evidence. *Centenary Proceedings of the Royal Meteorological Society* 1950, 188–195.

Faegri, K. (1963). Problems of immigration and dispersal of the Scandinavian flora. In *North Atlantic Biota and their History* (eds. A. Löve and D. Löve), Pergamon, Oxford.

Faegri, K. (1966). Some problems of representativity in pollen analysis. *Palaeobotanist* 15, 135–140.

Faegri, K. (1974). Quaternary pollen analysis—past, present and future. *Advances in Pollen Spore Research* 1, 62–69.

Faegri, K., and Deuse, P. (1960). Size variations in pollen grains with different treatments. *Pollen et Spores* 2, 293–298.

Faegri, K., and Iversen, J. (1975). Textbook of Pollen Analysis (Third edition revised by K. Faegri), Blackwell Scientific Publications, Oxford.

Faegri, K., and Ottestad, P. (1948). Statistical problems in pollen analysis. *Universitetet i Bergen Årbok 1948, Naturvitenskapelig Rekke* 3, 27 pp.

Fagerlind, F. (1952). The real signification of pollen diagrams. *Botaniska Notiser* 105, 185–224.

Falcon, R. M. S. (1976). Numerical methods in the definition of palynological assemblage zones in the Lower Karroo (Gondwana) of Rhodesia. *Palaeontographica Africa* 19, 1–20.

Fasham, M. J. R. (1977). A comparison of nonmetric multidimensional scaling, principal components, and reciprocal averaging for the ordination of simulated coenoclines and coenoplanes. *Ecology* 58, 551–561.

Ferguson, C. W. (1970). Concepts and techniques of dendrochronology. In *Scientific Methods in Medieval Archaeology* (ed. R. Berger), University of California Press, Berkeley, Los Angeles, and London.

Firbas, F. (1949). *Waldgeschichte Mitteleuropas*. Gustav Fischer, Jena.

Fisher, L., and Van Ness, J. W. (1971). Admissible clustering procedures. *Biometrika* 65, 91–104.

Fisher, W. D. (1958). On grouping for maximum homogeneity. *Journal of the American Statistical Association* 53, 789–798.

Flenley, J. R. (1973). The use of modern pollen rain samples in the study of the vegetational history of tropical regions. In *Quaternary Plant Ecology* (eds. H. J. B. Birks and R. G. West), Blackwell, Oxford.

Flenley, J. R. (1982). Book reviews. *Journal of Biogeography* 9, 91–94.

Fortier, J. J., and Solomon, H. (1966). Clustering procedures. In *Multivariate Analysis* (ed. P. R. Krishnaiah), Academic Press, London and New York.

Fox, W. T. (1968). Quantitative paleoecologic analysis of fossil communities in the Richmond Group. *Journal of Geology* 76, 613–640.

Frederiksen, N. O. (1974). Statistics in stratigraphic palynology: an annotated and indexed bibliography. *American Association of Stratigraphic Palynologists Contribution Series* No. 3, 65 pp.

Friedman, H. P., and Rubin, J. (1967). Some invariant criteria for grouping data. *Journal of the American Statistical Association* 62, 1159–1178.

Fries, M. (1967). Lennart von Post's pollen diagram series of 1916. *Review of Palaeobotany and Palynology* 4, 9–13.

Fritts, H. C. (1971). Dendroclimatology and dendroecology. *Quaternary Research* 1, 419–449.

Fritts, H. C. (1976). *Tree Rings and Climate*. Academic Press, London and New York.

Fritts, H. C., Blasing, T. J., Hayden, B. P., and Kutzbach, J. E. (1971). Multivariate techniques for specifying tree-growth and climate relationships and for reconstructing anomalies in paleoclimate. *Journal of Applied Meteorology* 10, 845–864.

Gabriel, K. R. (1971). The biplot graphic display of matrices with application to principal component analysis. *Biometrika* 58, 453–467.

Galloway, P. (1978). Restoring the map of Medieval Trondheim: a computer-aided investigation into the nightwatchmen's itinerary. *Journal of Archaeological Science* 5, 153–165.

Gauch, H. G. (1982). *Multivariate Analysis in Community Ecology*. Cambridge University Press, Cambridge.

Gauch, H. G., Whittaker, R. H., and Singer, S. B. (1981). A comparative study of nonmetric ordinations. *Journal of Ecology* 69, 135–152.

Gibbs, A. J., and MacIntyre, G. A. (1970). The diagram, a method for comparing sequences. *European Journal of Biochemistry* 16, 1–11.

Gill, D. (1970). Application of a statistical zonation method to reservoir evaluation and digitized-log analysis. *Bulletin of the American Association of Petroleum Geologists* 54, 719–729.

Gilmour, J. S. L., and Walters, S. M. (1963). Philosophy and classification. In *Recent Researches in Plant Taxonomy, Vistas in Botany* Vol. IV (ed. W. B. Turrill), Pergamon Press, Oxford.

Gittins, R. (1968). Trend-surface analysis of ecological data. *Journal of Ecology* 56, 845–869.

Gittins, R. (1979). Ecological applications of canonical analysis. In *Multivariate Methods in Ecological Work* (eds. L. Orloci, C. R. Rao, and W. M. Stiteler), International Co-operative Publishing House, Burtonsville, Maryland.

Gittins, R. (1981). Towards the analysis of vegetational succession. *Vegetatio* 46, 37–59.

Glahn, H. R. (1968). Canonical correlation and its relationship to discriminant analysis and multiple regression. *Journal of Atmospheric Sciences* 25, 23–31.

Gleason, H. A. (1939). The individualistic concept of the plant association. *American Midland Naturalist* 21, 92–110.

Gnanadesikan, R. (1977). *Methods for Statistical Data Analysis of Multivariate Observations*. J. Wiley and Sons, London and New York.

Godwin, H. (1934). Pollen analysis. An outline of the problems and potentialities of the method. I. Technique and interpretation. II. General applications of pollen analysis. *New Phytologist* 33, 278–305, 325–358.

Godwin, H. (1940). Pollen analysis and forest history of England and Wales. *New Phytologist* 39, 370–400.

Godwin, H. (1969). The value of plant materials for radiocarbon dating. *American Journal of Botany* 56, 723–731.

Goodman, D. (1975). The theory of diversity–stability relationships in ecology. *Quarterly Review of Biology* 50, 237–266.

Goodman, D. K. (1979). Dinoflagellate "communities" from the Lower Eocene Nanjemoy Formation of Maryland, U.S.A. *Palynology* 3, 169–190.

Göransson, H. (1977). The Flandrian vegetational history of southern Östergötland. Ph.D. Thesis, University of Lund.

Gordon, A. D. (1973a). Classification in the presence of constraints. *Biometrics* 29, 821–827.

Gordon, A. D. (1973b). A sequence-comparison statistic and algorithm. *Biometrika* 60, 197–200.

Gordon, A. D. (1974). Numerical methods in Quaternary palaeoecology. III. Sequential sampling strategies. *New Phytologist* 73, 781–792.

Gordon, A. D. (1980a). On the assessment and comparison of classifications. In *Analyse de Données et Informatique* (ed. R. Tomassone), I.N.R.I.A., Le Chesnay.

Gordon, A. D. (1980b). Methods of constrained classification. In *Analyse de Données et Informatique* (ed. R. Tomassone), I.N.R.I.A., Le Chesnay.

Gordon, A. D. (1980c). SLOTSEQ: a FORTRAN IV program for comparing two sequences of observations. *Computers and Geosciences* 6, 7–20.

Gordon, A. D. (1981). *Classification: Methods for the Exploratory Analysis of Multivariate Data*. Chapman and Hall, London and New York.

Gordon, A. D. (1982a). An investigation of two sequence-comparison statistics. *Australian Journal of Statistics* 24, 332–342.

Gordon, A. D. (1982b). On measuring and modelling the relationship between two stratigraphically-recorded variables. In *Quantitative Stratigraphic Correlation* (eds. J. M. Cubitt and R. A. Reyment), J. Wiley and Sons, Chichester and New York.

Gordon, A. D. (1982c). Numerical methods in Quaternary palaeoecology. V. Simultaneous graphical representation of the levels and taxa in a pollen diagram. *Review of Palaeobotany and Palynology* 37, 155–183.

Gordon, A. D. (1982d). Some observations on methods of estimating the proportions of morphologically similar pollen types in fossil samples. *Canadian Journal of Botany* 60, 1888–1894.

Gordon, A. D., and Birks, H. J. B. (1972). Numerical methods in Quaternary palaeoecology. I. Zonation of pollen diagrams. *New Phytologist* 71, 961–979.

Gordon, A. D., and Birks, H. J. B. (1974). Numerical methods in Quaternary palaeoecology. II. Comparison of pollen diagrams. *New Phytologist* 73, 221–249.

Gordon, A. D., and Henderson, J. T. (1977). An algorithm for Euclidean sum of squares classification. *Biometrics* 33, 355–362.

Gordon, A. D., and Prentice, I. C. (1977). Numerical methods in Quaternary palaeoecology. IV. Separating mixtures of morphologically similar pollen taxa. *Review of Palaeobotany and Palynology* 23, 359–372.

Gordon, A. D., and Reyment, R. A. (1979). Slotting of borehole sequences. *Journal of Mathematical Geology* 11, 309–327.

Gould, S. J. (1965). Is uniformitarianism necessary? *American Journal of Science* 263, 223–228.

Gower, J. C. (1966a). A Q-technique for the calculation of canonical variates. *Biometrika* 53, 588–590.

Gower, J. C. (1966b). Some distance properties of latent root and vector methods used in multivariate analysis. *Biometrika* 53, 325–338.

Gower, J. C. (1967a). Multivariate analysis and multidimensional geometry. *The Statistician* 17, 13–28.

Gower, J. C. (1967b). A survey of numerical methods useful in taxonomy. *Acarologia* 11, 357–375.

Gower, J. C. (1968). Adding a point to vector diagrams in multivariate analysis. *Biometrika* 55, 582–585.

Gower, J. C. (1970). Classification and geology. *Bulletin International Statistical Institute* 38, 35–41.

Gower, J. C. (1971). A general coefficient of similarity and some of its properties. *Biometrics* 27, 857–874.

Gower, J. C. (1972). In discussion after Sibson (1972b). *Journal of the Royal Statistical Society Series B* 34, 340–343.

Grabandt, R. A. J. (1980). Pollen rain in relation to arboreal vegetation in the Colombian Cordillera Oriental. *Review of Palaeobotany and Palynology* 29, 65–147.

Green, D. G. (1981). Time series and postglacial forest ecology. *Quaternary Research* 15, 265–277.

Green, D. G. (1982). Fire and stability in the postglacial forests of southwest Nova Scotia. *Journal of Biogeography* 9, 29–40.

Green, D. G. (1983). The ecological interpretation of fine resolution pollen records. *New Phytologist* 94, 459–477.

Green, P. E., and Rao, V. R. (1969). A note on proximity measures and cluster analyses. *Journal of Marketing Research* 6, 359–364.

Green, R. H. (1971). A multivariate statistical approach to the Hutchinsonian niche: bivalve molluscs of central Canada. *Ecology* 52, 543–556.

Green, R. H. (1974). Multivariate niche analysis with temporally varying environmental factors. *Ecology* 55, 73–83.

Greig-Smith, P. (1964). *Quantitative Plant Ecology* (Second edition). Butterworths, London.

Greig-Smith, P. (1980). The development of numerical classification and ordination. *Vegetatio* 42, 1–9.

Greig-Smith, P. (1983). *Quantitative Plant Ecology* (Third edition). Blackwell Scientific Publications, Oxford.

Grigal, D. F., and Arneman, H. F. (1969). Numerical classification of some forested Minnesota soils. *Proceedings of the Soil Science Society of America* 33, 433–438.

Grootes, P. M. (1978). Carbon-14 time scale extended: comparison of chronologies. *Science* 200, 11–15.

Hansen, B. S., and Cushing, E. J. (1973). Identification of pine pollen of late Quaternary age from the Chuska Mountains, New Mexico. *Geological Society of America Bulletin* 84, 1181–1200.

Harbaugh, J. W., and Bonham-Carter, G. (1970). *Computer Simulation in Geology*. J. Wiley and Sons, London and New York.

Harbaugh, J. W., and Merriam, D. F. (1968). *Computer Applications in Stratigraphic Analysis*. J. Wiley and Sons, London and New York.

Hare, F. K. (1976). Late Pleistocene and Holocene climates: some persistent problems. *Quaternary Research* 6, 507–517.

Hare, F. K., and Ritchie, J. C. (1972). The Boreal bioclimates. *Geographic Review* 62, 333–365.

Harris, W. F., and Norris, G. (1972). Ecologic significance of recurrent groups of pollen and spores in Quaternary sequences from New Zealand. *Palaeogeography, Palaeoclimatology, Palaeoecology* 11, 107–124.

Havinga, A. J. (1967). Palynology and pollen preservation. *Review of Palaeobotany and Palynology* 2, 81–98.

Hawkins, D. M. (1976a). FORTRAN IV program to segment multivariate sequences of data. *Computers and Geosciences* 1, 339–351.

Hawkins, D. M. (1976b). Point estimation of the parameters of piecewise regression models. *Applied Statistics* 25, 51–57.

Hawkins, D. M., and Merriam, D. F. (1973). Optimal zonation of digitized sequential data. *Journal of Mathematical Geology* 5, 389–395.

Hawkins, D. M., and Merriam, D. F. (1974). Zonation of multivariate sequences of digitized geologic data. *Journal of Mathematical Geology* 6, 263–269.

Hawkins, D. M., and Merriam, D. F. (1975). Segmentation of discrete sequences of geologic data. In *Quantitative Studies in the Geological Sciences, W. H. Krumbein Volume* (ed. E. T. H. Whitten), *Geological Society of America Memoir* 142.

Hawkins, D. M., and ten Krooden, J. A. (1979). A review of several methods of segmentation. In *Geomathematical and Petrophysical Studies in Sedimentology* (eds. D. Gill and D. F. Merriam), Pergamon Press, Oxford.

Haworth, E. Y. (1980). Comparison of continuous phytoplankton records with the diatom stratigraphy in the recent sediments of Blelham Tarn. *Limnology and Oceanography* 25, 1093–1603.

Hay, W. W., and Southam, J. R. (1978). Quantifying biostratigraphic correlation. *Annual Review of Earth and Planetary Sciences* 6, 353–375.

Hayes, J. G. (1974). Numerical methods for curve and surface fitting. *Bulletin of the Institute of Mathematics and its Applications* 10, 144–152.

Hazel, J. E. (1977) Use of certain multivariate and other techniques in assemblage zonal biostratigraphy: examples utilizing Cambrian, Cretaceous, and Tertiary Benthic Invertebrates. In *Concepts and Methods of Biostratigraphy* (eds. E. G. Kauffman and J. E. Hazel), Dowden, Hutchinson and Ross, Stroudsburg, Pennsylvania.

Healy, M. J. R. (1965). Descriptive uses of discriminant functions. In *Mathematics and Computer Science in Biology and Medicine*. Medical Research Council, H.M.S.O., London.

Hecht, A. D. (1973). A model for determining Pleistocene paleotemperatures from planktonic foraminiferal assemblages. *Micropaleontology* 19, 68–77.

Hedberg, H. D. (1972a). Introduction to an International Guide to Stratigraphic Classification, Terminology, and Usage. *Boreas* 1, 199–211.

Hedberg, H. D. (1972b). Summary of an International Guide to Stratigraphic Classification, Terminology, and Usage. *Boreas* 1, 213–239.

Hedberg, H. D. (Ed.) (1976). *International Stratigraphic Guide. A Guide to Stratigraphic Classification, Terminology, and Procedure.* J. Wiley and Sons, London and New York.

Hedgpeth, J. W. (1977). Models and muddles. Some philosophical observations. *Helgoländer wissenchaftliche Meeresunterssuchungen* 30, 92–104.

Heide, K. M. (1981). Late Quaternary vegetational history of north-central Wisconsin, U.S.A.: Estimating forest composition from pollen data. Ph.D. thesis, Brown University.

Heide, K. M., and Bradshaw, R. H. W. (1982). The pollen-tree relationship within forests of Wisconsin and Upper Michigan, U.S.A. *Review of Palaeobotany and Palynology* 36, 1–23.

Heim, J. (1962). Recherches sur les relations entre la végétation actuelle et le spectre pollinique récent dans les Ardennes Belges. *Bulletin Société royale de botanique de Belgique* 96, 5–92.

Heim, J. (1970). Les relations entre les spectres polliniques récents et la végétation actuelle en Europe Occidentale. Thèse, Université de Louvain.

Hemmings, S. K., and Rostron, J. (1972). A multivariate analysis of measurements on the Scottish Middle Old Red Sandstone antiarch fish genus *Pterichthyodes* Bleeker. *Biological Journal of the Linnean Society* 4, 15–28.

Hengeveld, R. (1982). Problems of scale in ecological research. Ph.D. thesis, University of Leiden.

Hengeveld, R., and Haeck, J. (1981). The distribution of abundance II. Models and implications. *Proceedings of the Koninklijke Nederlandse Akademie van Wetenschappen* C 84, 257–284.

Hengeveld, R., and Haeck, J. (1982). The distribution of abundance I. Measurements. *Journal of Biogeography* 9, 303–316.

Heusser, C. J. (1977). Quaternary palynology of the Pacific Slope of Washington. *Quaternary Research* 8, 282–306.

Heusser, C. J., and Heusser, L. E. (1981). Palynology and paleotemperature analysis of the Whidbay Formation, Puget Lowland, Washington. *Canadian Journal of Earth Sciences* 18, 136–149.

Heusser, C. J., Heusser, L. E., and Streeter, S. S. (1980). Quaternary temperatures and precipitation for the north-west coast of North America. *Nature, London* 286, 702–704.

Heusser, C. J., and Streeter, S. S. (1980). A temperature and precipitation record of the past 16,000 years in southern Chile. *Science* 210, 1345–1347.

Heusser, C. J., Streeter, S. S., and Stuiver, M. (1981). Temperature and precipitation record in southern Chile extended to ~43,000 yr ago. *Nature, London* 294, 65–67.

Hideux, M. J., and Ferguson, I. K. (1976). The stereostructure of the exine and its evolutionary significance in Saxifragaceae sensu lato. In *The Evolutionary Significance of the Exine* (eds. I. K. Ferguson and J. Muller), Academic Press, London and New York.

Hill, M. O. (1973). Reciprocal averaging: an eigenvector method of ordination. *Journal of Ecology* 61, 237–249.

Hill, M. O. (1974). Correspondence analysis: a neglected multivariate method. *Applied Statistics* 23, 340–354.

Hills, L. V., Klovan, J. E., and Sweet, A. R. (1974). *Juglans eocinerea* n. sp., Beaufort Formation (Tertiary), southwestern Banks Island, Arctic Canada. *Canadian Journal of Botany* 52, 65–90.

Hodge, M. C. (1983). Analysis of pollen data: mixture of normals model incorporating a scale change parameter. Honours dissertation, University of St. Andrews.

Hofmann, W. (1978). Analysis of animal microfossils from the Segeberger See (F.R.G.). *Archiv für Hydrobiologie* 82, 316–346.

Holmes, O. W. (1858). *The Autocrat of the Breakfast-Table*. Phillips, Samson, and Co., Boston.

Horn, H. S. (1976). Succession. In *Theoretical Ecology—Principles and Applications* (ed. R. M. May), Blackwell Scientific Publications, Oxford.

Howe, S., and Webb, T. (1977). Testing the statistical assumptions of paleoclimatic calibration functions. *Preprint Volume Fifth Conference on Probability and Statistics in Atmospheric Sciences*, 152–157.

Howe, S., and Webb, T., III. (1983). Calibrating pollen data in climatic terms: improving the methods. *Quaternary Science Reviews* 2, 17–51.

Hubbard, R. N. L. B., and Boutler, M. C. (1983). Reconstruction of Palaeogene climate from palynological evidence. *Nature, London* 301, 147–150.

Hughes, M. K., Kelly, P. M., Pilcher, J. R., and LaMarche, V. C., Jr. (1982). *Climate from Tree Rings*. Cambridge University Press, Cambridge.

Hunt, T. C., and Birks, H. J. B (1982). Devensian late-glacial vegetational history at Sea Mere, Norfolk. *Journal of Biogeography* 9, 517–538.

Huntley, B. (1976). The past and present vegetation of the Morrone Birkwoods and Caenlochan National Nature Reserves. Ph.D. thesis, University of Cambridge.

Huntley, B. (1981). The past and present vegetation of the Caenlochan National Nature Reserve, Scotland. II. Palaeoecological investigations. *New Phytologist* 87, 189–222.

Huntley, B., and Birks, H. J. B. (1979). The past and present vegetation of the Morrone Birkwoods National Nature Reserve, Scotland. II. Woodland vegetation and soils. *Journal of Ecology* 67, 447–467.

Huntley, B., and Birks, H. J. B. (1983). *An Atlas of Past and Present Pollen Maps for Europe: 0–13,000 years ago*. Cambridge University Press, Cambridge.
Hutson, W. H. (1977). Transfer functions under no-analog conditions: experiments with Indian Ocean planktonic foraminifera. *Quaternary Research* 8, 355–367.
Hutson, W. H. (1978). Application of transfer functions to Indian Ocean planktonic foraminifera. *Quaternary Research* 9, 87–112.
Huttunen, P. (1980). Early land use, especially the slash-and-burn cultivation in the commune of Lammi, southern Finland, interpreted mainly using pollen and charcoal analyses. *Acta Botanica Fennica* 113, 1–45.
Hyvärinen, H. (1972). Flandrian regional pollen assemblage zones in eastern Finland. *Commentationes Biologicae, Societas Scientiarum Fennica* 59, 3–25.
Hyvärinen, H. (1975). Absolute and relative pollen diagrams from northenmost Fennoscandia. *Fennia* 142, 23 pp.
Hyvärinen, H. (1976). Flandrian pollen deposition rates and treeline history in northern Fennoscandia. *Boreas* 5, 163–75.
Imbrie, J., and Kipp, N. G. (1971). A new micropaleontological method for quantitative paleoclimatology: application to a late Pleistocene Caribbean core. In *The Late Cenozoic Glacial Ages* (ed. K. K. Turekian), Yale University Press, New Haven and London.
Imbrie, J., Van Donk, J., and Kipp, N. G. (1973). Paleoclimatic investigation of a late Pleistocene Caribbean deep-sea core: comparison of isotopic and faunal methods. *Quaternary Research* 3, 10–38.
Imbrie, J., and Webb, T., III. (1981). Transfer functions: calibrating micropaleontological data in climatic terms. In *Climatic Variations and Variability: Facts and Theories* (ed. A. Berger), D. Reidel Publishing Company, Dordrecht and Boston.
International Study Group (1982). An inter-laboratory comparison of radiocarbon measurements in tree rings. *Nature, London* 298, 619–623.
Isebrands, J. G., and Crow, T. R. (1975). Introduction to uses and interpretation of principal component analysis in forest biology. *U.S.D.A. Forest Service General Technical Report* NC-17, 19 pp.
Iversen, J. (1936). Sekundäres pollen als Fehlerquelle. *Danmarks geologiske Undersøgelse* Series IV (2), 15, 24 pp.
Iversen, J. (1944). *Viscum, Hedera* and *Ilex* as climatic indicators. A contribution to the study of the post-glacial temperature climate. *Geologiska Föreningens i Stockholm Förhandlingar* 66, 463–483.
Iversen, J. (1947). In: Nordiskt kvärtargeologiskt möte den 5–9 november 1945. *Geologiska Föreningens i Stockholm Förhandlingar* 69, 241–242.
Iversen, J. (1949). The influence of prehistoric man on vegetation. *Danmarks geologiske Undersøgelse* Series IV 3(6), 25 pp.
Iversen, J. (1952–1953). Origin of the flora of western Greenland in the light of pollen analysis. *Oikos* 4, 86–103.
Iversen, J. (1954). The late-glacial flora of Denmark and its relation to climate and soil. *Danmarks geologiske Undersøgelse* Series II 80, 87–119.
Iversen, J. (1960). Problems of the early post-glacial forest development in Denmark. *Danmarks geologiske Undersøgelse* Series IV 4(3), 32 pp.
Iversen, J. (1964). Retrogressive vegetational succession in the post-glacial. *Journal of Ecology* 52 (Supplement), 59–70.
Iversen, J. (1969). Retrogressive development of a forest ecosystem demonstrated by pollen diagrams from fossil mor. *Oikos Supplement* 12, 35–49.
Iversen, J. (1973). The development of Denmark's nature since the last glacial. *Danmarks geologiske Undersøgelse* Series V 7C, 126 pp.
Jacobson, G. L. (1975). A palynological study of the history and ecology of white pine in Minnesota. Ph.D. thesis, University of Minnesota.
Jacobson, G. L. (1979). The palaeoecology of white pine (*Pinus strobus*) in Minnesota. *Journal of Ecology* 67, 697–726.

Jacobson, G. L., and Bradshaw, R. H. W. (1981). The selection of sites for paleoenvironmental studies. *Quaternary Research* 16, 80–96.

Janssen, C. R. (1966). Recent pollen spectra from the deciduous and coniferous–deciduous forests of northeastern Minnesota: a study in pollen dispersal. *Ecology* 47, 804–825.

Janssen, C. R. (1967a). A comparison between the recent regional pollen rain and the subrecent vegetation in four major vegetation types in Minnesota (U.S.A.). *Review of Palaeobotany and Palynology* 2, 331–342.

Janssen, C. R. (1967b). A post-glacial pollen diagram from a small *Typha* swamp in northwestern Minnesota, interpreted from pollen indicators and surface samples. *Ecological Monographs* 37, 145–172.

Janssen, C. R. (1970). Problems in the recognition of plant communities in pollen diagrams. *Vegetatio* 20, 187–198.

Janssen, C. R. (1972). The palaeoecology of plant communities in the Dommel Valley, North Brabant, Netherlands. *Journal of Ecology* 60, 411–437.

Janssen, C. R. (1973). Local and regional pollen deposition. In *Quaternary Plant Ecology* (eds. H. J. B. Birks and R. G. West), Blackwell Scientific Publications, Oxford.

Janssen, C. R. (1979). The development of palynology in relation to vegetation science, especially in The Netherlands. In *The Study of Vegetation* (ed. M. J. A. Werger), Dr. Junk, The Hague.

Janssen, C. R. (1980). Some remarks on facts and interpretation in Quaternary palyno-stratigraphy. *Bulletin de l'Association française pour l'Etude du Quaternaire* 1980–1984, 171–176.

Janssen, C. R. (1981a). Contemporary pollen assemblages from the Vosges (France). *Review of Palaeobotany and Palynology* 33, 183–313.

Janssen, C. R. (1981b). On the reconstruction of past vegetation by pollen analysis: a review. *Proceedings of the IVth International Palynological Conference, Lucknow* (1976–1977) 3, 163–172.

Járai-Komlódi, M. (1970). Studies on the vegetational history of *Picea omorika* Panc. on the Great Hungarian Plain. *Annales Universitatis scientiarum budapestinensis de Rolando Eötvös nominatae. I. Sectio biologica* 12, 143–156.

Jardine, N. (1972). Computational methods in the study of plant distributions. In *Taxonomy, Phytogeography, and Evolution* (ed. D. H. Valentine), Academic Press, London and New York.

Jatkar, S. A., Rushforth, S. R., and Brotherson, J. D. (1979). Diatom floristics and succession in a peat bog near Lily Lake, Summit County, Utah. *Great Basin Naturalist* 39, 15–43.

Jeffers, J. N. R. (1967). Two case studies in the application of principal component analysis. *Applied Statistics* 16, 225–236.

Jeffers, J. N. R. (1978). *An Introduction to Systems Analysis: with Ecological Applications*. Edward Arnold, London.

Johansen, J. (1975). Pollen diagrams from the Shetland and Faroe Islands. *New Phytologist* 75, 369–387.

Johnson, E. A. (1981). Vegetation organization and dynamics of lichen woodland communities in the Northwest Territories, Canada. *Ecology* 62, 200–215.

Johnson, L. A. S. (1970). Rainbow's end: the quest for an optimal taxonomy. *Systematic Zoology* 19, 203–239.

Johnson, N. L., and Kotz, S. (1969). *Distributions in Statistics: Discrete Distributions*. Houghton Mifflin Company, Boston.

Johnson, R. G. (1962). Interspecific associations in Pennsylvanian fossil assemblages. *Journal of Geology* 72, 32–55.

Johnson, R. H., Tallis, J. H., and Pearson, M. (1972). A temporary section through Late Devensian sediments at Green Lane, Dalton-in-Furness, Lancashire. *New Phytologist* 71, 533–544.

Jolicoeur, P. (1959). Multivariate geographical variation in the wolf *Canis lupus* L. *Evolution* 13, 283–299.

Jonassen, H. (1950). Recent pollen sedimentation and Jutland heath diagrams. *Dansk Botanisk Arkiv* 13(7), 1–168.

Jöreskog, K. G., Klovan, J. E., and Reyment, R. A. (1976). *Geological Factor Analysis*. Elsevier, Amsterdam.

Jørgensen, S. (1967). A method of absolute pollen counting. *New Phytologist* 66, 489–493.
Kabailiene, M. V. (1969). On formation of pollen spectra and restoration of vegetation. *Ministry of Geology of the SSR, Institute of Geology* (*Vil'nyus*), *Transactions* 11, 1–148.
Kac, M. (1969). Some mathematical models in science. *Science* 166, 695–699.
Kaesler, R. L. (1966). Quantitative re-evaluation of ecology and distribution of recent foraminifera and ostracoda of Todos Santos Bay, Baja California, Mexico. *University of Kansas Palaeontological Contributions* 10, 50 pp.
Kay, P. A. (1979). Multivariate statistical estimates of Holocene vegetation and climate change, forest–tundra transition zone, N.W.T., Canada. *Quaternary Research* 11, 125–140.
Kay, P. A., and Andrews, J. T. (1983). Re-evaluation of pollen-climate transfer functions in Keewatin, northern Canada. *Annals of the Association of American Geographers* 73, 550–559.
Kendall, D. G. (1971a). A mathematical approach to seriation. *Philosophical Transactions of the Royal Society of London Series A* 269, 125–135.
Kendall, D. G. (1971b). Seriation from abundance matrices. In *Mathematics in the Archaeological and Historical Sciences* (eds. F. R. Hodson, D. G. Kendall, and P. Tăutu), Edinburgh University Press, Edinburgh.
Kendall, D. G. (1975). The recovery of structure from fragmentary information. *Philosophical Transactions of the Royal Society of London Series A* 279, 547–582.
Kendall, M. G. (1970). *Rank Correlation Methods* (Fourth edition). Charles Griffin, London.
Kendall, R. L. (1969). An ecological history of the Lake Victoria basin. *Ecological Monographs* 39, 121–176.
Kershaw, A. P. (1970). A pollen diagram from Lake Euramoo, North-East Queensland, Australia. *New Phytologist* 69, 785–805.
Kershaw, A. P. (1971). A pollen diagram from Quincan Crater, North-East Queensland, Australia. *New Phytologist* 70, 669–681.
Kershaw, A. P. (1973). The numerical analysis of modern pollen spectra from northeast Queensland rain forests. *Geological Society of Australia Special Publication* 4, 191–199.
Kershaw, A. P. (1978). Record of last interglacial-glacial cycle from northeastern Queensland. *Nature, London* 272, 159–161.
Kershaw, A. P. (1979). Local pollen deposition in aquatic sediments on the Atherton Tableland, North-Eastern Australia. *Australian Journal of Ecology* 4, 253–263.
Kershaw, A. P., and Hyland, B. P. M. (1975). Pollen transfer and periodicity in a rain-forest situation. *Review of Palaeobotany and Palynology* 19, 129–138.
Kerslake, P. D. (1982). Vegetational history of wooded islands in Scottish lochs. Ph.D. thesis, University of Cambridge.
King, L. (1976). Pollenanalyse und Computer: Erfahrungen mit PALYNO, Programme zur Berechnung und Darstellung Pollenanalytischer Daten. *Pollen et Spores* 18, 93–104.
Kipp, N. G. (1976). New transfer function for estimating past sea-surface conditions from sea-bed distribution of planktonic foraminiferal assemblages in the North Atlantic. *Geological Society of America Memoir* 145, 3–41.
Kitchell, J. A., and Clark, D. L. (1979). A multivariate approach to biofacies analysis of deep-sea traces from the Central Arctic. *Journal of Paleontology* 53, 1045–1067.
Klovan, J. E., and Billings, G. K. (1967). Classification of geological samples by discriminant-function analysis. *Bulletin of the Canadian Petroleum Geologists* 15, 313–330.
Kohut, J. J. (1969). Determination, statistical analysis, and interpretation of recurrent conodont groups in Middle and Upper Ordovician strata of the Cincinnati region (Ohio, Kentucky, and Indiana). *Journal of Paleontology* 43, 392–412.
Krige, D. G. (1966). Two-dimensional weighted moving average trend surfaces for ore evaluation. *Journal of the South African Institute of Mining Metallurgy* 66, 13–38.
Krumbein, W. C., and Graybill, F. A. (1965). *An Introduction to Statistical Models in Geology*. McGraw-Hill, New York.
Kruskal, J. B. (1964a). Multidimensional scaling by optimizing goodness of fit to a nonmetric hypothesis. *Psychometrika* 29, 1–27.

Kruskal, J. B. (1964b). Nonmetric multidimensional scaling: a numerical method. *Psychometrika* 29, 115–129.

Kruskal, J. B. (1971). Comments on "A nonlinear mapping for data structure analysis". *IEEE Transactions on Computers* C-20, 1614.

Kruskal, J. B., and Carroll, J. (1969). Geometric models and badness-of-fit functions. In *Multivariate Analysis II* (ed. P. R. Krishnaiah), Academic Press, London and New York.

Kutzbach, J. E. (1981). Monsoon climate of the early Holocene: climate experiment with the earth's orbital parameters for 9000 years ago. *Science* 214, 59–61.

Kutzbach, J. E., and Bryson, R. A. (1974). Variance spectrum of Holocene climatic fluctuation in the North Atlantic sector. *Journal of Atmospheric Sciences* 31, 1958–1963.

Lamb, H. F. (1980). Late Quaternary vegetational history of southeastern Labrador. *Arctic and Alpine Research* 12, 117–135.

Lamb, H. F. (1982). Late Quaternary vegetation history of the forest–tundra ecotone in north central Labrador. Ph.D. thesis, University of Cambridge.

Lamb, H. F. (1984). Modern pollen spectra from Labrador and their use in reconstructing Holocene vegetational history. *Journal of Ecology* 72, 37–59.

Larsen, J. A. (1971). Vegetational relationships with air mass frequencies: boreal forest and tundra. *Arctic* 24, 177–194.

Lefkovitch, L. P. (1976). Hierarchical clustering from principal coordinates: an efficient method for small to very large numbers of objects. *Mathematical Biosciences* 31, 154–174.

Lehman, J. T. (1975). Reconstructing the rate of accumulation of lake sediment: the effect of sediment focusing. *Quaternary Research* 5, 541–550.

Leopold, E. B. (1964). Reconstruction of Quaternary environments using palynology. In *The Reconstruction of Past Environments* (ed. F. Wendorf), Fort Burgwin Research Center, Taos, New Mexico Publication 3, 43–50.

Lichti-Federovich, S., and Ritchie, J. C. (1965). Contemporary pollen spectra in central Canada. II. The forest–grassland transition in Manitoba. *Pollen et Spores* 7, 63–87.

Lichti-Federovich, S., and Ritchie, J. C. (1968). Recent pollen assemblages from the western interior of Canada. *Review of Palaeobotany and Palynology* 7, 297–344.

Likens, G. E., and Davis, M. B. (1975). Post-glacial history of Mirror Lake and its watershed in New Hampshire, U.S.A.: an initial report. *Verhandlungen Internationalen Vereiningung Limnologie* 19, 982–993.

Livingstone, D. A. (1968). Some interstadial and postglacial pollen diagrams from eastern Canada. *Ecological Monographs* 38, 87–125.

Livingstone, D. A. (1969). Communities of the past. In *Essays in Plant Geography and Ecology* (ed. K. N. H. Greenidge), Nova Scotia Museum, Halifax, Nova Scotia.

Livingstone, D. A., and Estes, A. H. (1967). A carbon-dated pollen diagram from the Cape Breton Plateau, Nova Scotia. *Canadian Journal of Botany* 45, 339–359.

Loubère, P. (1982). Plankton ecology and the paleoceanographic-climatic record. *Quaternary Research* 17, 314–324.

McAndrews, J. H. (1966). Postglacial history of prairie, savanna, and forest in northwestern Minnesota. *Memoirs of the Torrey Botanical Club* 22(2), 72 pp.

McAndrews, J. H. (1967). Pollen analysis and vegetation history of the Itasca region, Minnesota. In *Quaternary Paleoecology* (eds. E. J. Cushing and H. E. Wright, Jr.), Yale University Press, New Haven and London.

McAndrews, J. H. (1976). Fossil history of man's impact on the Canadian flora: an example from southern Ontario. *Canadian Botanical Association Bulletin Supplement* 9, 1–6.

McAndrews, J. H. (1982). Holocene environment of a fossil bison from Kenora, Ontario. *Ontario Archaeology* 37, 41–51.

McAndrews, J. H., and Adams, R. J. (1974). Modern pollen deposition in Ontario: a trend surface analysis. *American Journal of Botany* 61, 59.

McAndrews, J. H., Berti, A. A., and Norris, G. (1973). Key to the Quaternary pollen and spores of the

Great Lakes Region. *Royal Ontario Museum Life Sciences Miscellaneous Publication,* University of Toronto Press, Toronto.

McAndrews, J. H., and Manville, G. C. (1980). Modern pollen rain of Canada. *Abstracts 5th International Palynological Conference,* 238.

McAndrews, J. H., and Power, D. M. (1973). Palynology of the Great Lakes: the surface sediments of Lake Ontario. *Canadian Journal of Earth Sciences* 10, 777–792.

McAndrews, J. H., Riley, J. L., and Davis, A. M. (1982). Vegetation history of the Hudson Bay Lowland: a postglacial pollen diagram from the Sutton Ridge. *Naturaliste canadiense* 109, 597–608.

McAndrews, J. H., and Wright, H. E., Jr., (1969). Modern pollen rain across the Wyoming basins and the northern Great Plains (U.S.A). *Review of Palaeobotany and Palynology* 9, 17–43.

MacDonald, G. M. (1983). Holocene vegetation history of the Upper Natla River Area, Northwest Territories, Canada. *Arctic and Alpine Research* 15, 169–180.

McGee, V. E., and Carleton, W. T. (1970). Piecewise regression. *Journal of the American Statistical Association* 65, 1109–1124.

McIntosh, R. P. (1973). Matrix and plexus techniques. In *Handbook of Vegetation Science. V. Ordination and Classification of Vegetation* (ed. R. H. Whittaker), Dr. Junk, The Hague.

McIntosh, R. P. (1974). Plant ecology 1947–1972. *Annals of the Missouri Botanical Garden* 61, 132–165.

McIntosh, R. P. (1975). H. A. Gleason—"Individualistic Ecologist" 1882–1975: his contributions to ecological theory. *Bulletin of the Torrey Botanical Club* 102, 253–273.

McIntosh, R. P. (1980). The background and some current problems of theoretical ecology. *Synthese* 43, 195–225.

McNamara, K. J., and Fordham, B. G. (1981). Mid-Cautleyan (Ashgill Series) trilobites and facies in the English Lake District. *Palaeogeography, Palaeoclimatology, Palaeoecology* 34, 137–161.

Mack, R. N., and Bryant, V. M. (1974). Modern pollen spectra from the Columbia Basin, Washington. *Northwest Science* 48, 183–194.

Maddocks, R. F. (1966). Distribution patterns of living and subfossil Podocopid Ostracods in the Nosy Bé area, Northern Madagascar. *University of Kansas Paleontological Contributions* 12, 72 pp.

Maher, L. J. (1972a). Absolute pollen diagram of Redrock Lake, Boulder County, Colorado. *Quaternary Research* 2, 531–553.

Maher, L. J. (1972b). Nomograms for computing 0.95 confidence limits of pollen data. *Review of Palaeobotany and Palynology* 13, 85–93.

Maher, L. J. (1977). Palynological studies in the western arm of Lake Superior. *Quaternary Research* 7, 14–44.

Maher, L. J. (1980). The confidence limit is a necessary statistic for relative and absolute pollen data. *Proceedings of the IVth International Palynological Conference, Lucknow* (1976–1977) 3, 152–162.

Maher, L. J. (1981). Statistics for microfossil concentration measurements employing samples spiked with marker grains. *Review of Palaeobotany and Palynology* 32, 153–191.

Maher, L. J. (1982). The palynology of Devils Lake, Sauk County, Wisconsin. In *Quaternary History of the Driftless Area* (eds. J. C. Knox, L. Clayton, and D. M. Mickelson), Wisconsin Geological and Natural History Survey Field Trip Guide Book 5.

Malmgren, B. A. (1974). Morphometric studies of planktonic foraminifers from the type Danian of southern Scandinavia. *Stockholm Contributions in Geology* 29, 126 pp.

Malmgren, B. A., and Kennett, J. P. (1973). Recent planktonic foraminiferal distribution in high latitudes of the South Pacific: a multivariate statistical study. *Palaeogeography, Palaeoclimatology, Palaeoecology* 14, 127–136.

Malmgren, B. A., and Kennett, J. P. (1978a). Late Quaternary paleoclimatic applications of mean size variations in *Globigerina bulloides* D'Orbigny in the southern Indian Ocean. *Journal of Paleontology* 52, 1195–1207.

Malmgren, B. A., and Kennett, J. P. (1978b). Test size variation in *Globigerina bulloides* in response to Quaternary palaeoceanographic changes. *Nature, London* 275, 123–124.

Malmgren, B., Oviatt, C., Gerber, R., and Jeffries, H. P. (1978). Correspondence analysis: applications to biological oceanographic data. *Estuarine and Coastal Marine Science* 6, 429–437.

Mangerud, J. (1970). Late Weichselian vegetation and ice-front oscillations in the Bergen District, Western Norway. *Norsk geografisk Tidsskrift* 24, 121–148.

Mangerud, J., Andersen, S. Th., Berglund, B. E., and Donner, J. J. (1974). Quaternary stratigraphy of Norden, a proposal for terminology and classification. *Boreas* 3, 109–128.

Mann, C. J. (1979). Obstacles to quantitative lithostratigraphic correlation. In *Geomathematical and Petrophysical Studies in Sedimentology* (eds. D. Gill and D. F. Merriam), Pergamon Press, Oxford.

Manten, A. A. (1967). Lennart von Post and the foundation of modern palynology. *Review of Palaeobotany and Palynology* 1, 11–22.

Markgraf, V., D'Antoni, H. L., and Ager, T. A. (1981). Modern pollen dispersal in Argentina. *Palynology* 5, 43–63.

Martin, A. R. H. (1959). South African palynological studies. I. Statistical and morphological variation in the pollen of the South African species of *Podocarpus*. *Grana Palynologica* 2, 40–68.

Martin, A. R. H. (1968). Pollen analysis of Groenvlei lake sediments, Knysna (South Africa). *Review of Palaeobotany and Palynology* 7, 107–144.

Martin, P. S. (1973). The discovery of America. *Science* 179, 969–974.

Martin, P. S., and Gray, J. (1962). Pollen analysis and the Cenozoic. *Science* 137, 103–111.

Martin, P. S., and Mosimann, J. E. (1965). Geochronology of Pluvial Lake Cochise, Southern Arizona, III. Pollen statistics and Pleistocene metastability. *American Journal of Science* 263, 313–358.

Mastrogiuseppe, J. D., Cridland, A. A., and Bogyo, T. P. (1970). Multivariate comparison of fossil and recent *Ginkgo* wood. *Lethaia* 3, 271–277.

Mather, P. M. (1976). *Computational Methods of Multivariate Analysis in Physical Geography*. J. Wiley and Sons, London and New York.

Mathewes, R. W. and Heusser, L. E. (1981). A 12,000 year palynological record of temperature and precipitation trends in southwestern British Columbia. *Canadian Journal of Botany* 59, 707–710.

Matthews, J. (1969). The assessment of a method for the determination of absolute pollen frequencies. *New Phytologist* 68, 161–166.

Maugh, T. H. (1978a). Enrichment: an alternative way to extend dating range. *Science* 200, 638.

Maugh, T. H. (1978b). Radiodating: direct detection extends range of the technique. *Science* 200, 635, 637.

Maxwell, A. E. (1977). *Multivariate Analysis in Behavioural Research*. Chapman and Hall, London and New York.

Médus, J., and Ipert, C. (1977). Introduction de l'analyse multifactorielle dans le traitement de données palynofloristiques Santoniennes (Crétacé Supérieur). *Review of Palaeobotany and Palynology* 24, 141–154.

Mehringer, P. J., Arno, S. F., and Petersen, K. L. (1977). Postglacial history of Lost Trail Pass Bog, Bitterroot Mountains, Montana. *Arctic and Alpine Research* 9, 345–368.

Melguen, M. (1973). Correspondence analysis for recognition of facies in homogeneous sediments off an Iranian river mouth. In *The Persian Gulf* (ed. B. H. Purser), Springer Verlag, New York.

Melguen, M. (1974). Facies analysis by correspondence analysis: numerous advantages of this new statistical technique. *Marine Geology* 17, 165–182.

Mello, J. F., and Buzas, M. A. (1968). An application of cluster analysis as a method of determining biofacies. *Journal of Paleontology* 42, 747–758.

Meredith, W. (1964). Canonical correlations with fallible data. *Psychometrika* 29, 55–65.

Merriam, D. F., and Sneath, P. H. A. (1967). Comparison of cyclic rock sequences using cross-association. In *Essays in Paleontology and Stratigraphy: Raymond C. Moore Commemorative Volume* (eds. C. Teichert and E. L. Yockelson), University of Kansas Press, Lawrence.

Mikkelsen, V. M. (1949). Praestø Fjord. The development of the post-glacial vegetation and a contribution to the history of the Baltic Sea. *Dansk Botanisk Arkiv* 13(5), 171 pp.

Mikkelsen, V. M. (1963). Beech as a natural forest tree in Bornholm. *Botanisk Tidsskrift* 58, 253–280.

Millendorf, S. A., Brower, J. C., and Dyman, T. S. (1978). A comparison of methods for the quantification of assemblage zones. *Computers and Geosciences* 4, 229–242.

Miller, N. G. (1973). Late-glacial and postglacial vegetation change in southwestern New York State. *New York Museum and Science Service Bulletin* 420, 102 pp.

Molfino, B., Kipp, N. G., and Morley, J. J. (1982). Comparison of foraminiferal, coccolithophorid, and radiolarian paleotemperature equations: assemblage coherency and estimate concordancy. *Quaternary Research* 17, 279–313.

Monmonier, M. S., and Finn, F. E. (1973). Improving the interpretation of geographical canonical correlation models, *Professional Geographer* 25, 140–142.

Mood, A. M., Graybill, F. A., and Boes, D. C. (1974). *Introduction to the Theory of Statistics* (Third edition). McGraw-Hill, Tokyo.

Moore, A. W., Russell, J. S., and Ward, W. T. (1972). Numerical analysis of soils: a comparison of three profile models with field classification. *Journal of Soil Science* 23, 193–209.

Moore, P. D. (1973). Objective classification of peats on the basis of their macrofossil content. *Symposium of the International Peat Society Glasgow 1973*, 1–9.

Moore, P. D., and Webb, J. A. (1978). *An Illustrated Guide to Pollen Analysis*. Hodder and Stoughton, London.

Morrison, D. F. (1976). *Multivariate Statistical Methods* (Second edition). McGraw-Hill, Tokyo.

Mosimann, J. E. (1962). On the compound multinomial distribution, the multivariate β-distribution, and correlations among proportions. *Biometrika* 49, 65–82.

Mosimann, J. E. (1963). On the compound negative multinomial distribution and correlations among inversely sampled pollen counts. *Biometrika* 50, 47–54.

Mosimann, J. E. (1965). Statistical methods for the pollen analyst: multinomial and negative multinomial techniques. In *Handbook of Paleontological Techniques* (eds. B. Kummel and D. Raup), W. H. Freeman, San Francisco.

Mosimann, J. E. (1970). Discrete distribution models arising in pollen studies. In *Random Counts in Physical Science, Geoscience, and Business* (ed. G. P. Patil), Pennsylvania State University Press, University Park.

Mosimann, J. E., and Greenstreet, R. L. (1971). Representation-insensitive methods for palaeoecological pollen studies. In *Statistical Ecology Vol. 1, Spatial Patterns and Statistical Distributions* (eds. G. P. Patil, E. C. Pielou and W. E. Waters), Pennsylvania State University Press, University Park.

Mosimann, J. E., and Martin, P. S. (1975). Simulating overkill by paleoindians. *American Scientist* 63, 304–313.

Mott, R. J. (1969). Palynological studies in central Saskatchewan. Contemporary pollen spectra from surface samples. *Geological Survey of Canada Paper* 69–32, 13 pp.

Muller, R. A. (1977). Radioisotope dating with a cyclotron. *Science* 196, 489–494.

Müller, P. (1937). Das Hochmoor von Etzelwil. *Bericht über das Geobotanische Forschungsinstitut Rübel in Zurich* 1936, 85–106.

Needleman, S. B., and Wunsch, C. D. (1970). A general method applicable to the search for similarities in the amino acid sequences of two proteins. *Journal of Molecular Biology* 48, 443–453.

Nichols, H. (1975). Palynological and paleoclimatic study of the late Quaternary displacement of the Boreal forest–tundra ecotone in Keewatin and MacKenzie, N.W.T., Canada. *Institute of Arctic and Alpine Research, University of Colorado, Occasional Paper* 15, 87 pp.

Nichols, H., Kelly, P. M., and Andrews, J. T. (1978). Holocene palaeo-wind evidence from palynology in Baffin Island. *Nature, London* 273, 140–142.

Norris, J. M. (1971). Comparison of different methods in the transition matrix approach to the numerical classification of soil profiles. *Proceedings of the Soil Science Society of America* 35, 965–968.

Norris, J. M., and Dale, M. B. (1971). Transition matrix approach to numerical classification of soil profiles. *Proceedings of the Soil Science Society of America* 35, 487–491.

Norton, S. A., Davis. R. B., and Brakke, D. F. (1981). Responses of northern New England lakes to atmospheric inputs of acids and heavy metals. *Report to Land and Water Resources Center, University of Maine, Project A-048-ME,* 90 pp.

Noy-Meir, I. (1973). Data transformations in ecological ordination. I. Some advantages of non-centering. *Journal of Ecology* 61, 329–341.

Noy-Meir, I., Walker, D., and Williams, W. T. (1975). Data transformations in ecological ordination. II. On the meaning of data standardization. *Journal of Ecology* 63, 779–800.

Noy-Meir, I., and Whittaker, R. H. (1977). Continuous multivariate methods in community analysis: some problems and developments. *Vegetatio* 33, 79–98.

Noy-Meir, I., and Whittaker, R. H. (1978). Recent developments in continuous multivariate techniques. In *Ordination of Plant Communities* (ed. R. H. Whittaker), Dr. Junk, The Hague.

Ogden, J. G. (1964). Problems in vegetational interpretation of surface pollen spectra from eastern North America. Paper read at the American Association for the Advancement of Science meeting. Montreal, December 27, 1964.

Ogden, J. G. (1967). Radiocarbon determinations of sedimentation rates from hard and soft-water lakes in northeastern North America. In *Quaternary Paleoecology* (eds. E. J. Cushing and H. E. Wright, Jr.), Yale University Press, New Haven and London.

Ogden, J. G. (1969). Correlation of contemporary and Late Pleistocene pollen records in the reconstruction of postglacial environments in northeastern North America. *Mitteilungen Internationalen Vereinigung Limnologie* 17, 64–77.

Ogden, J. G. (1977a). Limiting factors in paleoenvironmental reconstruction. In *Geobotany* (ed. R. C. Romans), Plenum Press, New York.

Ogden, J. G. (1977b). Pollen analysis: state of the art. *Géographie physique et Quaternaire* 37, 151–159.

Oldfield, F. (1970). Some aspects of scale and complexity in pollen-analytically based palaeoecology. *Pollen et Spores* 12, 163–171.

Oldfield, F., Brown, A., and Thompson, R. (1979). The effect of microtopography and vegetation on the catchment of airborne particles measured by remanent magnetism. *Quaternary Research* 12, 326–332.

Olsson, I. U. (1979). A warning against radiocarbon dating of samples containing little carbon. *Boreas* 8, 203–207.

Oltz, D. F. (1969). Numerical analyses of palynological data from Cretaceous and early Tertiary sediments in east central Montana. *Palaeontographica* B 128, 166 pp.

Oltz, D. F. (1971). Cluster analyses of Late Cretaceous–Early Tertiary pollen and spore data. *Micropaleontology* 17, 221–232.

Ord, J. K. (1979). Time-series and spatial patterns in ecology. In *Spatial and Temporal Analysis in Ecology* (eds. R. M. Cormack and J. K. Ord), International Co-operative Publishing House, Burtonsville, Maryland.

Orloci, L. (1967a). An agglomerative method for classification of plant communities. *Journal of Ecology* 55, 193–206.

Orloci, L. (1967b). Data centering: a review and evaluation with reference to component analysis. *Systematic Zoology* 16, 208–212.

Orloci, L. (1972). On objective functions of phytosociological resemblance. *American Midland Naturalist* 88, 28–55.

Orloci, L. (1973). Ordination by resemblance matrices. In *Handbook of Vegetation Science V. Ordination and Classification of Vegetation* (ed. R. H. Whittaker), Dr. Junk, The Hague.

Orloci, L. (1978). *Multivariate Analysis in Vegetation Research* (Second edition). Dr. Junk, The Hague.

O'Sullivan, P. E. (1973). Contemporary pollen studies in a native Scots pine ecosystem. *Oikos* 24, 143–150.

O'Sullivan, P. E., and Riley, D. H. (1974). Multivariate numerical analysis of surface pollen spectra from a native Scots pine forest. *Pollen et Spores* 16, 239–264.

Ovenden, L. (1982). Vegetation history of a polygonal peatland, northern Yukon. *Boreas* 11, 209–224.

Overpeck, J. T., and Webb, T., III. (1983). Calibration of numerical dissimilarity measures for matching modern and fossil pollen spectra. *Bulletin of the Ecological Society of America* 64, 155

Oxnard, C. E. (1972). Some African fossil foot bones: a note on the interpolation of fossils into a matrix of extant species. *American Journal of Physical Anthropology* 37, 3–12.

Oxnard, C. E. (1973). *Form and Pattern in Human Evolution.* University of Chicago Press, Chicago and London.

Pardi, R., and Marcus, L. (1977). Non-counting errors in ^{14}C dating. *Annals of the New York Academy of Sciences* 288, 174–180.

Park, R. A. (1968). Paleoecology of *Venericardia* sensu lato (Pelecypoda) in the Atlantic and Gulf Coastal Province: an application of paleosynecologic methods. *Journal of Paleontology* 42, 955–986.

Parsons, R. W., Gordon, A. D., and Prentice, I. C. (1983). Statistical uncertainty in forest composition estimates obtained from fossil pollen spectra via the R-value model. *Review of Palaeobotany and Palynology* 40, 177–189.

Parsons, R. W., Prentice, I. C., and Saarnisto, M. (1980). Statistical studies on pollen representation in Finnish lake sediments in relation to forest inventory data. *Annales Botanici Fennici* 17, 379–393.

Parsons, R. W., and Prentice, I. C. (1981). Statistical approaches to R-values and the pollen-vegetation relationship. *Review of Palaeobotany and Palynology* 32, 127–152.

Pawlikowski, M., Ralska-Jasiewiczowa, M., Schönborn, W., Stupnicka, E., and Szeroczynska, K. (1982). Woryty near Gietrzwałd, Olsztyn Lake District, NE Poland—vegetational history and lake development during the last 12000 years. *Acta Palaeobotanica* 22, 85–116.

Peck, R. M. (1973). Pollen budget studies in a small Yorkshire catchment. In *Quaternary Plant Ecology* (eds. H. J. B. Birks and R. G. West), Blackwell Scientific Publications, Oxford.

Peck, R. M. (1974). A comparison of four absolute pollen preparation techniques. *New Phytologist* 73, 567–587.

Peglar, S. M., Fritz, S. C., Alapieti, T., Saarnisto, M., and Birks, H. J. B. (1984). Composition and formation of laminated sediments in Diss Mere, Norfolk, England. *Boreas* 13, 13–28.

Pennington, W. (1964). Pollen analyses from the deposits of six upland tarns in the Lake District. *Philosophical Transactions of the Royal Society of London Series B* 248, 205–244.

Pennington, W. (1970). Vegetation history in the North-West of England: a regional synthesis. In *Studies in the Vegetational History of the British Isles* (eds. D. Walker and R. G. West), Cambridge University Press, London.

Pennington, W. (1973). Absolute pollen frequencies in sediments of lakes of different morphometry. In *Quaternary Plant Ecology* (eds. H. J. B. Birks and R. G. West), Blackwell Scientific Publications, Oxford.

Pennington, W. (1977). The Late Devensian flora and vegetation of Britain. *Philosophical Transactions of the Royal Society of London Series B* 280, 247–271.

Pennington, W. (1979). The origin of pollen in lake sediments: an enclosed lake compared with one receiving inflow streams. *New Phytologist* 83, 189–213.

Pennington, W., and Bonny, A. P. (1970). Absolute pollen diagram from the British Late-Glacial. *Nature, London* 226, 871–873.

Pennington, W., Cambray, R. S., Eakins, J. D., and Harkness, D. D. (1976). Radionuclide dating of the recent sediments of Blelham Tarn. *Freshwater Biology* 6, 317–331.

Pennington, W., and Sackin, M. J. (1975). An application of principal components analysis to the zonation of two Late-Devensian profiles. *New Phytologist* 75, 419–453.

Peters, R. H. (1980). Useful concepts for predictive ecology. *Synthese* 43, 257–269.

Pielou, E. C. (1977). *Mathematical Ecology.* J. Wiley and Sons, London and New York.

Pielou, E. C. (1981). The usefulness of ecological models: a stocktaking. *Quarterly Review of Biology* 56, 17–31.

Pigott, C. D., and Huntley, J. P. (1981). Factors controlling the distribution of *Tilia cordata* at the northern limits of its geographical range. III. Nature and causes of seed sterility. *New Phytologist* 87, 817–839.

Pimm, S. L. (1984). The complexity and stability of ecosystems. *Nature, London* 307, 321–326.

Platt, T., and Denman, K. (1975). Spectral analysis in ecology. *Annual Review Ecology and Systematics* 6, 189–210.

Pohl, F. (1937). Die Pollenerzeugung der Windblütler. *Beiheflen zum Botanischen Centralblatt* 56, 365–470.

Poore, M. E. D. (1955). The use of phytosociological methods in ecological investigations. II. Practical issues involved in an attempt to apply the Braun-Blanquet system. *Journal of Ecology* 43, 245–269.

Prentice, H. C. (1982). Late Weichselian and early Flandrian vegetational history of the Varanger peninsula, north-east Norway. *Boreas* 11, 187–208.

Prentice, I. C. (1977). Non-metric ordination methods in ecology. *Journal of Ecology* 65, 85–94.

Prentice, I. C. (1978). Modern pollen spectra from lake sediments in Finland and Finnmark, north Norway. *Boreas* 7, 131–153.

Prentice, I. C. (1980). Multidimensional scaling as a research tool in Quaternary palynology: a review of theory and methods. *Review of Palaeobotany and Palynology* 31, 71–104.

Prentice, I. C. (1981). Quantitative birch (*Betula* L.) pollen separation by analysis of size frequency data. *New Phytologist* 89, 145–157.

Prentice, I. C. (1982a). Calibration of pollen spectra in terms of species abundance. In *Palaeohydrological Changes in the Temperate Zone in the last 15000 years. Subproject B. Lake and Mire Environments. Volume III. Specific Methods* (ed. B. E. Berglund), University of Lund, Lund.

Prentice, I. C. (1982b). Multivariate methods for the presentation and analysis of data. In *Palaeohydrological Changes in the Temperate Zone in the last 15000 years. Subproject B. Lake and Mire Environments. Volume III. Specific Methods* (ed. B. E. Berglund), University of Lund, Lund.

Prentice, I. C. (1983a). Pollen mapping of regional vegetation patterns in south and central Sweden. *Journal of Biogeography* 10, 441–454.

Prentice, I. C. (1983b). Postglacial climatic change: vegetation dynamics and the pollen record. *Progress in Physical Geography* 7, 273–286.

Prentice, I. C., and Parsons, R. W. (1983). Maximum likelihood linear calibration of pollen spectra in terms of forest composition. *Biometrics* 39, 1051–1057.

Pritchard, N. M., and Anderson, A. J. B. (1971). Observations on the use of cluster analysis in botany with an ecological example. *Journal of Ecology* 59, 727–747.

Ralska-Jasiewiczowa, M. (1983). Isopollen maps for Poland: 0–11000 years B. P. *New Phytologist* 94, 133–175.

Rampton, V. (1971). Late Quaternary vegetational and climatic history of the Snag-Klutlan area, south-western Yukon Territory, Canada. *Geological Society of America Bulletin* 82, 959–978.

Rand, W. M. (1971). Objective criteria for the evaluation of clustering methods. *Journal of the American Statistical Association* 66, 846–850.

Rao, C. R. (1970). *Advanced Statistical Methods in Biometric Research*. Hafner, Darien, Connecticut.

Rayner, J. H. (1966). Classification of soils by numerical methods. *Journal of Soil Science* 17, 79–82.

Read, W. A., and Sackin, M. J. (1971). A quantitative comparison, using cross-association of vertical sections of Namurian (E_1) paralic sediments in the Kincardine Basin, Scotland. *Institute of Geological Sciences Report* 71/14, 21 pp., London.

Regal, R. R., and Cushing, E. J. (1979). Confidence intervals for absolute pollen counts. *Biometrics* 35, 557–565.

Reitsma, Tj. (1969). Size modifications of recent pollen grains under different treatments. *Review of Palaeobotany and Palynology* 9, 175–202.

Rempe, U., and Weber, E. E. (1972). An illustration of the principal ideas of MANOVA. *Biometrics* 28, 235–238.

Reyment, R. A. (1963a). Multivariate analytical treatment of quantitative species associations: an example from palaeoecology. *Journal of Animal Ecology* 32, 535–547.

Reyment, R. A. (1963b). Studies on Nigerian Upper Cretaceous and Lower Tertiary Ostracoda. 2. Danian, Paleocene and Eocene Ostracoda. *Stockholm Contributions in Geology* 10, 1–286.

Reyment, R. A. (1966). Studies on Nigerian Upper Cretaceous and Lower Tertiary Ostracoda. 3.

Stratigraphical, paleoecological, and biometrical conclusions. *Stockholm Contributions in Geology* 14, 151 pp.
Reyment, R. A. (1967). Systems analysis in paleoecology. *Geologiska Föreningens i Stockholm Förhandlingar* 89, 440–447.
Reyment, R. A. (1969a). Some case studies of the statistical analysis of sexual dimorphism. *Bulletin of the Geological Institution of the University of Upsala* N.S.1, 97–119.
Reyment, R. A. (1969b). Statistical analysis of some volcanologic data regarded as series of point events. *Pure and Applied Geophysics* 74, 57–77.
Reyment, R. A. (1971). *Introduction to Quantitative Paleoecology*. Elsevier, Amsterdam.
Reyment, R. A. (1972). Models for studying the occurrence of lead and zinc in a deltaic environment. In *Mathematical Models of Sedimentary Processes* (ed. D. F. Merriam), Plenum Press, New York.
Reyment, R. A. (1975). Analysis of generic level transition in Cretaceous ammonites. *Evolution* 28, 665–676.
Reyment, R. A. (1976a). Chemical components of the environment and Late Campanian microfossil frequencies. *Geologiska Föreningens i Stockholm Förhandlingar* 98, 322–328.
Reyment, R. A. (1976b). Some applications of point processes in geology. *Journal of Mathematical Geology* 8, 95–97.
Reyment, R. A. (1978a). Biostratigraphical logging methods. *Computers and Geosciences* 4, 261–268.
Reyment, R. A. (1978b). Quantitative biostratigraphical analysis exemplified by Moroccan Cretaceous ostracods. *Micropaleontology* 24, 24–43.
Reyment, R. A. (1979). Multivariate analysis in statistical paleoecology. In *Multivariate Methods in Ecological Work* (eds. L. Orloci, C. R. Rao, and W. M. Stiteler), International Co-operative Publishing House, Burtonsville, Maryland.
Reyment, R. A. (1980a). *Morphometric Methods in Biostratigraphy*. Academic Press, London and New York.
Reyment, R. A. (1980b). Trends in Cretaceous and Tertiary geomagnetic reversals. *Cretaceous Research* 1, 27–48.
Reyment, R. A. (1982). Phenotypic evolution in a Cretaceous foraminifer. *Evolution* 36, 1182–1199.
Reyment, R. A., and Brännstrom, B. (1962). Certain aspects of the physiology of *Cypridopsis* (Ostracoda, Crustacea). *Stockholm Contributions in Geology* 9, 207–242.
Reyment, R. A., Hayami, I., and Carbonnel, G. (1977). Variation of discrete morphological characters in *Cytheridea* (Crustacea:Ostracoda). *Bulletin of the Geological Institution of the University of Upsala* N.S.7, 23–36.
Reyment, R. A., and Ramden, H-A. (1970). FORTRAN IV program for canonical variates analysis for the CDC 3600 computer. *Kansas Geological Survey Computer Contribution* 47, 40 pp.
Richard, P. (1976). Relations entre la végétation actuelle et le spectre pollinique au Québec. *Naturaliste canadiense* 103, 53–66.
Riggs, D. S., Guarnieri, J. A., and Addelman, S. (1978). Fitting straight lines when both variables are subject to error. *Life Sciences* 22, 1305–1360.
Ritchie, J. C. (1959). The vegetation of northern Manitoba. III. Studies in the Subarctic. *Arctic Institute of North America Technical Paper* 3, 56 pp.
Ritchie, J. C. (1960). The vegetation of northern Manitoba. V. Establishing the major zonation. *Arctic* 13, 211–229.
Ritchie, J. C. (1967). Holocene vegetation of the northwestern precincts of the Glacial Lake Agassiz basin. In *Life, Land, and Water* (ed. W. J. Mayer-Oakes), University of Manitoba Press, Winnipeg.
Ritchie, J. C. (1974). Modern pollen assemblages near the arctic tree line, Mackenzie Delta region, Northwest Territories. *Canadian Journal of Botany* 52, 381–396.
Ritchie, J. C. (1976). The late-Quaternary vegetational history of the Western Interior of Canada. *Canadian Journal of Botany* 54, 1793–1818.
Ritchie, J. C. (1977). The modern and late Quaternary vegetation of the Campbell-Dolomite Uplands, near Inuvik, N.W.T., Canada. *Ecological Monographs* 47, 401–423.

Ritchie, J. C. (1981). Problems of interpretation of the pollen stratigraphy of northwest North America. In *Quaternary Paleoclimate* (ed. W. C. Mahaney), GeoAbstracts Ltd, Norwich.

Ritchie, J. C. (1982). The modern and late-Quaternary vegetation of the Doll Creek area, North Yukon, Canada. *New Phytologist* 90, 563–603.

Ritchie, J. C., and Hare, F. K. (1971). Late Quaternary vegetation and climate near the arctic tree line of northwestern North America. *Quaternary Research* 1, 331–342.

Ritchie, J. C., and Lichti-Federovich, S. (1967). Pollen dispersal phenomena in Arctic-Subarctic Canada. *Review of Palaeobotany and Palynology* 3, 255–266.

Ritchie, J. C., and Yarranton, G. A. (1978a). The late-Quaternary history of the Boreal Forest of central Canada, based on standard pollen stratigraphy and principal components analysis. *Journal of Ecology* 66, 199–212.

Ritchie, J. C., and Yarranton, G. A. (1978b). Patterns of change in the late-Quaternary vegetation of the western interior of Canada. *Canadian Journal of Botany* 56, 2177–2183.

Robichaux, R. H., and Taylor, D. W. (1977). Vegetation-analysis techniques applied to late Tertiary fossil floras from western United States. *Journal of Ecology* 65, 643–660.

Robinson, A. H. (1962). Mapping the correspondence of isarithmic maps. *Annals of the Association of American Geographers* 52, 414–425.

Rohlf, F. J. (1972). An empirical comparison of three ordination techniques in numerical taxonomy. *Systematic Zoology* 21, 271–280.

Rohlf, F. J. (1974). Methods of comparing classifications. *Annual Review of Ecology and Systematics* 5, 101–113.

Rowe, J. S. (1966). Phytogeographic zonation: an ecological appreciation. In *The Evolution of Canada's Flora* (eds. R. L. Taylor and R. A. Ludwig), University of Toronto Press, Toronto.

Rowell, A. J., McBride, D. J., and Palmer, A. R. (1973). Quantitative study of Trempealeauian (Latest Cambrian) trilobite distribution in North America. *Geological Society of America Bulletin* 84, 3429–3442.

Rudman, A. J., and Lankston, R. W. (1973). Stratigraphic correlation of well logs by computer techniques. *Bulletin of the American Association of Petroleum Geologists* 97, 577–588.

Rybniček, K., and Rybničkova, E. (1968). The history of flora and vegetation on the Blato Mire in southeastern Bohemia, Czechoslovakia. *Folia Geobotanica Phytotaxonomica* 3, 117–142.

Rymer, L. (1974). The palaeoecology and historical ecology of the Parish of North Knapdale, Argyllshire. Ph.D. thesis, University of Cambridge.

Rymer, L. (1977). A late-glacial and early post-glacial pollen diagram from Drimnagall, North Knapdale, Argyllshire. *New Phytologist* 79, 211–221.

Rymer, L. (1978). The use of uniformitarianism and analogy in palaeoecology, particularly pollen analysis. In *Biology and Quaternary Environments* (eds. D. Walker and J. C. Guppy), Australian Academy of Sciences, Canberra.

Saarnisto, M. (1979a). Application of annually laminated lake sediments: a review. *Acta Universitatis Ouluensis Series A* 82, *Geologica* 3, 97–108.

Saarnisto, M. (1979b). Studies of annually laminated lake sediments. In *Palaeohydrological Changes in the Temperate Zone in the last 15000 years. Subproject B. Lake and Mire Environments. Volume III. Specific Methods* (ed. B. E. Berglund), University of Lund, Lund.

Saarnisto, M., Huttunen, P., and Tolonen, K. (1977). Annual lamination of sediments in Lake Lovojärvi, southern Finland, during the past 600 years. *Annales Botanici Fennici* 14, 35–45.

Sachs, H. M., Webb, T., III, and Clark, D. R. (1977). Palaeoecological transfer functions. *Annual Review of Earth and Planetary Sciences* 5, 159–178.

Sackin, M. J. (1971). Cross-association: a method for comparing protein sequences. *Biochemical Genetics* 5, 287–313.

Sackin, M. J., Sneath, P. H. A., and Merriam, D. F. (1965). ALGOL program for cross-association of nonmetric sequences using a medium-size computer. *Kansas Geological Survey Special Distribution Publication* 23, 36 pp.

Sammon, J. W. (1969). A nonlinear mapping for data structure analysis. *IEEE Transactions on Computers* C-18, 401–409.

Schaffer, W. M., and Leigh, E. G. (1976). The prospective role of mathematical theory in plant ecology. *Systematic Botany* 1, 209–232.
Schindel, D. E. (1980). Microstratigraphic sampling and the limits of paleontologic resolution. *Paleobiology* 6, 408–426.
Scott, E. L. (1979). Correlation and suggestions of causality: spurious correlation. In *Multivariate Methods in Ecological Work* (eds. L. Orloci, C. R. Rao, and W. M. Stiteler), International Cooperative Publishing House, Burtonsville, Maryland.
Seal, H. (1964). *Multivariate Statistical Analysis for Biologists.* Methuen, London.
Sears, P. B., and Clisby, K. H. (1955). Palynology in southern North America. IV. Pleistocene climate in New Mexico. *Geological Society of America Bulletin* 66, 521–530.
Sercelj, A., and Adam, D. P. (1975). A late Holocene pollen diagram from near Lake Tahoe, El Dorado County, California. *Journal of Research of the United States Geological Survey* 3, 737–745.
Shackelton, J. (1982). Environmental histories from Whitefish and Imruk Lakes, Seward Peninsula, Alaska. *Ohio State University of Polar Studies Report* 76, 50 pp.
Shaffer, B. L., and Wilke, S. C. (1965). The ordination of fossil communities: an approach to the study of species interrelationships and communal structure. *Papers of the Michigan Academy of Science, Arts, and Letters* 50, 199–214.
Shaw, B. R., and Cubitt, J. M. (1979). Stratigraphic correlation of well logs: an automated approach. In *Geomathematical and Petrophysical Studies in Sedimentology* (eds. D. Gill and D. F. Merriam), Pergamon Press, Oxford.
Shepard, R. N. (1962a). The analysis of proximities: multidimensional scaling with an unknown distance function I. *Psychometrika* 27, 125–140.
Shepard, R. N. (1962b). The analysis of proximities: multidimensional scaling with an unknown distance function II. *Psychometrika* 27, 219–246.
Shotton, F. W. (1967). The problems and contributions of methods of absolute dating within the Pleistocene period. *Quarterly Journal of Geological Society of London* 122, 356–383.
Sibson, R. (1969). Information radius. *Zeitschrift für Wahrscheinlichkeitstheorie und verwandte Gebiete* 14, 149–160.
Sibson, R. (1972a). Multidimensional scaling in theory and practice. *Les Méthodes Mathématiques de l'Archéologie* (ed. M. Borillo), C.N.R.S., Paris.
Sibson, R. (1972b). Order invariant methods for data analysis. *Journal of the Royal Statistical Society Series B* 34, 311–349.
Sibson, R. (1981). A brief description of natural neighbour interpolation. In *Interpreting Multivariate Data* (ed. V. Barnett), J. Wiley and Sons, Chichester and New York.
Simberloff, D. (1982). The status of competition theory in ecology. *Annales Zoologici Fennici* 19, 241–253.
Singh, G., Kershaw, A. P., and Clark, R. (1981). Quaternary vegetation and fire history in Australia. In *Fire and Australian Biota* (eds. A. M. Gill, R. A. Groves, and I. R. Noble), Australian Academy of Sciences, Canberra.
Sissons, J. B. (1979a). The Loch Lomond stadial in the British Isles. *Nature, London* 280, 199–203.
Sissons, J. B. (1979b). Palaeoclimatic inferences from former glaciers in Scotland and the Lake District. *Nature, London* 278, 518–521.
Sluiter, I. R., and Kershaw, A. P. (1982). The nature of Late Tertiary vegetation in Australia. *Alcheringa* 6, 211–222.
Smith, A. G. (1958). Pollen analytical investigations of the mire at Fallahogy Td, Co. Derry. *Proceedings of the Royal Irish Academy* 59B, 329–343.
Smith, A. G. (1961). Cannons Lough, Kilrea, Co. Derry: stratigraphy and pollen analysis. *Proceedings of the Royal Irish Academy* 61B, 369–383.
Smith, A. G., and Willis E. H. (1961–1962). Radiocarbon dating of the Fallahogy Landnam Phase. *Ulster Journal of Archaeology* 25, 16–24.
Sneath, P. H. A., and Sokal, R. R. (1973). *Numerical Taxonomy*. W. H. Freeman, San Francisco.
Solomon, A. M., Blasing, T. J., and Solomon, J. A. (1982). Interpretation of floodplain pollen in alluvial sediments from an arid region. *Quaternary Research* 18, 52–71.

Solomon, A. M., Delcourt, H. R., West, D. C., and Blasing, T. J. (1980). Testing a simulation model for reconstruction of prehistoric forest-stand dynamics. *Quaternary Research* 14, 275–293.

Solomon, A. M., and Harrington, J. B. (1979). Palynology models. In *Aerobiology—The Ecological Systems Approach* (ed. R. L. Edmonds), Dowden, Hutchinson and Ross Inc., Stroudsburg, Pennsylvania.

Solomon, A. M., West, D. C., and Solomon, J. A. (1981). The role of climatic change and species immigration in forest succession. In *Forest Succession-Concepts and Applications* (eds. D. C. West, H. H. Shugart, and D. B. Botkin), Springer-Verlag, New York.

Solomon, D. L. (1979). On a paradigm for mathematical modeling. In *Contemporary Quantitative Ecology and Related Ecometrics* (eds. G. P. Patil and M. Rosenzweig), International Co-operative Publishing House, Burtonsville, Maryland.

Sønstegaard, E., and Mangerud, J. (1977). Stratigraphy and dating of Holocene gully sediments in Os, Western Norway. *Norsk geologisk Tidsskrift* 57, 313–346.

Southam, J. R., and Hay, W. W. (1978). Correlation of stratigraphic sections by continuous variables. *Computers and Geosciences* 4, 257–260.

Spicer, R. A. (1981). The sorting and deposition of allochthonous plant material in a modern environment at Silwood Lake, Silwood Park, Berkshire, England. *United States Geological Survey Professional Paper* 1143, 77 pp.

Spicer, R. A., and Hill, C. R. (1979). Principal components and correspondence analyses of quantitative data from a Jurassic plant bed. *Review of Palaeobotany and Palynology* 28, 273–299.

Squires, R. H. (1970). A computer program for the presentation of pollen data. *Department of Geography, University of Durham Occasional Papers* 11, 53 pp.

Squires, R. H., and Holder, A. P. (1970). The use of computers in the presentation of pollen data. *New Phytologist* 69, 875–883.

Starling, R. N., and Crowder, A. (1981). Pollen in the Salmon River System, Ontario. *Review of Palaeobotany and Palynology* 31, 311–334.

Stearns, S. C. (1976). Life-history tactics: a review of the ideas. *Quarterly Review of Biology* 51, 3–47.

Steinberg, K. (1944). Zur spät und nacheiszeitlichen Vegetationsgeschichte des Untereichsfeldes. *Hercynia* 3, 529–587.

Stephenson, W. (1978). Analyses of periodicity in macrobenthos using constructed and real data. *Australian Journal of Ecology* 3, 321–336.

Stewart, D., and Love, W. (1968). A general canonical correlation index. *Psychological Bulletin* 70, 160–163.

Stockmarr, J. (1971). Tablets with spores used in absolute pollen analysis. *Pollen et Spores* 13, 615–621.

Stockmarr, J. (1972). Determination of spore concentration with an electronic particle counter. *Danmarks geologiske Undersøgelse Årbog* 1972, 87–89.

Stockmarr, J. (1975). Retrogressive forest development, as reflected in a mor pollen diagram from Mantingerbos, Drenthe, The Netherlands. *Palaeohistoria* 17, 37–51.

Strong, D. R. (1980). Null hypotheses in ecology. *Synthese* 43, 271–285.

Strong, W. L. (1977). Pre- and post-settlement palynology of southern Alberta. *Review of Palaeobotany and Palynology* 23, 373–387.

Sudia, T. W. (1952). A statistical analysis of twenty pollen spectra from a single stratum of Amanda Bog (Ohio). *Ohio Journal of Science* 52, 213–215.

Sutherland, D. G. (1980). Problems of radiocarbon dating deposits from newly deglaciated terrain: examples from the Scottish lateglacial. In *Studies in the Lateglacial of North-West Europe* (eds. J. J. Lowe, J. M. Gray, and J. E. Robinson), Pergamon Press, Oxford.

Swain, A. M. (1973). A history of fire and vegetation in northeastern Minnesota as recorded in lake sediments. *Quaternary Research* 3, 383–396.

Swain, A. M. (1978). Environmental changes during the past 2000 years in north-central Wisconsin: analysis of pollen, charcoal, and seeds from varved lake sediments. *Quaternary Research* 10, 55–68.

Swain, A. M. (1980). Landscape patterns and forest history in the Boundary Waters Canoe Area, Minnesota: a pollen study from Hug Lake. *Ecology* 61, 747–754.

Swain, A. M., Kutzbach, J. E., and Hastenrath, S. (1983). Estimates of Holocene precipitation for Rajasthan, India, based on pollen and lake-level data. *Quaternary Research* 19, 1–17.

Symons, F., and De Meuter, F. (1974). Foraminiferal associations of the Mid-Tertiary Edegem Sands at Terhagen, Belgium. *Journal of Mathematical Geology* 6, 1–15.

Synerholm, C. C. (1979). The chydorid cladocera from surface lake sediments in Minnesota, North America. *Archiv Hydrobiologie* 86, 137–151.

Szafer, W. (1935). The significance of isopollen lines for the investigations of the geographical distribution of trees in the post-glacial period. *Bulletin de l'Académie Polonaise des Sciences et des Lettres B* 1935, 235–239.

Taillie, C., Ord, J. K., Mosimann, J. E., and Patil, G. P. (1979). Chance mechanisms underlying multivariate distributions. In *Statistical Distributions in Ecological Work* (eds. J. K. Ord, G. P. Patil, and C. Taillie), International Co-operative Publishing House, Burtonsville, Maryland.

Tallis, J. H., and Johnson, R. H. (1980). The dating of landslides in Longdendale, north Derbyshire, using pollen-analytical techniques. In *Timescales in Geomorphology* (eds. R. A. Cullingford, D. A. Davidson, and J. Lewin), J. Wiley and Sons, London and New York.

Tamboer-Van den Heuvel, G., and Janssen, C. R. (1976). Recent pollen assemblages from the crest region of the Vosges Mountains (France). *Review of Palaeobotany and Palynology* 21, 219–240.

Tauber, H. (1965). Differential pollen dispersion and the interpretation of pollen diagrams. *Danmarks geologiske Undersøgelse* Series II 89, 69 pp.

Tauber, H. (1967). Investigations of the mode of pollen transfer in forested areas. *Review of Palaeobotany and Palynology* 3, 277–286.

Tauber, H. (1977). Investigations of aerial pollen transport in a forested area. *Dansk Botanisk Arkiv* 32(1), 121 pp.

Thompson, R. (1979). Palaeomagnetic correlation and dating. In *Palaeohydrological Changes in the Temperate Zone in the last 15000 years. Subproject B. Lake and Mire Environments. Volume III. Specific Methods* (ed. B. E. Berglund), University of Lund, Lund.

Thompson, R. (1980). Use of the word "influx" in palaeolimnological studies. *Quaternary Research* 14, 269–270.

Tinsley, H. M., and Smith, R. T. (1974). Surface pollen studies across a woodland/heath transition and their application to the interpretation of pollen diagrams. *New Phytologist* 73, 547–565.

Tipper, J. C. (1975). Lower Silurian animal communities—three case histories. *Lethaia* 8, 287–299.

Tipper, J. C. (1980). Some distributional models for fossil animals. *Paleobiology* 6, 77–95.

Tolonen, M. (1978). Palaeoecology of annually laminated sediments in Lake Ahvenainen, S. Finland I. Pollen and charcoal analyses, and their relation to human impact. *Annales Botanici Fennici* 15, 177–208.

Torgerson, W. S. (1952). Multidimensional scaling: I. Theory and method. *Psychometrika* 17, 401–419.

Traverse, A., and Ginsburg, R. N. (1967). Pollen and associated microfossils in the marine surface sediments of the Great Bahama Bank. *Review of Palaeobotany and Palynology* 3, 243–254.

Troels-Smith, J. (1955). Karakterisering af løse jordater. Characterisation of unconsolidated sediments. *Danmarks geologiske Undersøgelse* Series IV 3(10), 73 pp.

Tsukada, M. (1957). Pollen analytical studies of postglacial age in Japan. I. Hyotan-ike Ponds on Shiga Heights, Nagano Prefecture. *Journal of the Institute of Polytechnics, Osaka City University Series D* 8, 203–216.

Tsukada, M. (1958). Untersuchungen über das Verhältnis zwischen dem Pollengehalt der Oberflachenproben und der Vegetation des Hochlandes Shiga. *Journal of the Institute of Polytechnics, Osaka City University Series D* 9, 217–234.

Tsukada, M. (1981). *Cryptomeria japonica* D. Don. I. Pollen dispersal and logistic forest expansion. *Japanese Journal of Ecology* 31, 371–383.

Tsukada, M. (1982a). *Cryptomeria japonica:* glacial refugia and late-glacial and postglacial migration. *Ecology* 63, 1091–1105.

Tsukada, M. (1982b). Late-Quaternary development of the *Fagus* forest in the Japanese Archipelago. *Japanese Journal of Ecology* 32, 113–118.

Tsukada, M. (1982c). *Pseudotsuga menziesii* (Mirb.) Franco: its pollen dispersal and the late Quaternary history in the Pacific northwest. *Japanese Journal of Ecology* 32, 159–187.

Tsukada, M. (1983). Late-Quaternary spruce decline and rise in Japan and Sakhalin. *Botanical Magazine, Tokyo* 96, 127–133.

Tsukada, M., and Sugita, S. (1982). Late Quaternary dynamics of pollen influx at Mineral Lake, Washington. *Botanical Magazine, Tokyo* 95, 401–418.

Tukey, J. W. (1977). *Exploratory Data Analysis*. Addison-Wesley Company, Reading, Massachusetts.

Turner, J. (1964). Surface sample analyses from Ayrshire, Scotland. *Pollen et Spores* 6, 583–592.

Turner, J. (1965). A contribution to the history of forest clearance. *Proceedings of the Royal Society of London Series B* 161, 343–354.

Turner, J. (1970). Post-Neolithic disturbance of British vegetation. In *Studies in the Vegetational History of the British Isles* (eds. D. Walker and R. G. West), Cambridge University Press, London.

Turner, J. (1975). The evidence for land use by prehistoric farming communities: the use of three-dimensional pollen diagrams. In *The Effect of Man on the Landscape: The Highland Zone. CBA Research Report* 11, 86–95.

Turner, J., and Hodgson, J. (1979). Studies in the vegetational history of the Northern Pennines. I. Variations in the composition of the early Flandrian forests. *Journal of Ecology* 67, 629–646.

Turner, J., and Hodgson, J. (1981). Studies in the vegetational history of the Northern Pennines. II. An atypical pollen diagram from Pow Hill, Co. Durham. *Journal of Ecology* 69, 171–188.

Turner, J., and Hodgson, J. (1983). Studies in the vegetational history of the Northern Pennines. III. Variations in the composition of the mid-Flandrian forests. *Journal of Ecology* 71, 95–118.

Unwin, D. (1975). An introduction to trend surface analysis. *Concepts and Techniques in Modern Geography* 5, 40 pp.

Unwin, D. J., and Hepple, L. W. (1974). The statistical analysis of spatial series. *The Statistician* 23, 211–227.

Usher, M. B. (1973). *Biological Management and Conservation*. Chapman and Hall, London.

Usinger, H. (1975). Pollenanalytische und stratigraphische Untersuchungen an zwei Spätglazial-Vorkommen in Schleswig-Holstein. *Mitterlungen Arbeitsgemeinschaft Geobotanik in Schleswig-Holstein und Hamburg* 25, 183 pp.

Usinger, H. (1978a). Bölling-Interstadial und Laacher Bimstuff in einem neuen Spätglazial-Profil aus dem Vallensgard Mose/Bornholm. Mit pollengrossenstatistischer Trennung der Birken. *Danmarks geologiske Undersøgelse Arbog* 1977, 5–29.

Usinger, H. (1978b). Pollen und grossrestanalytische Untersuchengen zur Frage des Bölling-Interstadials und der Spätglazialen Baumbirken-Einwanderung in Schleswig-Holstein. *Schriften des Naturwissenschaftlichen Vereins fur Schleswig-Holstein*. 48, 41–61.

Valentine, J. W., and Mallory, B. (1965). Recurrent groups of bonded species in mixed death assemblages. *Journal of Geology* 73, 683–701.

Valentine, J. W., and Peddicord, R. G. (1967). Evaluation of fossil assemblages by cluster analysis. *Journal of Paleontology* 41, 502–507.

Van der Maarel, E. (1975). The Braun-Blanquet approach in perspective. *Vegetatio* 30, 213–219.

Van der Pluym, A., and Hideux, M. (1977a). Numerical analysis study of pollen grain populations of *Eryngium maritimum* L. (Umbelliferae). *Review of Palaeobotany and Palynology* 24, 119–139.

Van der Pluym, A., and Hideux, M. J. (1977b). Application d'une méthodologie quantitative à la palynologie d'*Eryngium maritimum* (Umbelliferae). *Plant Systematics and Evolution* 127, 55–85.

Van Hulst, R. (1979). On the dynamics of vegetation: Markov chains as models of succession. *Vegetatio* 40, 3–14.

Van Hulst, R. (1980). Vegetation dynamics of ecosystem dynamics: dynamic sufficiency in succession theory. *Vegetatio* 43, 147–151.

Van Ness, J. W. (1973). Admissible clustering procedures. *Biometrika* 60, 422–424.

Veldkamp, A. C., Hagen, T., and Van der Woude, J. D. (1981). Laser plotting of pollen diagrams. *Review of Palaeobotany and Palynology* 32, 441–443.

Vilks, G., Anthony, E. H., and Williams, W. T. (1970). Application of association analysis to distribution studies of recent foraminifera. *Canadian Journal of Earth Science* 7, 1462–1469.

Von Post, L. (1918). Skogasträd pollen i sydvenska torvmosselagerföldjer. *Förhandlingar Skandinavika Naturforskeres* 16, möte 1916, 432–465.

Von Post, L. (1946). The prospect for pollen analysis in the study of the earth's climatic history. *New Phytologist* 45, 193–217.

Von Post, L. (1967). Forest tree pollen in south Swedish peat bog deposits (Translation by M. B. Davis and K. Faegri). *Pollen et Spores* 9, 375–401.

Voorrips, A. (1973). An ALGOL-60 program for computation and graphical representation of pollen analytical data. *Acta Botanica Neerlandica* 22, 645–654.

Voorrips, A. (1974). An ALGOL-60 program for pollen analytical data; the CDC version. *Acta Botanica Neerlandica* 23, 701–704.

Vuilleumier, F. (1979). La niche de certains modelisateurs: paramètres d'un monde réal ou d'un univers fictif? *Terre Vie* 33, 375–423.

Vuorela, I. (1977). Pollen grains indicating culture in peat, mud and till. *Grana* 1, 211–214.

Vuorela, I. (1980). Old organic material as a source of dating errors in sediments from Vanajavesi and its manifestation in the pollen stratigraphy. *Annales Botanici Fennici* 17, 244–257.

Vuorinen, J., and Huttunen, P. (1981). The desktop computer in processing biostratigraphic data. *Pollen et Spores* 23, 165–172.

Vuorinen, J., and Tolonen, K. (1975). Flandrian pollen deposition in Lake Pappilanlampi, eastern Finland. *Publications of the University of Joensuu Series B* 3, 12 pp.

Waddington, J. C. B. (1969). A stratigraphic record of the pollen influx to a lake in the Big Woods of Minnesota. *Geological Society of America Special Paper* 123, 263–282.

Walker, D. (1966). The late Quaternary history of the Cumberland Lowland. *Philosophical Transactions of the Royal Society of London Series B* 251, 1–210.

Walker, D. (1970). Direction and rate in some British post-glacial hydroseres. In *Studies of the Vegetational History of the British Isles* (eds. D. Walker and R. G. West), Cambridge University Press, London.

Walker, D. (1972). Quantification in historical plant ecology. *Proceedings of the Ecological Society of Australia* 6, 91–104.

Walker, D. (1978). Envoi. In *Biology and Quaternary Environments* (eds. D. Walker and J. C. Guppy), Australian Academy of Science, Canberra.

Walker, D. (1982a). The development of resilience in burned vegetation. In *The Plant Community as a Working Mechanism* (ed. E. I. Newman), Blackwell Scientific Publications, Oxford.

Walker, D. (1982b). Vegetation's fourth dimension. *New Phytologist* 90, 419–429.

Walker, D., and Flenley, J. R. (1979). Late Quaternary vegetational history of the Enga Province of upland Papua New Guinea. *Philosophical Transactions of the Royal Society of London Series B* 286, 265–344.

Walker, D., and Pittelkow, Y. (1981). Some applications of the independent treatment of taxa in pollen analysis. *Journal of Biogeography* 8, 37–51.

Walker, D., and Wilson, S. R. (1978). A statistical alternative to the zoning of pollen diagrams. *Journal of Biogeography* 5, 1–21.

Walker, M. J. C., and Lowe, J. J. (1977). Postglacial environmental history of Rannoch Moor, Scotland. I. Three pollen diagrams from the Kingshouse area. *Journal of Biogeography* 4, 333–351.

Ward, J. H. (1963). Hierarchical grouping to optimize an objective function. *Journal of the American Statistical Association* 58, 236–244.

Watts, W. A. (1961). Post-Atlantic forest in Ireland. *Proceedings of the Linnean Society of London* 172, 33–38.

Watts, W. A. (1973). Rates of change and stability in vegetation in the perspective of long periods of time. In *Quaternary Plant Ecology* (eds. H. J. B. Birks and R. G. West), Blackwell Scientific Publications, Oxford.

Watts, W. A. (1977). The Late Devensian vegetation of Ireland. *Philosophical Transactions of the Royal Society of London Series B* 280, 273–293.
Watts, W. A. (1982). Response of biotic populations to rapid environmental and climatic change. *Abstracts Seventh Biennial AMQUA Conference,* 19–21.
Watts, W. A., and Winter, T. C. (1966). Plant macrofossils from Kirchner Marsh, Minnesota—a paleoecological study. *Geological Society of America Bulletin* 77, 1339–1360.
Webb, D. A. (1954). Is the classification of plant communities either possible or desirable? *Botanisk Tidsskrift* 5, 362–370.
Webb, J. A., and Moore, P. D. (1982). The Late Devensian vegetational history of the Whitlaw Mosses, southeast Scotland. *New Phytologist* 91, 341–398.
Webb, T., III. (1971). The late- and post-glacial sequence of climatic events in Wisconsin and east-central Minnesota: quantitative estimates derived from fossil pollen spectra by multivariate statistical analysis. Ph.D. thesis, University of Wisconsin.
Webb, T., III. (1973). A comparison of modern and presettlement pollen from southern Michigan (U.S.A.). *Review of Palaeobotany and Palynology* 16, 137–156.
Webb, T., III. (1974a). Corresponding patterns of pollen and vegetation in Lower Michigan: a comparison of quantitative data. *Ecology* 55, 17–28.
Webb, T., III. (1974b). The pollen-vegetation relationship in southern Michigan: an application of isopolls and principal components analysis. *Geoscience and Man* 9, 7–14.
Webb, T., III. (1974c). A vegetational history from northern Wisconsin: evidence from modern and fossil pollen. *American Midland Naturalist* 92, 12–34.
Webb, T., III. (1980). The reconstruction of climatic sequences from botanical data. *Journal of Interdisciplinary History* 10, 749–772.
Webb, T., III. (1981). The past 11,000 years of vegetational change in eastern North America. *Bioscience* 31, 501–506.
Webb, T., III. (1982). Temporal resolution in Holocene pollen data. *Proceedings Third North American Paleontological Convention* 2, 569–572.
Webb, T., III. (1983). Calibration of Holocene pollen data in climatic terms. *Quaternary Studies in Poland* 4, 107–113.
Webb, T., III., and Bryson, R. A. (1972). Late- and post-glacial climatic change in the northern Midwest, USA: quantitative estimates derived from fossil pollen spectra by multivariate statistical analysis. *Quaternary Research* 2, 70–115.
Webb, T., III., and Clark, D. R. (1977). Calibrating micropaleontological data in climatic terms: a critical review. *Annals of the New York Academy of Sciences* 288, 93–118.
Webb, T., III., Howe, S., Bradshaw, R. H. W., and Heide, K. (1981). Estimating plant abundances from pollen percentages: the use of regression analysis. *Review of Palaeobotany and Palynology* 34, 269–300.
Webb, T., III., Laseski, R. A., and Bernabo, J. C. (1978). Sensing vegetational patterns with pollen data: choosing the data. *Ecology* 59, 1151–1163.
Webb, T., III., and McAndrews, J. H. (1976). Corresponding patterns of contemporary pollen and vegetation in central North America. *Geological Society of America Memoir* 145, 267–299.
Webb, T., Richard, P. J. H., and Mott, R. J. (1983). A mapped history of Holocene vegetation in southern Quebec. *Syllogeus* 49, 273–336.
Webb, T., III., Yeracaris, G. Y., and Richard, P. (1978). Mapped patterns in sediment samples of modern pollen from southeastern Canada and northeastern United States. *Géographie physique et Quaternaire* 32, 163–176.
Webster, R. (1973). Automatic soil-boundary location from transect data. *Journal of Mathematical Geology* 5, 27–37.
Webster, R. (1977). *Quantitative and Numerical Methods in Soil Classification and Survey*. Clarendon Press, Oxford.
Webster, R. (1978). Optimally partitioning soil transects. *Journal of Soil Science* 29, 388–402.
West, R. G. (1964). Inter-relations of ecology and Quaternary palaeobotany. *Journal of Ecology* 52 (Supplement), 47–57.

West, R. G. (1970). Pollen zones in the Pleistocene of Great Britain and their correlation. *New Phytologist* 69, 1179–1183.
West, R. G. (1971). Studying the past by pollen analysis. *Oxford Biology Readers* 10, 16 pp.
West, R. G. (1977). *Pleistocene Geology and Biology* (Second edition), Longmans, London.
Westenberg, J. (1947a). Mathematics of pollen diagrams. I. *Koninklijke Nederlande Akademie van Wetenschapen* 50, 509–520.
Westenberg, J. (1947b). Mathematics of pollen diagrams. II. *Koninklijke Nederlande Akadamie van Wetenschapen* 50, 640–648.
Westenberg, J. (1964). Nomograms for significance in a 2 × 2 contingency table for sample sizes of 10 to 500. *Koninklijke Nederlande Akademie van Wetenschapen* 67, 441–467.
Westenberg, J. (1967). Testing significance of difference in a pair of relative frequencies in pollen analysis. *Review of Palaeobotany and Palynology* 3, 359–369.
Whitehead, D. R. (1964). Fossil pine pollen and full-glacial vegetation in southeastern North Carolina. *Ecology* 45, 767–777.
Whitehead, D. R. (1979). Late-glacial and post-glacial vegetational history of the Berkshires, western Massachusetts. *Quaternary Research* 12, 333–357.
Whitehead, D. R. (1981). Late-Pleistocene vegetational changes in northeastern North Carolina. *Ecological Monographs* 51, 451–471.
Whitehead, D. R., and Langham, E. J. (1965). Measurement as a means of identifying fossil maize pollen. *Bulletin of the Torrey Botanical Club* 92, 7–20.
Whitehead, D. R., and Sheehan, M. C. (1971). Measurement as a means of identifying fossil maize pollen, II: The effect of slide thickness. *Bulletin of the Torrey Botanical Club* 98, 268–271.
Whitehead, D. R., and Tan, K. W. (1969). Modern vegetation and pollen rain in Bladen county, North Carolina. *Ecology* 50, 235–248.
Whiteside, M. C. (1970). Danish chydorid cladocera: modern ecology and core studies. *Ecological Monographs* 40, 79–118.
Whittaker, R. H. (1962). Classification of natural communities. *Botanical Review* 28, 1–239.
Whittaker, R. H. (Ed.) (1978a). *Classification of Plant Communities*. Dr. Junk, The Hague.
Whittaker, R. H. (Ed.) (1978b). *Ordination of Plant Communities*. Dr. Junk, The Hague.
Whitten, E. H. T. (1975). The practical use of trend-surface analysis in the geological sciences. In *Display and Analysis of Spatial Data* (eds. J. C. Davis and M. J. McCullagh), J. Wiley and Sons, London and New York.
Wiens, J. A. (1977). On competition and variable environments. *American Scientist* 65, 590–597.
Wijmstra, T. A. (1969). Palynology of the first 30 metres of a 120 m deep section in northern Greece. *Acta Botanica Neerlandica* 18, 511–527.
Wijmstra, T. A., and Smit, A. (1976). Palynology of the middle part (30–78 metres) of the 120 m deep section in northern Greece (Macedonia). *Acta Botanica Neerlandica* 25, 297–312.
Wilkinson, C. (1970). Adding a point to a principal coordinates analysis. *Systematic Zoology* 19, 258–263.
Williams, B. K. (1983). Some observations on the use of discriminant analysis in ecology. *Ecology* 64, 1283–1291.
Williams, W. (1977). The Flandrian vegetational history of the Isle of Skye and the Morar Peninsula. Ph.D. thesis, University of Cambridge.
Williams, W. T. (Ed.) (1976). *Pattern Analysis in Agricultural Science*. CSIRO and Elsevier, Melbourne and Amsterdam.
Williamson, I. (1981). Seismic data processing, sampled Quaternary data and pollen analysis. *Quaternary Studies* 1, 95–113.
Williamson, M. (1975). The biological interpretation of time-series analysis. *Bulletin of the Institute of Mathematics and its Applications* 11, 67–69.
Wilson, M. V. H. (1980). Eocene lake environments: depth and distance-from-shore variation in fish, insect, and plant assemblages. *Palaeogeography, Palaeoclimatology, Palaeoecology* 32, 21–44.
Wiltshire, P. E. J., and Moore, P. D. (1983). Palaeovegetation and palaeohydrology in upland Britain. In *Background to Palaeohydrology* (ed. K. J. Gregory), J. Wiley & Sons, Chichester.

Wishart, D. (1969). An algorithm for hierarchical classifications. *Biometrics* 22, 165–170.
Woillard, G. M. (1978). Grande Pile Peat Bog: a continuous pollen record for the last 140,000 years. *Quaternary Research* 9, 1–21.
Woillard, G. M., and Mook, W. G. (1982). Carbon-14 dates at Grande Pile: correlation of land and sea chronologies. *Science* 215, 159–161.
Woodhead, N., and Hodgson, L. M. (1935). A preliminary study of some Snowdonian peats. *New Phytologist* 34, 263–282.
Wright, H. E., Jr. (1967). The use of surface samples in Quaternary pollen analysis. *Review of Palaeobotany and Palynology* 2, 321–330.
Wright, H. E., Jr. (1974). Deciphering vegetation history. *Science* 186, 728.
Wright, H. E., Jr. (1976). The dynamic nature of Holocene vegetation. A problem in paleoclimatology, biogeography, and stratigraphic nomenclature. *Quaternary Research* 6, 581–596.
Wright, H. E., Jr., Bent, A. M., Hansen, B. S., and Maher, L. J. (1973). Present and past vegetation of the Chuska Mountains, northwestern New Mexico. *Geological Society of America Bulletin* 84, 1155–1180.
Wright, H. E., Jr., McAndrews, J. H., and Van Zeist, W. (1967). Modern pollen rain in western Iran, and its relation to plant geography and Quaternary vegetational history. *Journal of Ecology* 55, 415–443.
Wright, H. E., Jr., and Patten, H. L. (1963). The pollen sum. *Pollen et Spores* 5, 445–450.
Wright, H. E., Jr., Winter, T. C., and Patten, H. L. (1963). Two pollen diagrams from southeastern Minnesota: problems in the regional lateglacial and postglacial vegetational history. *Geological Society of America Bulletin* 74, 1371–1396.
Yarranton, G. A., and Ritchie, J. C. (1972). Sequential correlations as an aid in placing pollen zone boundaries. *Pollen et Spores* 14, 213–223.
Yule, G. U. (1926). Why do we sometimes get nonsense-correlations between time-series?—a study in sampling and the nature of time series. *Journal of the Royal Statistical Society* 89, 1–69.

Index